Essentials of Physiology

There is a mask of theory over the whole face of nature.

William Whewell (1847) quoted in
Medawar, P. B. (1967) *The Art of the Soluble*,
Methuen & Co

We are more likely to reach the truth through error than through confusion.

Francis Bacon

Essentials of Physiology

J. F. LAMB
MB, PhD, FRCP(E), FRS(E)
Chandos Professor

C. G. INGRAM
MB, ChB
Senior Lecturer

I. A. JOHNSTON
BSc, PhD
Professor of Comparative Physiology

R. M. PITMAN
BSc, PhD
Lecturer

All of the Department of Physiology and Pharmacology
University of St Andrews, Fife

Second Edition

BLACKWELL
SCIENTIFIC PUBLICATIONS
OXFORD · LONDON · EDINBURGH
BOSTON · PALO ALTO · MELBOURNE

First published 1980
Reprinted 1981, 1982, 1983
Second Edition 1984
Reprinted with corrections 1986, 1988

Printed and bound in Great Britain by
Butler & Tanner Ltd,
Frome and London

DISTRIBUTORS

USA
  Year Book Medical Publishers
  200 North LaSalle Street
  Chicago, Illinois 60601

Canada
  The C. V. Mosby Company
  5240 Finch Avenue East
  Scarborough, Ontario

Australia
  Blackwell Scientific Publications
  (Australia) Pty Ltd
  107 Barry Street
  Carlton, Victoria 3053

British Library
Cataloguing in Publication Data

Essentials of physiology.—2nd ed
  1. Human physiology
  I. Lamb, J. F.
  612   QP34.5

ISBN 0–632–01259–5

Contents

Preface

This book is aimed at those students of science, medicine, dentistry and allied professions who are studying a first year course of Physiology at University or equivalent level.

We have kept the book short so as to encourage the student to feel that he can grasp and understand the whole subject; a point especially important at a time when other subjects are making large demands on his time. We have chosen those aspects of the subject which we think are essential for the student to understand, and arranged them in such a way that students can use them in future as pegs to hang more information on. In order to achieve this, we have predigested most of the usual information given to students in larger text books, and then used graphs and diagrams to present it as simply and concisely as possible. This book is, however, concerned more with current ideas in Physiology than with a list of the currently known facts of the subject; for we feel that this approach is more likely to interest students who are starting the subject and, in consequence, is more likely to lead to their remembering the subject. Historically it is also true that the importance, and sometimes even the validity of 'facts', varies with the current theories about the subject.

A major difficulty in teaching physiology is to give a simple explanation of the complex mechanisms involved in the control of various bodily functions. Part of this difficulty is due to the problem of showing dynamic feedback loops in static diagrams. In our teaching, therefore, we have started to adopt a convention devised by Allweis for constructing such diagrams; in the book we have included some examples of these diagrams which our own students have found useful. (The convention is illustrated in Fig. 5.48.)

In choosing the material to be included in the book we have been guided by a set of Objectives for a course on Physiology produced by the Physiology Department of the University of Aarhus in Denmark (translated and published by the Physiological Society), and by *Learning Objectives in Medical Physiology* (1976) published by the University of London Board of Studies in Physiology. The reference list included is generally that which we found useful, and usually consists of other secondary references, i.e. larger texts, monographs, etc. We have occasionally referred to recent original papers where these are relevant.

We are grateful to Mr N. W. Palmer of Blackwell Scientific Publications who gave us a detailed specification for the book, and to colleagues who read various sections.

J.F.L., C.G.I.,
I.A.J., R.M.P.

Preface to the Second Edition

Our objectives in this edition are as before: to present an account of modern ideas about Physiology as clearly and as concisely as possible. For the former we have relied on the articles listed later, on seminars and particularily on colleagues familiar with fields foreign to us. For the latter we have tried to follow the adage that 'if a thing can be described it can be described simply'.

Those unfamiliar with this book will find that the text is generally simpler to understand than some of the illustrations. We often use the illustrations, particularly the captions, to 'amplify' rather than 'illustrate' the text. We suggest that on a first reading you read the text by itself, leaving the more conceptual Figures outlined in boxes for a second reading.

The chapters on the Gastrointestinal tract, Nutrition, and Kidney have been extensively revised, a short chapter on Circadian Rhythms and a section on Homeostasis added. Most of the other sections have been updated. Those familiar with the First Edition will find that the book has moved slightly 'up market', it has also become a little longer and acquired a second colour.

Many people were kind enough to send us helpful comments on the First Edition; we hope that readers will continue to point out our errors, stupidities and misconceptions in this Edition.

We are most indebted to Bridget Cook, of Blackwell Scientific Publications for much skilful, dedicated help in the preparation of this Edition.

Part 1
Basic Properties

Chapter 1
Introduction and
Cells

INTRODUCTORY IDEAS

The structure and function of cells

There are three generalisations in biology:

1   The theory of evolution—all species living at present are a subset of all those which have lived and all are related to each other.

2   The cell theory—all living organisms are composed of cells: other life forms may have existed once but they have died out. Cells are a closed domain, this allows the concentrations of essential materials to be kept high enough so that the chemical reactions needed for life can proceed at fast rates. This is possible because there are membranes round and within cells and these act as selective barriers.

3   The unity of biochemical and physiological processes—all organisms now living share certain basic biochemical reactions. The two main things in common are:

a   The chemistry of the tape, i.e. the genetic code is the same, and always based on nucleic acids.

b   The machinery—proteins and devices for making proteins—are generally the same.

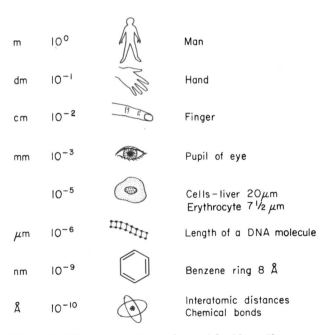

m	10^0		Man
dm	10^{-1}		Hand
cm	10^{-2}		Finger
mm	10^{-3}		Pupil of eye
	10^{-5}		Cells – liver $20\,\mu m$ Erythrocyte $7\frac{1}{2}\,\mu m$
μm	10^{-6}		Length of a DNA molecule
nm	10^{-9}		Benzene ring 8 Å
Å	10^{-10}		Interatomic distances Chemical bonds

Fig. 1.1. Relative scales of the component parts of a man (after D'Arcy Thomson, see Source of Figures).

Because of this common basis we can study certain processes of life in whichever animal is convenient, e.g. the process of nerve conduction was worked out on the squid giant axon, that of blood pressure control in dogs, reflexes in cats and dogs and the genetic code in bacteria and so on. The data on which any 'human' course is based has been obtained largely from other animals.

Cells could not be seen until microscopes were invented in the 17th century, for they are small on our scale, although large compared to the atoms of the chemist. In size they lie about half way (on a logarithmic scale) between ourselves and the atoms (Fig. 1.1) and contain about as many atoms as we contain cells (more than 10^{14}). In the evolution of

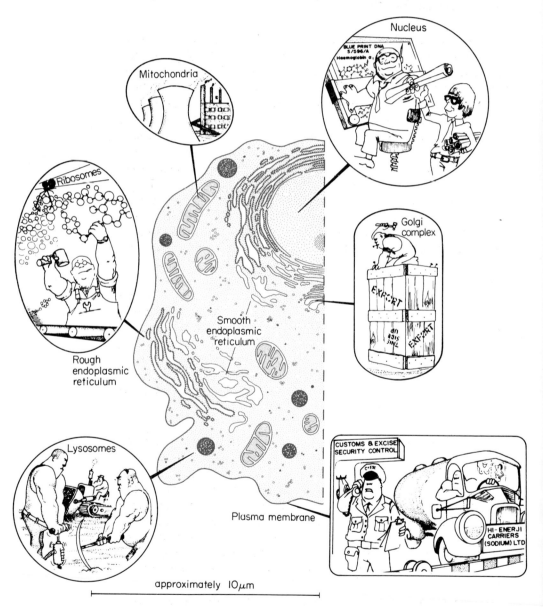

Nucleus

Mitochondria

Ribosomes

Golgi complex

Smooth endoplasmic reticulum

Rough endoplasmic reticulum

Lysosomes

Plasma membrane

CUSTOMS & EXCISE SECURITY CONTROL

HI-ENERJI CARRIERS (SODIUM) LTD

approximately 10 μm

Fig. 1.2. Cellular components with an indication of their function (after Finean).

multicellular organisms, cells became specialised and organised into different systems, cardiovascular, muscle, kidney and so on, but generally retained common features (Fig. 1.2) and indeed have the same genetic information (in any one organism).

Such diagrams tend to give you the misleading impression that cells are static and fixed, whereas in life all parts of each cell are wearing out, being replaced or modified so that our cells of today do not contain the same molecules as our cells of yesterday. 'Biological structure is not like that of a rock or a crystal, but rather like that of a flame or a waterfall' (Robinson). The rate of molecular replacement varies from minutes to months; that of cells from the few hours of the alimentary tract to the years of our bones.

Membranes in cells. The membranes of animal cells make up some 40–90 percent of the total dry mass in various cell types. They are of major importance in biological systems, because they divide cells into various compartments; this allows high concentrations of substrates to be created in parts of the cell with the possibility of high reaction rates, and allows control to be exercised by altering the rates of movement between compartments. Although their detailed structure varies from one location in the cell to another, all membranes have certain features in common. The present view is of a double layer of phospholipid some 10 nm thick) with proteins studded in and on each side of it (Fig. 1.3). The phospholipid molecules can be thought of as a clothes-pin (Ganong) with a water-soluble, charged phosphate head and a water-insoluble, non-charged, lipid tail; these are arranged in a double layer with the heads in the water at each side of the membrane, and the tails in the centre. Proteins in globular form are embedded in the inner or outer membranes; in a more elongated form they span the membrane. There are both lipoproteins (lipid containing) and glycoproteins (carbohydrate containing) in the membrane. Some of the proteins which span the outer cell membrane perhaps act as receptors for hormones or neurotransmitters, and others may act as ion channels. It is now clear that membrane constituents (proteins and some lipids) can move laterally within the membrane at quite fast rates. So the membrane can be thought of as a fluid mosaic, largely composed of phospholipids but with many embedded proteins.

The nucleus. The individual active units of heredity—the genes—are strung together along the chromosomes, which are thread-like bodies in the nucleus of each cell. These genes contain all the information needed by the cell to reproduce itself; they do this by bearing within themselves, in coded form, the detailed specifications for the many thousands of protein molecules required by the cell for its moment-to-moment existence. The information in the gene is stored as a one-dimensional message, which is translated by the cellular machinery into the one-dimensional sequence of amino acids in the final protein molecule.

The genetic material in all cells is the giant chain-like molecule of deoxyribonucleic acid (DNA); in mammals dozens of these are clustered

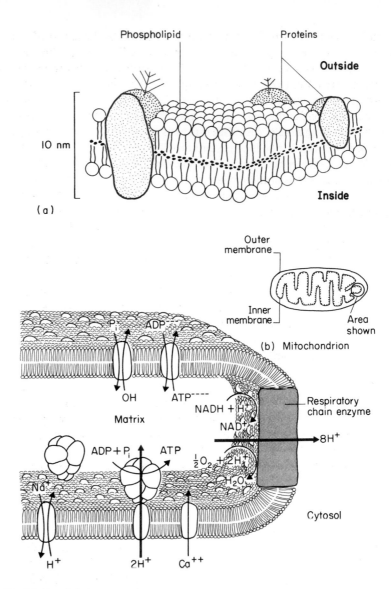

Fig. 1.3. Detailed structures of membranes. (a) Plasma membrane. (b) Mitochondrial membrane. Electrons and protons from carbohydrates and fats are carried by molecules of the hydrogen carrier, NADH, to a system of enzymes embedded in the mitochondrial membrane. These enzymes convey protons across the mitochondrial membrane to the cell cytoplasm. The gradient of proton and electrical potential so formed forces protons back through the membrane driving the process of ATP synthesis and the various carrier systems in the mitochondrial membrane (after Hinkle and McCarty).

together in the chromosomes. The DNA molecules are arranged like a ladder that has been twisted into a helix; the sides of the ladder are formed by alternating units of 5-carbon sugar and phosphate groups. The rungs, which join two sugar units, are made of pairs of bases; either adenine (A) paired with thymine (T) or guanine (G) paired with cytosine (C). These are the four 'letters' which spell out the genetic message; the exact sequence of bases along the sides of the 'ladder' determines the particular protein molecule.

Proteins are made from a standard set of 20 amino acids, uniform throughout nature, which are joined end to end to form the long polypeptide chains of protein molecules. Each protein has its own characteristic sequence of amino acids. Each polypeptide may have from 100 to more than 300 amino acids in it.

The genetic code is the 'dictionary' used by the cell to translate from the four-letter language of the nucleic acid to the twenty-letter language of the proteins. In doing so the cell uses a variety of intermediate molecules and mechanisms. The DNA message is first transcribed into a similar molecule called messenger RNA (ribonucleic acid), which has the bases adenine, guanine, cytosine, and uracil (U) instead of thymine. This moves out into the cytoplasm of the cell where the ribosomes travel along it, 'reading' the code contained in its base sequence and synthesising the polypeptide chain of the protein, starting at the amino end (NH$_2$). Each amino acid is recognised by a transfer RNA, which carries it to the growing polypeptide chain to which it is added.

The genetic code is a triplet code, i.e. the bases on the DNA are read three at a time rather than singly, and each group—called a codon— corresponds to a particular amino acid. Mathematically, the four letters of the DNA code can be combined into 64 distinct triplets so that there are more codons available than amino acids; it appears that some amino acids are represented by several codons.

The endoplasmic reticulum is a series of tubules in the cytoplasm of the cell. They are either smooth or rough in appearance. The roughness comes from the presence of ribosomes; globules some 10 nm in diameter made up of two subunits, a larger 50S and a smaller 30S (called after their rate of sedimentation in the centrifuge). Ribosomes may also be found in clumps of three to five as polysomes (polyribosomes) or free in the cytoplasm. Ribosomes manufacture protein on templates (messenger RNA) produced by the cell nucleus, using amino acids free in the cytoplasm. Cells which export* protein (pancreas, etc.) have their ribosomes organised on the endoplasmic reticulum; those which produce protein for replacement purposes within the cell have free ribosomes. Ribosomes are attached to membranes in pairs, probably with a pore present in the centre of the four subunits through which proteins are exported.

Smooth endoplasmic reticulum in muscle (= sarcoplasmic reticulum) is important in contraction; in steroid secretory cells it is the place where steroid hormones are made and in other cells is a site for detoxification.

The Golgi complex is a filamentous or plate-like collection of smooth membranes. Although present in the cytoplasm of all cells it is largest and most important in exporting cells and is believed to be responsible for 'packaging' the protein (for export). The most characteristic enzymes found here are those responsible for linking sugars with proteins to form glycoproteins.

---

* Manufacture protein for use outside the producing cell

The lysosomes are made in the Golgi apparatus, and are large and irregular membrane-bound structures. They contain a variety of enzymes which can break down proteins, carbohydrates, phosphate esters and RNA and DNA. They can be thought of as the digestive system of the cell, whose function is to break down unwanted parts of the cell, release hormones by breaking down precursors (for example, thyroid), kill engulfed bacteria, etc. Several of these functions are brought about by the lysosome merging with another vesicle containing the 'substrate' material. The resultant products are either excreted to the exterior of the cell or absorbed into the cell.

Microfilaments and microtubules are elongated structures present in many cells; the microfilaments are solid rods some 4–6 nm in diameter, whereas the microtubules are hollow cylinders some 25 nm in diameter with walls 5 nm thick. Microfilaments are made up of the same filaments—actin and myosin—that give muscles their contractile properties; if they contain much tropomyosin they are readily seen in electron micrographs (EM), if not they are not easily seen. Microtubules are made up of various proteins—the main one being tubulin—sometimes with side arms. Various drugs—for example, vinblastin and colchicine—and cold break up the labile variety of microtubules.

These filamentous structures are thought to be involved in cellular functions such as movement, ingestion of food, spindle formation during cell division, controlling the shape of cells, sensory transduction and perhaps the organisation of proteins in the cell membrane. The precise way in which this occurs is, at present, unclear; one view is that the microfilaments are the 'muscles' of the cell whereas the microtubules provide a framework for the organisation and co-ordination of the power so produced. The microtubules are the railway lines, the microfilaments the railway engine.

The mitochondria are sausage-shaped structures in the cytoplasm of cells. They have an outer membrane and an inner membrane folded into a variable number of cristae. The mitochondria are the 'power-house' of the cell where ATP* is made by a process called oxidative phosphorylation. It is found that the number of mitochondria and the number of cristae per mitochondrion are related to the energy requirements of the cell.

The current view is that mitochondria make ATP by a 'chemi-osmotic' method. It is thought that oxidation drives protons from the inside of the mitochondrion (the matrix) out into the cytoplasm; as the protons flow back the complex enzymes in the mitochondrial membrane form ATP (Fig. 1.3b).

The mitochondria may have originated as independent bodies which progressed through a symbiotic relationship with primeval cells to their present state. They have their own DNA and can make protein. Their DNA, however, is unlike that in the nucleus in that it is ring-shaped and attached to the mitochondrial boundary membrane and the coding is also different in mitochondria from that in the nucleus.

* Adenosine triphosphate—the universal energy currency of cells

This section contains a summary of those aspects of physiology 'that show some immediate prospect of being described in terms of the known laws of physics and chemistry' (Davson).

Water is important in the body, for our cells are largely composed of it, are suspended in it, our food is dissolved in it and we excrete into it.

The three atoms of the water molecule are arranged at an angle of about 104°, with the central oxygen electronegative and the hydrogen positive (Fig. 1.4). The properties of water which are important for our present purpose follow from this structure.

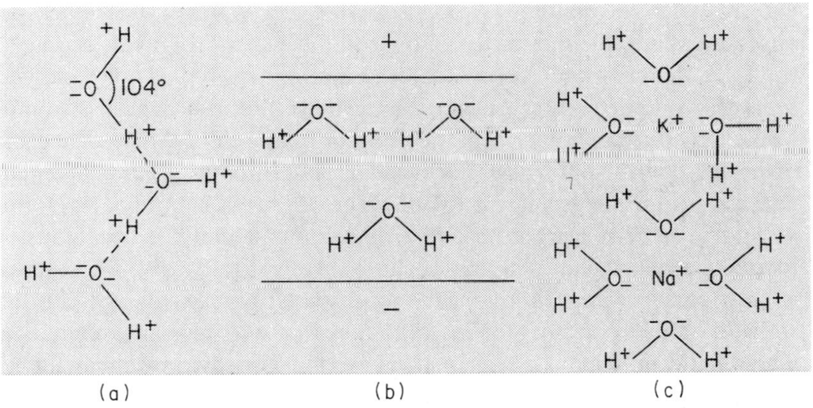

(a)   (b)   (c)

Fig. 1.4. (a) The clusters of dipolar molecules in liquid water; (b) its dielectric effect on electric forces between charged plates; and (c) the formation of hydration shells around Na^+ and K^+ ions.

Na^+ and K^+ have the same charge but the atomic radius of Na^+ is smaller and so it has a more intense field at its surface, leading to a greater hydration shell. The molecules of water in free clusters (a) or in shells (c) exchange rapidly with other water molecules, perhaps at rates of 10^{11} and 10^9 times per second.

1 Liquid water is formed of clusters or chains of molecules held together by hydrogen bonds between the oxygen of one molecule and the hydrogen of another. When water freezes, a regular structure of water molecules is formed with each molecule bound to four others by hydrogen bonds. On melting, some bonds are broken; further increases in temperature lead to more and more bonds being broken until only single molecules are left in water vapour. So, change in temperature is associated with chemical as well as physical changes in the structure of water, and requires much more heat than would be expected on simple grounds. Water, compared to say chloroform, has a very high specific heat and a high latent heat of evaporation. This means that water can readily absorb local heat produced by chemical reactions and that sweating in the tropics is a very effective way of losing heat. If we used chloroform instead of water as the solvent in our bodies, we would need to sweat 200 litres/day instead of the 10–15 litres of water we can sweat in the tropics to lose the same heat.

2   Water molecules, because of their charge, become orientated in an electric field. This means (a) that if an electric field exists across a membrane then the water molecules all line up and so reduce the electrostatic forces across this membrane (water is said to have a large dielectric* property, due to its dipolar nature); and (b) water molecules 'cluster' like a shell around a charged particle. These hydration shells (Fig. 1.4c) increase the size of the ions and partly obscure the chemical nature of the central ion, which only has direct access to its surroundings when the water molecules surrounding it exchange with those in free solution.

Water distribution in the body

Some two-thirds of our body consists of water, the actual amount varying with fat content (which contains less water). This water is, anatomically, split into many compartments rather than existing as a single pool. Fortunately, however, it can be considered as a few simple separate compartments. About two-thirds of the total water is found inside cells (intracellular), and the remaining one-third outside the cells (extracellular). The intracellular water is split into a multitude of small packets—each corresponding to one of the 10^{14} cells in our bodies—and only forms a compartment because its behaviour and dissolved constituents are all similar. The extracellular water again exists in several spaces: cerebrospinal fluid in the brain; joints; blood; etc., but can be considered as consisting of interstitial water and plasma water, the water between the cells and in the vascular system respectively. The main difference between intra- and extracellular fluids is that intracellular fluid contains potassium and organic anions (mainly organic phosphate) whereas extracellular fluid contains sodium chloride (Fig. 1.5). The extracellular fluid can be considered as the descendant of the seas our ancestral cells inhabited, although compared to the present-day sea it is only one-third as concentrated. Thus even though our remote ancestors left the sea to colonise the land, our cells are still tiny aquatic organisms floating in their own sea.

Homeostasis and control system theory

Claude Bernard introduced the idea that our lives are free and independent because our internal sea is fixed and constant: most of our cells do not have to share the same rigours that our whole body has to. Much of physiology is concerned with the homeostatic mechanisms which keep this internal sea constant. The magnitude of the problem can be appreciated by reflecting that our cellular mass has to live in a fluid environment of roughly one-quarter of its own volume† (Fig. 1.5). A real pond with this proportion of organisms to pond water would have a consistency like porridge and would soon become desperately polluted. Our bodies remain in a constant state because our extracellular space is very well mixed, has its gases replenished, is supplied with food and has its waste products removed.

---

* Transmitting electric effects without conducting
† About a 'legful' of water for our bodies

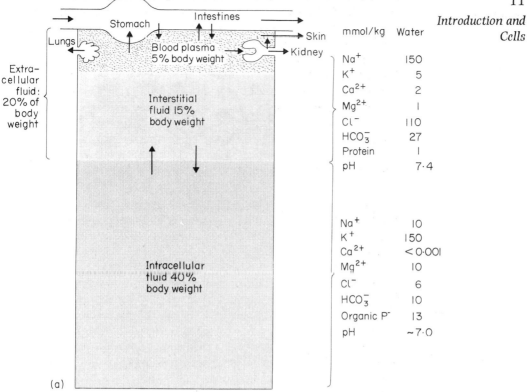

Fig. 1.5. The sizes and dissolved substances in the body fluid compartments. The arrows represent fluid exchanges.

The intracellular compartment can be considered as 'the discontinuous dispersed phase of an emulsion in the continuous extracellular fluid' (Robinson). All compartments have the same osmotic pressure, i.e. are isosmotic. (after Gamble)

Many conditions in the body are in fact so constant that minor departures from the normal can be used in the diagnosis of disease, if they persist. Thus, in man, the body temperature, and in the blood, the amount of sugar, the slight alkalinity, the concentration of various salts, the number of red or white cells etc., are constant or vary only within narrow limits.

In order to keep the output of a system constant (or to vary it in some controlled way) the inputs need to be manipulated. We will describe this by using a system familiar to you; keeping a room heated to a constant temperature. The essentials of such a control system consist of a heater, a sensor, i.e. thermometer, a controller and connections between them (Fig. 1.6). The temperature of the room is measured, compared with a desired temperature (set point) and the heater switched on or off. The sensor therefore detects the effects of the applied disturbance on the regulated variable, and *feeds back* a signal to the controller.

Additional points on the more complex arrangements present in physiological systems:

1   Most feedback systems are *negative*, i.e. the output of the sensor is subtracted from the set point in the controller to give an 'error' signal.

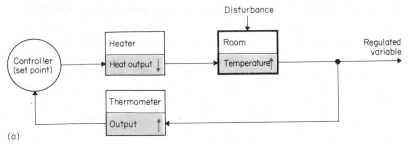

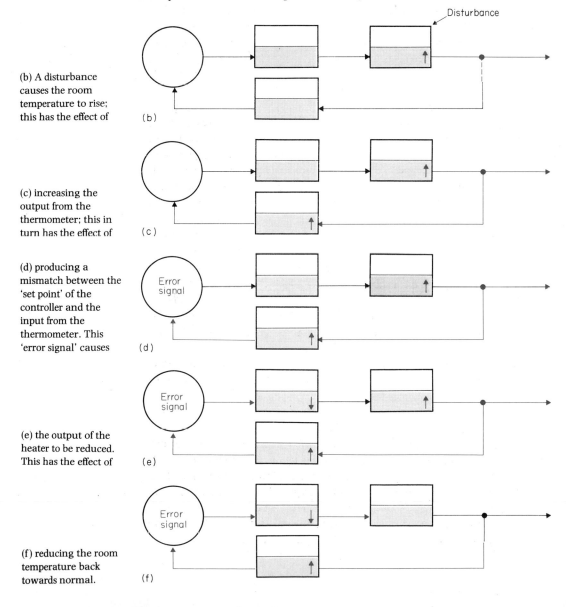

Fig. 1.6. (a) Basic components and signals of a control system to regulate room temperature. The upper name in each box identifies the component, the lower name (in red box) the variable. The box on the right, shown in bolder outline, contains the variable which is regulated. Lines joining boxes mean 'has an effect on' (after Allweiss).

A disturbance of temperature (shown as a rise) is followed by the consequences shown (red arrows) which corrects the disturbance. The following figures show the sequence of events in this negative feedback system:

(b) A disturbance causes the room temperature to rise; this has the effect of

(c) increasing the output from the thermometer; this in turn has the effect of

(d) producing a mismatch between the 'set point' of the controller and the input from the thermometer. This 'error signal' causes

(e) the output of the heater to be reduced. This has the effect of

(f) reducing the room temperature back towards normal.

2   Feedback systems have the disadvantage that the compensatory actions may be excessively slow, since they must await the development of the error that drives them. This problem can be reduced by introducing feedback sensitive to the rate of change of the regulated variable. The use of such a system allows a strong compensatory action to be taken in advance of the full error. Physiological systems have many such devices; in fact almost all receptors signal rate of change better than steady states, e.g. the baroreceptors for blood pressure control.

3   Feedforward regulation. Disturbances, of course, occur before any effects they produce; so if the disturbance can be detected before it has an effect then action can be taken before the effects appear, this is *feedforward* regulation. A familiar example of this is the 'mouth watering' which occurs at the prospect of tasty food. It will be noticed that in feedforward the controller has to calculate what it expects to happen; for this reason feedforward arrangements are inherently less reliable.

4   Adaptive control systems are those which change to meet changing needs; they differ from the others in showing long-term changes, rather than the moment-to-moment changes of the others. Examples are the hypertrophy of a muscle on training, learning to ride a bicycle etc. Good adaptive systems 'learn' to improve their performance by past experience; our learning processes in an adaptive process.

5   The set point in physiological regulators. Fig. 1.6 shows the set point as a single stage of the process, but the controller for a physiological regulator is usually a succession of neuronal stages that convert the feedback signals sent from sensors into appropriate controlling signals. In general, each neurone at each stage will receive other inputs as well as feedback signals, so the reference for a physiological regulator is a *distributed* rather than a localised signal. It follows that there is no localised error detector either; instead, error detection is an operation that is distributed among each of the neurones in the feedback chain.

6   A servosystem is a control system which operates like a negative feedback system, but the reference signal varies with time and so is called a *command signal*. The muscle spindle system works in this way.

7   If the control system is disabled an applied disturbance will lead to a maximum change in the regulated variable; the ratio of this change to that with the control intact gives the gain of the system. Gains of 4–10 are common in biological control systems.

In the physiological diagrams shown later, the boxes show the anatomic region, the red lines refer to hormones and the dashed lines to action potentials.

The final common path between capillaries and cells

Almost all cells are within 10 μm of a capillary. Food and oxygen are delivered to the capillaries and waste products removed from the capillaries by the action of the cardiovascular, alimentary, respiratory and kidney systems. Movement between the capillaries and the cells is by diffusion (Fig. 1.7). Much of this book is concerned with the way in which the body transports substances to and from the capillaries; the next section deals with movement from capillary to cell.

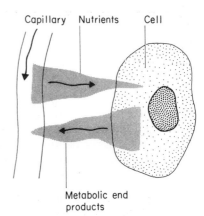

Capillary   Nutrients   Cell

Metabolic end
products

Fig. 1.7 The final common path by which
substances travel to and from cells is
that through the capillary wall, the
interstitial space and the cell membrane
(after Vander, Sherman & Luciani).

Movement from capillary to cell
Fluid movement in the gut and cardiovascular system is largely by bulk
flow, where water with its contents is physically moved from one place
to another by the action of an external force. The movement of substances from capillaries to cells occurs by diffusion, the process by which
matter is transported by spontaneous random movements of individual
molecules from areas of higher to those of lower concentration. In bulk
flow all the constituent materials move at the same rate; in diffusion
each has a flow whose rate and amount depends on the individual
circumstance of the substance.

Diffusion
The observed rates at which substances diffuse is well described by Fick's
law of diffusion (Fig. 1.8). This law is simply a mathematical expression
for something we know from experience. Imagine that a substance (e.g.
dye in water) can move from one place to another; the rate of movement
will increase as the driving force increases (concentration of the dye) but
decrease with increasing distance; the rate will also increase as the
aperture (i.e. the area) between the places increases. In Fick's law these
factors are collected together and a diffusion coefficient (D) added to
complete the equation. The diffusion coefficient is constant for a given
substance under specified conditions. It allows for molecular size, the
viscosity of the solution and temperature.

Brownian movement. The fundamental event in diffusion consists of the
random jumping of molecules (Brownian movement), the jumps having
a mean length depending on the circumstances. If there is a concentration difference between two areas there is a slow drift of molecules to
even this out. Einstein treated this situation theoretically by considering
that the movement of each particle was a 'random walk' (or 'drunkard's
walk'—Gamow) from its starting place. Einstein considered the rate at
which this dispersion would occur and concluded that the time taken
would be proportional to the square of the distance; so double the
distance takes four times as long but half the distance only a quarter of
the time. We are used to a 'linear world' and expect that if we double the
number of steps in any direction we would cover twice the distance; if

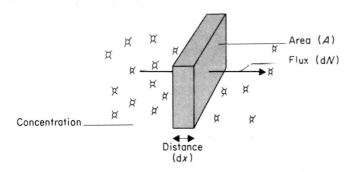

Fig. 1.8. Fick's law of diffusion illustrated. The flux per unit time increases with the concentration difference and area for diffusion and decreases with the distance for diffusion.

$$\frac{dN}{dt} = -DA \times \frac{dc}{dx}$$

where dN/dt is the amount of substance which moves per unit time (i.e. rate) passing through an area A under a gradient of dc/dx. D, the diffusion coefficient is negative as the movement is downhill; it depends on the size of the molecule and the medium in which it is moving and the temperature. This equation is similar to Ohm's law

$$I = V \times \frac{1}{r}$$

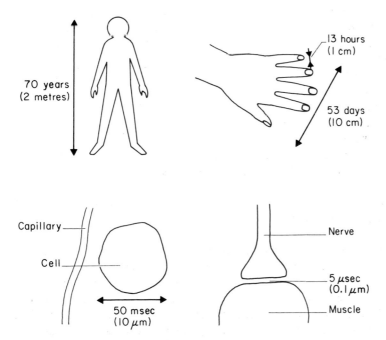

Fig. 1.9. *Diffusion times in water.* The times shown are for diffusion to proceed to a concentration of 99 percent of the starting value. Clearly diffusion is very adequate over cellular distances but is much too slow over bodily distances.

we lived in a 'random walk' world we would need to multiply the number of steps (and hence the time) by four to double the distance travelled, but conversely could cover half the original distance in a quarter of the number of steps or time. If actual diffusion times for typical distances are calculated (Fig. 1.9) it is clear that diffusion over cellular distances is fast but that over even a distance of 10 cm is impossibly slow. Diffusion is very adequate for short distances but totally inadequate for distances much greater than the diameter of a cell. This circumstance has probably led to body cells being about 10 μm in diameter and has determined the density of capillaries in tissues. The rates of diffusion therefore have important consequences for the design of biological systems.

Diffusion in gases is much faster than in liquids, as we all know (e.g. a perfume soon pervades a room but dyes only move slowly through water). The reason for this is that gas molecules are much less tightly packed than liquid molecules so that the size of the random movements in a gas (mean free path) is greater than the intermolecular distance of the molecules, but in a liquid it is less than this distance. The mean free path is therefore much greater in a gas than in a liquid.

Osmosis

Living systems are characterised by the presence of membranes (Fig. 1.10). Such membranes often alter diffusion by slowing down the move-

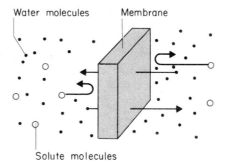

Water molecules   Membrane

Solute molecules

Fig. 1.10. Osmotic pressure created by the presence of a membrane which allows solvent molecules to pass but not the solute molecules (red).

ment of solute molecules more than that of solvent (usually water) molecules. This leads to a greater concentration of solute molecules on one side of the membrane than on the other; alternatively we can consider that the water on one side of the membrane is diluted with foreign molecules to a greater extent than the other side. This deficit of water means that a water movement occurs across the membrane to equalise the water concentration. The pressure built up by this movement is the osmotic pressure. The nature of the solute molecules is unimportant, merely their number.

The capillary wall

Substances of molecular weights up to about 70,000 can cross the capillary wall. They do so by moving through tortuous channels between the cells. These channels behave as water-filled pores some 10 nm in diameter occupying about 0·1 percent of the total available area. The spaces between the cells can best be visualised as that present in tangled undergrowth rather than a regular series of straight pores (Fig. 1.11).

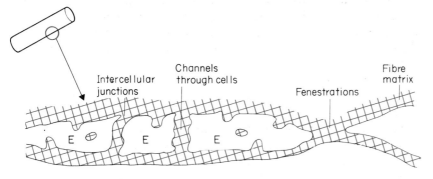

Fig. 1.11. The fibre matrix model of capillary permeability. The sizes and number of the various channels shown fix the area available for permeation but the molecular sieving properties are determined by the interstices of the fibre matrix which is present on the cell surface and in the channels, (E = cytoplasm) (after Michel).

Small molecules such as Na^+, Cl^-, glucose, amino acids and the hormones can cross the capillary wall readily; only slowed because the surface area available for diffusion is restricted. Large molecules, mainly the albumin of the plasma proteins, are held back and so lead to an osmotic pressure difference between the plasma and the interstitial fluid. The total size of this osmotic pressure—the oncotic pressure—is about 1·5 mosmol equivalent to a hydrostatic pressure of about 25 mmHg. This pressure, though very important, is small compared to the total osmotic pressure which would arise if the capillary wall was an ideal semipermeable membrane (i.e. also impermeable to the ions), say 300 mosmol. Starling proposed that this oncotic pressure was responsible for balancing the blood pressure in the capillaries and so determining the water distribution between the circulatory system and the interstitial fluid (see p. 19).

Calculation shows that only about 16 mmHg of the oncotic pressure can be accounted for by the direct presence of the plasma proteins; the other 9 mm arise because of an uneven distribution of diffusible ions (mainly Na^+ and Cl^-) across the capillary wall due to the negative charge on the plasma proteins. This is the Gibbs–Donnan effect.

Gibbs–Donnan distribution

Most students find the concepts involved in this distribution rather difficult. As it is mainly of importance in this section and perhaps in red cells we think that students should start by knowing the consequences of this distribution and (if they wish) progress to its derivation later (given for example in Robinson).

A Gibbs-Donnan distribution arises (in biology) when water containing charged protein molecules and salts is separated from one containing only salts, by a membrane which allows the salts to cross easily but which is impermeable to the protein. The consequence (Fig. 1.12) is that the diffusable ions take up a distribution such that: (a) a potential occurs across the membrane, so that (at body pH) the protein side becomes negative; (b) there are more diffusible ions on the protein side of the membrane than on the other side and so the osmotic difference between the sides is due partly to the protein, and partly to the excess of diffusible ions on the protein side; and (c) the product of the diffusible ions on each side of the membrane is equal.

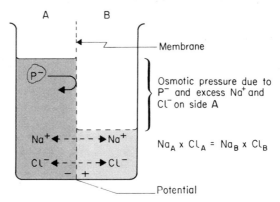

Fig. 1.12. Gibbs–Donnan distribution. Normal situation with a membrane permeable to small ions (Na^+, Cl^-, etc.) but impermeable to P^-, with sodium proteinate and NaCl present.

Capillary wall exchanges

Capillaries must have a pressure within them to drive blood onwards to the veins and yet also must be leaky to allow exchanges to occur with their surroundings. Capillary perfusion is therefore analogous to having leaky pipes in a central heating system. How is the fluid retained within them? The opposing force which does this is the oncotic pressure of the plasma proteins; this exactly balances the blood pressure and so stops any net fluid loss from the capillaries (Fig. 1.13). They are not, however, equally balanced all along a capillary; the blood pressure varies, in a typical capillary, from about 32 mm at the arterial end through 25 mm in the middle to 12 mm at the venous end, whereas the oncotic pressure is relatively constant at 25 mm. So water leaves the capillary at the arterial end and returns at the venous end (Fig. 1.13). About 20 litres of water circulate in this way every day in our bodies through the capillary walls—say, the area of two tennis courts—with 10 percent of it forming lymph (see p. 87). This bulk flow of water with contained solutes controls the volume distribution between the blood and the rest of the extracellular fluid. It has superimposed on it a diffusional flow of water and solutes due to the random thermal movement of the molecules; this latter flow equalises the distribution of these solutes and is much faster than the bulk flow. Thus bulk flow causes the interstitial fluid to be exchanged once per day whereas diffusional flow causes the water to exchange in 4 seconds, the glucose every 20 seconds and the O_2 and

CO_2 every few seconds. Bulk flow controls the volume distribution of water, diffusional flow evens out the concentration differences of the substances dissolved in the water and so is tailored to the requirements of individual substances such as O_2, CO_2, glucose, amino acids, etc.

Fig. 1.13 is oversimplified both as to the forces involved in bulk flow and the situation in real capillaries:

1   Other forces involved are tissue pressure which, if large, causes back flow into the capillary; protein does leave capillaries slowly—about one-third of the total per day—and so there is a tissue oncotic pressure as well as a blood oncotic pressure. In the lungs there is also a surface-tension effect due to the air–fluid interface (see p. 182). These forces are

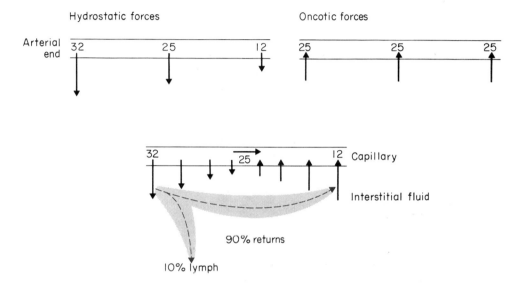

Fig. 1.13. The forces driving water across a capillary wall (Starling's hypothesis).

important in understanding fluid distribution in certain areas of the body (lungs, kidney). Posture affects fluid distribution in that feet swell during the day when standing upright due to increased blood pressure in feet, and shrink at night again.

2   Real capillaries, as opposed to the ideal capillary shown, are perfused intermittently so that in any one capillary there are periods when exudation occurs all along it; this is followed by reabsorption along it. This intermittent perfusion is due to the action of the precapillary sphincters (p. 114). The whole system obeys Starling's hypothesis but an individual capillary does not do so at any one time.

The system can be seen to be self-regulating in that if there is some loss of blood then there is a drop in the mean pressure in the capillary bed (partly directly and partly because the arterioles contract) and so fluid is reabsorbed from the interstitial space to expand the plasma volume. This is the explanation for the haemodilution seen after haemorrhage. The oedema associated with a low protein diet may be a consequence of low plasma protein concentration.

CELL MEMBRANE EXCHANGES

The membranes of animal cells are a much more substantial and selective barrier to the passage of substances than the capillary wall, but all cell membranes do not have the same properties. Their permeability is adjusted to their function, for example, erythrocytes have a high permeability to water and Cl^- and a low one to Na^+ and K^+, whereas nerve and muscle have a high K^+ permeability (Fig. 1.14).

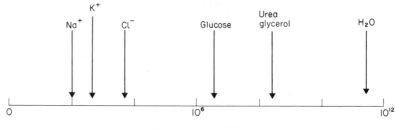

Relative permeability

Fig. 1.14. Relative permeabilities of a lipid bilayer membrane to various common substances. Biological membranes contain embedded proteins to increase the permeability to Na^+, K^+, Cl^- etc.

Note log scale. The permeability of water is around 10^{-2} cm/s (after Stryer).

Mechanisms by which substances cross the cell membrane

Cell membranes consist of a fatty sheet containing narrow, water-filled pores and specialised 'transport molecules'. Substances may enter or leave cells by any or all of these routes (Fig. 1.15). (In addition, substances may cross by membrane-bound vesicles containing extra- or intracellular fluid. This will not be considered further here.)

By solution in the membrane material

Molecules which are readily soluble in fat (i.e. have a high fat/water partition coefficient) enter cells by dissolving in the lipid of the membrane, diffusing within the membrane and then entering the intracellular fluid from the inside of the membrane. The rate at which such substances cross cell membranes depends on this coefficient, a relationship which holds good over a very wide range of molecular sizes. The respiratory gases, some anaesthetics and some drugs enter cells in this way.

Diffusion in water-filled pores

Small, non-fat soluble molecules enter cells by water-filled pores in the membrane. These pores are believed to be within a protein molecule and bear a net positive charge, for they allow the passage of small anions more readily than small cations (because of attraction rather than repulsion). The main factors determining the rate of entry of any particular molecule is the hydrated size and the charge on the particle. Water, urea, Cl^-, Na^+, K^+* and other small molecules enter cells in this way.

* A bare Na^+ is smaller than a bare K^+, but the hydrated Na^+ is larger than a hydrated K^+; this perhaps explains why the Na^+ permeability is generally lower than the K^+ permeability.

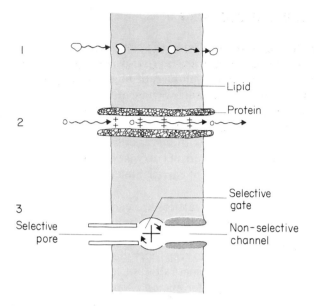

Fig. 1.15. Mechanisms by which substances cross cell membranes (1) dissolving in membrane substance; (2) via water-filled pores, perhaps within a protein molecule; (3) by interaction of the substance with specific molecules in the membrane. Current views suggest that these membrane components span the membrane and consist of various parts. The model shown consists of a selective outer pore which leads to a selective gate and then to a non-selective channel. In active transport it can be imagined that the gate is energised, thus applying a force to the particle.

It will be noticed that diffusion is important in all cases, either driven by a concentration gradient of the substance or substance–carrier complex. Carrier molecules operate about 30 times per second allowing 30 or so molecules to pass per second whereas 10^6 molecules enter a pore per second.

These pores can be imagined as long straight tunnels some 1 nm (10 Å) wide traversing the 10 nm of the membrane and occupying a small fraction of it; they cannot be seen in electron micrographs. Because they are so narrow the water in them is partly immobilised, or organised, by hydrogen bonding to the protein walls—as occurs at any such interface—so that it has properties more like ice than liquid water.

Temporary combination with a component of the membrane
Many molecules enter cells by a temporary combination with some membrane component; the formation of this complex is relatively slow but highly specific. Substances entering cells by this 'carrier-mediated' method are often identified because their transport into cells shows characteristics which are quite unlike those expected from the two methods just described. These characteristics are discussed below in some detail but in general they are features which one would not expect from a process of simple diffusion in a fatty layer or a water-filled pore (e.g. the rapid entry of large molecules, such as sugars or amino acids, which are not lipid-soluble). These carrier molecules are hypothetical so far because there are generally too few of them (hundreds to thousands) present in the membrane of a cell to be isolated and analysed chemically, so that they are defined in terms of their properties rather than their

structure. They can be thought of as large, complex lipoprotein molecules embedded in the membrane, able to combine specifically with the much smaller molecules entering the cell. A simple analogy is to compare carriers to turnstiles at sports grounds, there are a finite number of such turnstiles and a maximum rate of working of each. Most are 'specialised' in that they only allow entry of certain types of ticket holder.

These carrier molecules show some or all of the following characteristics:

1 They are saturable, i.e. with increasing concentration of transporting substance a maximum rate of entry is reached; this is because there are only a certain number of carrier molecules per cell and each can only work at a certain rate.

2 High specificity—many carriers are very specific for the substances they carry, e.g. Na^+ rather than K^+, one amino acid rather than another.

3 Competition and inhibition—similar molecules can compete for the same transport system and so keep other molecules out. If the competing molecule occupies the transport site but immobilises the carrier then specific inhibition occurs—for example, ouabain (see p. 24). Imagine a very fat man trying to get through a turnstile, he might stick and so immobilise it.

4 Usually such systems are very sensitive to temperature change, i.e. changing the temperature by 10°C changes the rate of uptake by a factor of 2–3 rather than the 1·3 or so one expects for diffusional processes. This has been taken to mean that the process requires a substantial energy of activation, as with enzymes.

For those familiar with the kinetics of chemical reactions it will be noticed that these characteristics are very similar to the combination of substrate molecules with the specific chemical groups of enzymes. Similar theories cover both cases.

Carrier-mediated transport may be divided into two main classes, passive and active. In passive transport, substances are carried down energy gradients, in active transport they are carried 'uphill' so that concentration occurs.

Active transport

In the economy of living cells many substances must be accumulated within them or kept out of them, often against an energy gradient. Thus it has been known for 100 years that the intracellular K^+ concentration is much higher than the extracellular K^+ concentration whereas the intracellular Na^+ concentration is lower than that outside. It was shown that the maintenance of this state depended on metabolism, for stopping metabolism led to these gradients running down and restoration of metabolism to their recovery.

So the first essential characteristic of active transport is that it requires a continuous supply of energy from metabolism, either from ATP or some other source. If this energy supply is removed, then active transport stops and in general the direction of the net movement of substances is reversed.

The second characteristic is that net transport must occur 'uphill', i.e. in the direction opposite to that which would occur spontaneously from the other driving forces. For uncharged molecules this means that active transport occurs against a concentration gradient; for charged molecules it is more difficult because ions are affected both by concentration gradients and by the electrical gradient across the membrane. These two gradients are combined in the electrochemical potential and active transport occurs 'uphill' against this.

In cells with a high membrane potential (nerve and muscle) there is a steep gradient inwards for Na^+ and a feeble outwards gradient for K^+ (Fig. 1.16). In cells with a low membrane potential (for example erythrocytes and most other body cells) the inward Na^+ gradient is somewhat less and the outward K^+ gradient somewhat greater than in muscle and nerve. The action of the 'sodium pump' balances this passive movement and so keeps the ionic concentration of cells in a steady state.

The other characteristics of active transport are very similar to those for all carrier-mediated transport and provide evidence for the current view that active-transport systems are carrier-mediated.

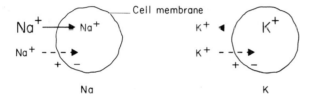

Fig. 1.16. The chemical and electrical forces acting on passive Na^+ and K^+ movements in a cell with a high membrane potential.

Both forces cause the cell to gain Na^+, giving a large inwards movement; the forces on K^+ are almost balanced giving a feeble outwards movement.
→ indicates the chemical gradient and −−→ the electrical gradient.

The sodium pump

It is now known that all the cellular Na^+ and K^+ turns over continuously, with Na^+ leaking into the cell all the time and being pumped out by an active process (the sodium pump); the situation is analogous to a leaky boat, water enters continuously and must be continually pumped out to stop the boat being swamped.

The current view of this pump is that it is a large phospholipid molecule in the surface membrane of the cell, which has ATPase (ability to split ATP) properties (Fig. 1.17). The pump molecule is thought to be a double molecule, each half consisting of two large fractions, one about 40,000 daltons, the other 90,000 daltons. At rest, it has a Na^+ binding site and a separate ATP binding site facing the interior of the cell and a K^+ binding site facing outwards. When all these sites are occupied the molecule undergoes a transformation which results in the ATP being split to release ADP and P_i into the cell, the Na^+ being released to the outside of the cell and the K^+ being released to the inside of the cell. The molecule then resumes its resting state. In erythrocytes under normal conditions two K^+ ions are transported inwards for each three Na^+ ions transported outwards and one ATP molecule is split but it seems likely that this ratio can be varied in different circumstances. It also seems possible that such pumps can produce a potential directly—the electro-

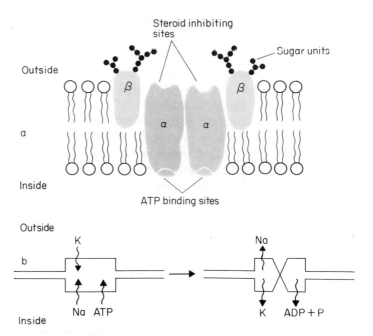

Fig. 1.17. (a) Schematic diagram of the subunit structure and orientation of the sodium pump. (b) the simplest model of its working Na$^+$, K$^+$ and ATP bind randomly, the pump molecule then 'flips over' to release Na$^+$ externally, K$^+$, ADP and P$_i$ internally. The energy for the molecular rearrangement comes from the splitting of the ATP.

The pump molecules are very large and so are unlikely to move readily in the membrane. Perhaps only part of them move, as suggested in Fig. 1.15 ([a] after Stryer).

genic potential—by the asymmetric transport of charge, in addition to the potentials produced indirectly as a consequence of the establishment of ionic gradients. The pump molecule also has another site on its external face which binds cardiac glycosides (e.g. ouabain); when this occurs the pump is unable to transport Na$^+$ or K$^+$. This may be the K$^+$ site, for ouabain and K$^+$ compete with each other.

The external potassium site has an affinity for K$^+$—K_m (see p. 448) of about 1 mM—such that it is normally saturated at the usual K$^+$ concentration (5 mM) found in extracellular fluids. The internal Na$^+$ site has an affinity for Na$^+$—K_m of 20 mM—such that it is less than half saturated at the normal Na$^+$ concentration (10 mM) found in intracellular fluids. This means that as internal Na$^+$ concentrations rise, the pumping rate increases greatly. A normal diameter (10 μm) body cell probably has about a million sodium pumps in its membrane, and each pump works about 30 times per second. Recent evidence suggests that the actual number of sodium pumps per cell is variable and can be adjusted to meet different conditions. Cells use a lot of energy to fuel their sodium pumps; sometimes as much as 30 percent of the total.

Consequences of having sodium pumps. All cells have sodium pumps at their membranes which keep their intracellular Na$^+$ lower and the intracellular K$^+$ higher than the surrounding medium. Several consequences follow from this:

1   Cells have a higher concentration of proteins and other large molecules than interstitial fluid and so tend to gain water; the action of the sodium pump keeps Na$^+$ out of the cell and so balances these macromolecules. If the sodium pump is stopped, cells (eventually) swell and burst.

2   Enzymes within cells have evolved which require K$^+$ and are inhibited by Na$^+$.

3   High intracellular K$^+$ and low intracellular Na$^+$ provide the basis for the electrical activity characteristic of excitable cells.

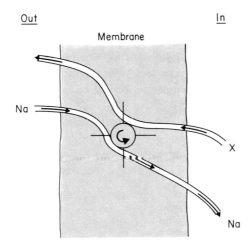

Fig. 1.18. Using the energy in the sodium gradient to drive another substance 'uphill'.

Fig. 1.19. Cartoon of receptor-mediated endocytosis. The substance binds to a surface receptor (R), several receptors clump together in a 'coated pit' which then enters the cell and fuses with a lysosome, where the substance is detached (after Chuck, 1980).

4   The inward Na$^+$ gradient is used as an energy source to drive other active transport processes, for example, amino-acid uptake, Ca^{2+} extrusion, etc. If the normal Na$^+$ leak into a cell is likened to that of water rushing into a leaking boat then this mode could be imagined as a waterwheel, driven by the leak, being used to do other work (Fig. 1.18).

5   Epithelial cells are often polarised by having more sodium pumps on one face than on the other. Sodium is then transported across the sheet of cells, followed passively by Cl$^-$ and water. This is the basis of fluid transport in kidney, frog skin, gut, bladder, etc. (Fig. 6.16, p. 162).

Receptor mediated endocytosis. This mechanism solves a number of cellular problems in a neat way. The problems are:

1   the uptake of large molecules by the cell through the plasma membrane, which as we have seen is very impermeable;

2   recovering membrane inserted during exocytosis, e.g. at the neuro-muscular junction (p. 38);

3   'cleaning-up receptors' which have various substances bound to them.

In the last few years a new mechanism has been described which does all these things (Fig. 1.19). This process requires energy at the internalisation step at least.

Chapter 2
Nerve and
Synapse

Most cells maintain an electrical potential difference across their plasma membranes, with the inside negative to the outside at rest; in excitable cells such as nerve and muscle this potential is -70 to $-90\,$mV, in non-excitable cells such as erythrocyte and liver, it is -10 to $-30\,$mV. In excitable cells this negative resting potential shows large transient changes during activity; changes not shown by non-excitable cells. These transient changes—the action potentials—last less than a millisecond (nerve and muscle) to a few hundred milliseconds (heart muscle) and cause the potential across the plasma membrane to become about $+30\,$mV—the overshoot. The current view (discussed below) is that these potentials arise as a consequence of the uneven distribution of ions across the plasma membrane, together with the permeability of the cell membrane to these ions.

Resting potential

The *resting potential* is thought to be a type of diffusion potential, mainly due to K^+ moving out of the cell down its large concentration gradient; whereas the *action potential* is a similar diffusion potential mainly set up by the diffusion of Na^+ (Fig. 2.1).

If a cell is perfectly permeable to K^+ and impermeable to all other ions it would have a membrane potential equal to E_K (i.e. $-100\,$mV) if perfectly permeable to Na^+ and impermeable to all other ions it would have a membrane potential equal to E_{Na} (i.e. $+50\,$mV). Real cells are somewhat permeable both to Na^+ and K^+ so that the potential across their membranes must lie somewhere in between the equilibrium potentials of Na^+ and K^+ (Fig. 2.2). The exact value of the membrane potential will depend on the relative values of the sodium and potassium permeabilities (P_{Na} and P_K, respectively); if they are equal then the potential will lie midway between the equilibrium potentials of K^+ and Na^+; this is the situation in the erythrocyte. If the potassium permeability greatly exceeds the sodium permeability (i.e. $P_K > P_{Na}$) then the membrane potential is near the potassium equilibrium potential; this is the resting potential of nerve and muscle. If the sodium permeability exceeds the potassium permeability (i.e. $P_{Na} > P_K$) then the membrane potential is near the sodium equilibrium potential; this is the potential at the height of the action potential in nerve and muscle.

The action potential

The current view of the action potential is that it is due to a sequence of permeability changes to sodium and potassium. At the start of the action potential the sodium permeability increases about a thousandfold, caus-

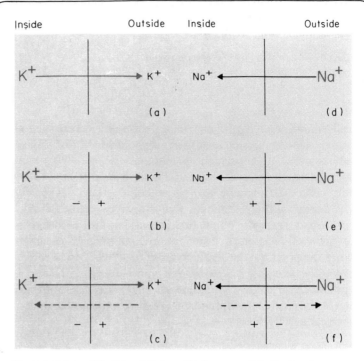

Inside   Outside   Inside   Outside

Fig. 2.1. Origin of the K^+ and Na^+ equilibrium potentials. During the development of the K^+ equilibrium potential it is supposed that it is the only ion present. The actual movement of charge necessary to cause these potentials is only some two millionths of a percent of the K^+ or Na^+ present and is therefore not detectable chemically.

The situation can best be understood if you consider a membrane, which is permeable only to K^+, and initially has a concentration gradient but no potential across it. Some K^+ diffuses out of the cell (a) carrying positive charges, these positive charges cause a small potential to build up across the membrane (b), this process continues (c) until the membrane potential is big enough to cause a reverse flow of K^+ equal to the K^+ flow down the concentration gradient. At this equilibrium position the K^+ flow driven by the concentration gradient equals that driven by the membrane potential; this potential is the equilibrium potential (E_K) for the K^+ ion. In normal animal cells this potential is about -100 mV. Similar potentials can be calculated for other ions across the cell membrane, assuming that only they can cross the membrane. For Na^+ (E_{Na}) this comes to about $+50$ mV.

ing the potential to swing towards a positive value; soon this change in P_{Na} starts to decrease (inactivation) so that the potential returns towards its resting value again (Fig. 2.3). In many tissues P_K increases during the action potential so that repolarisation is faster than it would otherwise be. The sodium and potassium permeability channels of the membrane can be visualised as water-filled pores crossing the plasma membrane with protein 'gates' able to swing across to close them off. On this basis the action potential is due to the sodium gates being opened for a millisecond or so, and then closed—perhaps by another molecule.

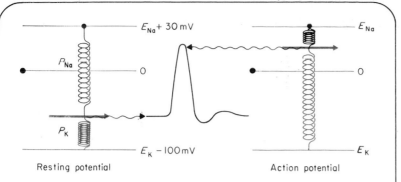

Fig. 2.2. Model of the membrane-potential mechanism. E_{Na} and E_K represent the extreme positions (i.e. voltages) which the pointer (i.e. the membrane potential) can take up. The actual position taken up depends on the relative strength of the springs (representing P_K and P_{Na}, the potassium and sodium permeabilities respectively). Normally the potassium spring is greater than the sodium spring (i.e. $P_K > P_{Na}$) and so the pointer sits near E_K; during the action potential the sodium spring greatly increases in strength, thus exceeding the potassium spring (i.e. $P_{Na} > P_K$) and so the pointer (i.e. E_M) swings towards E_{Na}. The sodium spring then declines in strength and the potassium spring increases somewhat so that the pointer returns towards E_K (i.e. repolarisation occurs).

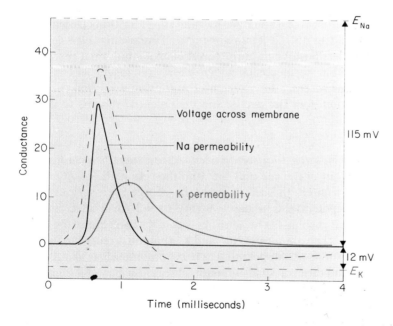

Fig. 2.3. Action potentials and permeability changes in nerve (after Hodgkin & Huxley 1952).

The inactivation of the sodium and potassium permeability changes are both time dependent. The conductance units are mmho/cm^2.

At the resting potential there are more potassium channels (p. 39) open than sodium channels; during the action potential many more sodium channels open. At rest there are more negative ions inside the cell than positive ions (though this difference is so small it cannot be measured chemically) which leads to a negative potential across the cell membrane.

Setting up an action potential. When the membrane potential of a resting cell is depolarised—making the potential across the cell membrane less—then there is a regenerative increase (positive feedback) in the sodium permeability. The relationship between depolarisation and increased sodium permeability is very steep so that small depolarisations lead to large increases in P_{Na}. This process can be thought of as an opening of the sodium gates by the potential across the membrane; the more positive the potential, the more gates open. In this context it should be appreciated that the normal electric field across a plasma membrane is very large so that small changes in potential might be expected to alter the orientation of gating molecules. (100 mV over a 10-nm membrane is equivalent to a million volts over 10 cm.) Action potentials are normally set up by a depolarisation of the membrane potential. When this depolarisation is large enough to exceed a threshold value (usually some 20 mV less than the resting potential) there is a large enough rise in the sodium permeability to allow Na^+ ions to enter faster than K^+ ions are leaving so that an action potential is set up. This action potential is of a standard size and shape for that cell and it either occurs or it does not, i.e. it is 'all-or-none'. If the resting cell is depolarised to a value less positive than the threshold potential, then there is a rise in the sodium permeability which allows an insufficient number of sodium ions to enter to overcome the continual K^+ leak so that the depolarisation does not persist and the potential returns to its resting value.

Part played by active transport in nerve. At rest there is a continual slow leak of K^+ out of the cell and Na^+ into the cell; with activity a further leak of Na^+ enters the cell to carry the charge necessary to alter the membrane potential from say -90 mV to $+30$ mV; after activity the membrane is repolarised by a further leak of K^+ to recharge the potential to -90 mV. The amount of Na^+ and K^+ that must cross the membrane to cause a single action potential in nerve and muscle is about 3 pmole/cm^2 of membrane; a small amount chemically but still representing the movement of 2×10^{12} ions per action potential. These large numbers of ions can enter through relatively few channels (say $2 \times 10^7/cm^2$, or $2/\mu m^2$) because each channel can carry about 1×10^6 ions/second. This extra leak of Na^+ and K^+ is small for a squid axon, with its large volume per surface area, but represents a change of about 1 percent of the K^+ in a 0.5 μm mammalian nerve fibre. Small mammalian cells can therefore only conduct a relatively few action potentials unless their 'ion batteries' are recharged by the sodium pump. When pressure is applied to one of our nerves—e.g. by the edge of a chair on our sciatic nerve—then our leg soon goes 'numb'. This is probably due to the ionic gradients running down, following stoppage of the blood supply.

Propagated action potential

So far, we have described the mechanism by which an action potential is set up at one part of a nerve or muscle cell membrane, a 'membrane action potential'. In the present section we describe the way in which this action potential is propagated within the boundaries of a single cell. This spread of an action potential may be likened to that of the flame of a gunpowder trail. If one end of the gunpowder is lit then the flame moves slowly down the trail. Essentially two processes happen, the heat from the flame at any one point warms the gunpowder ahead of it and then sets up a new flame. The two processes are thus conduction of the heat and then a regeneration of the heat by the flame. The propagation of an action potential in nerve is also essentially a two-stage process by which (1) current spreads from the region occupied by the action potential to the adjacent membrane, and (2) there the signal is boosted to full strength by a new action potential (Fig. 2.4).

The solutions inside and outside nerve and muscle cells contain ions and so can conduct electricity. When one part of the membrane of a cell becomes depolarised by an action potential, this potential 'leaks away' to the surrounding plasma membrane (Fig. 2.4a) due to current flowing

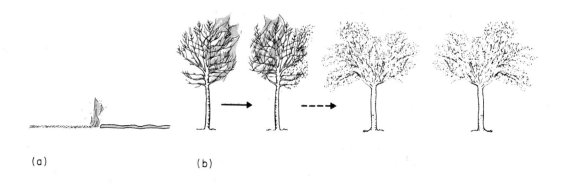

(a)   (b)

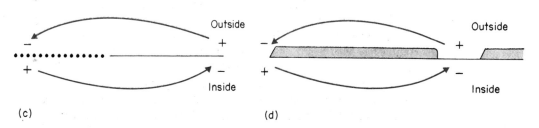

(c)   (d)

Fig. 2.4. Action potentials compared to the spread of fire, (a) along a gunpowder trail and (b) leaping from one tree to the next in a forest fire. Unmyelinated fibres (c) behave like a, and myelinated ones (d) like b.

inside and outside the cell. This flow of current depolarises the adjacent membrane of the cell and, when the threshold is reached, sets up an action potential. As the size of the action potential is some 120 mV there is no problem in causing the depolarisation of 20 mV in the adjacent membrane sufficient to start an action potential. The passive spread of current down an axon is decremental, i.e. the voltage induced decreases with distance. The action potential mechanism is like a relay mechanism which boosts the signal to full strength at each point on the membrane.

In unmyelinated axons and muscle cells this mechanism causes an action potential to sweep along the plasma membrane in all directions from the initiating point. The rate of this propagation—conduction velocity—is determined by the resistance of the axoplasm. Bigger axons have a larger cross-sectional area and hence a smaller longitudinal resistance, so that the conduction velocity of an axon increases with its diameter. For a large squid axon (1 mm) it is about 25 m/s but only a few metres per second for our small unmyelinated axons.

Large mammalian axons are covered with myelin, which consists of a wrapping of many layers of cell membrane, interrupted every few millimetres by the gaps of the nodes of Ranvier. The myelin insulates the axon from the extracellular fluid, which can only reach the plasma membrane at these nodes. It is found that propagation of the action potential occurs from one node to the next, so that action potentials are only set up at the nodes of Ranvier (Fig. 2.4b). The action potential therefore jumps from one node to the next, hence the name saltatory (jumping) conduction. Myelination confers two advantages on an axon compared to a bare axon of similar diameter:

1   The conduction velocity is greatly increased, for the 'jumping' process is relatively fast compared to the action potential mechanism as sodium entry is relatively slow. Thus 20-μm human axons have a conduction velocity of 100 m/s; a velocity only approached by the enormous (1 mm) unmyelinated axons of invertebrates.

2   The nerve impulse is less costly in terms of ion 'run down', for ion exchanges only occur at the nodes of Ranvier and not over the whole nerve membrane. Myelinated fibres probably use less than 1 percent of the equivalent energy of unmyelinated ones. By the process of myelination a mammal can have several hundred axons in the space of one squid axon, achieving high conduction velocities at low cost.

Quantitative aspects

Quantitatively the equilibrium potential for an ion is described by the Nernst equation which, at 37°C for K+, is

$$E_K = -61 \log \frac{[K^+]_1}{[K^+]_0} \, mV$$

The more complex situation which occurs in real cells can be described by the more complex formulae derived by Goldman and Hodgkin—the constant-field equation—which describes the membrane potential in terms of the relative sodium and potassium permeabilities:

$$E = -61 \log \frac{[K^+]_1 + b[Na^+]_1}{[K^+]_0 + b[Na^+]_0} \text{ mV}$$

where $b = P_{Na}/P_K$; if $b = 0$ then this becomes the Nernst equation.

In muscle or nerve at rest $b = 1/100$ whereas during activity it becomes about 10, due to a large increase in P_{Na}. Substitution of the appropriate values in these equations gives resting and action potentials close to these observed. Chloride is often (but not always, p. 238) passively distributed across the membrane so that

$$E_{Cl} = -61 \log \frac{[Cl^-]_0}{[Cl^-]_1} = \text{resting potential mV}$$

In skeletal muscle the chloride permeability is high—about twice that of potassium—so that chloride tends to clamp the potential at the resting potential. Once an action potential is set up, Cl⁻ carries a substantial part of the current required for repolarisation of the membrane. Measurements of absolute permeabilities of various excitable and non-excitable cells suggest that the resting values of the sodium permeabilities are similar in all cells but that the potassium permeability of excitable cells is some 100 times that of non-excitable cells; this accounts for the high resting potentials of such cells.

Compound action potential

Action potentials from a single axon recorded with extracellular electrodes show small (1 mV) potentials. If such a recording is made from one end of a nerve trunk and the other end is stimulated maximally, then a compound action potential is obtained (Fig. 2.5). This record is composed of the sum of the potentials produced in each nerve fibre within that trunk; as these nerve fibres are all of different sizes they all have different conduction velocities and so each action potential takes a different time to traverse the distance between the electrodes. The large nerve trunks conduct faster than the smaller ones, and so contribute to

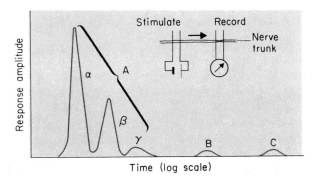

Fig. 2.5. Diagram of a compound action potential obtained from a nerve trunk arranged as shown in the inset. These peaks are used to classify nerve fibres as shown in Table 2.1. (After Ehrlanger & Gasser.)

the first peak of the compound action potential; the very small fibres (unmyelinated) conduct slowly and give the last peak, intermediate fibres give intermediate peaks.

Nerve fibres have been classified according to their conduction velocity and axon diameter. Table 2.1 shows one such classification of nerve fibres and lists the functions of the different nerve fibre types.

Table 2.1. Classification of nerve fibres

Fibre type	Conduction velocity (m/s)	Diameter (μm)	Function
A α	70–120	12–20	Proprioception, somatic motor.
β	30–70	5–12	Touch pressure.
γ	15–30	3–6	Motor to muscle spindles.
δ	12–30	2–5	Pain, temperature.
B	3–15	1–3	Preganglionic autonomic.
C	0·5–2·0	0·2–2	Postganglionic sympathetic, pain, and possibly heat, cold and pressure.

Sensory nerve fibres can also be classified by an alternative system based upon the type of sensory receptor to which they are connected (Table 2.2). Unfortunately, both systems of classification are frequently used together in descriptions which include both motor and sensory nerve fibres.

Table 2.2. Alternative classification for sensory nerves

Number	Origin	Equivalent
IA	Muscle spindle.	A α
IB	Golgi tendon organ.	A α
II	Muscle spindle; touch, pressure receptors.	A β and γ
III	Pain and temperature receptors.	A δ
IV	Pain and other receptors.	C (dorsal root)

CONNECTIONS BETWEEN CELLS

It is clear that cells in a tissue must be connected together mechanically; recently it has also become clear that the cells in many tissues are also connected together chemically.

These *mechanical connections* may conveniently be described by considering the epithelial cells which form the lining of the gut. Functionally, these cells transport substances from the lumen into the gut wall with a small amount of 'back leakage' between the cells. The intercell connections ensure this low leakage, as well as mechanical strength between cells (Fig. 2.6). Adjacent cells are connected together by: (1) a continuous belt-like region of intimate contact between adjacent plasma

membranes near their luminal surface. This band—the tight junction—
is formed by fusion of the outer layers of adjacent cell membranes with
some modification of the remaining membrane. It stops diffusion in the
cleft between the cells; (2) a series of spot connections—the desmosomes— connect the basal ends of the clefts together. These connections
are mechanically very strong, due to the presence of membrane deposits
and strengthening fibrils; they act as 'spot welds' holding the cells
together. They have no effects on diffusion.

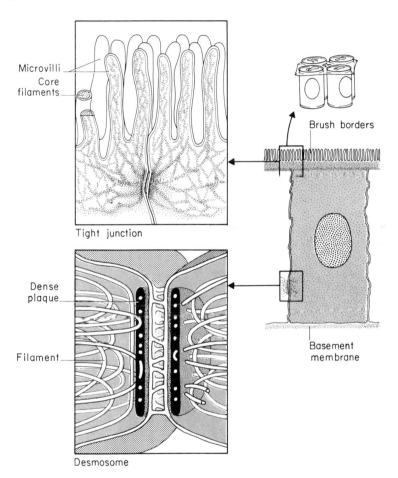

Microvilli
Core
filaments

Brush borders

Tight junction

Dense
plaque

Filament

Basement
membrane

Desmosome

Fig. 2.6. Diagram of the junctions between adjacent epithelial cells. The tight junction
encircles each cell (see inset showing beer can analogy) the desmosome is a spot
connection.

The tight junction stops or slows diffusion in the cleft between the cells. The
desmosomes have no effect on diffusion in the cleft and probably only hold the cells
together mechanically.

Types of epithelia

Epithelial sheets can be divided into 'tight' and 'loose' groups depending
on their electrical resistance. The tight group include frog skin, bladder
and distal tubules of the kidney with a resistance of about 1 KΩ/cm^2 and
the loose group intestine, gall-bladder and proximal tubule with a re-

sistance about 20 times less ($50 \, \Omega/cm^2$). Microscopically there is little to distinguish these groups of membranes, for the junctions look similar. The current view is that the areas of membrane fusion in these tight junctions ('kissing in the dark'—Diamond) are larger in the high resistance membranes than in the low.

Cytoplasmic connections occur between cells in most tissues (except skeletal muscle) by 'gap-junctions'. In these junctions the plasma membranes of two cells are separated by a gap of about 2 nm, with well-marked pores connecting the interior of one cell with that of the other (Fig. 2.7). These pores are arranged in clusters and are thought to

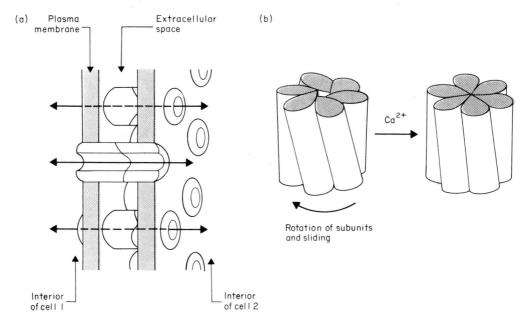

Fig. 2.7. (a) Schematic view of a gap junction; the cell membranes are separated by some 2 nm at the junction. (b) A model for the graded closure of gap junctions by Ca^{2+} (after Stryer).

provide an aqueous channel between the cells, allowing the passage of nucleotides and other molecules up to a molecular weight of about 500 between the cells. They form within a few minutes when cells come in contact; they probably allow about a million nucleotides to pass between cells per second. This means that most cells are not isolated from each other but can share all their small molecules with their neighbours; only large control molecules cannot be shared.

Gap junctions are the basis of electrical transmission between cells (pp. 58, 232), and may be involved in the processes of growth and differentiation of cells.

JUNCTIONAL AND SYNAPTIC TRANSMISSION

When an action potential is set up in a nerve cell it spreads throughout that cell; to reach another nerve or muscle cell it must cross a synapse or junction. These junctions generally function by releasing a chemical substance from one cell, which then diffuses across to the next cell where it combines with receptor molecules and causes a characteristic response in that second cell. Such synaptic transmission is in one direction only and so makes transmission in the nervous system unidirectional. The simplest and most studied junction is the neuromuscular junction, so this will be considered first.

Transmission from nerve to muscle

The function of the neuromuscular junction is to transfer nerve impulses from the small motor neurones to the large muscle fibre and cause it to contract. At the normal neuromuscular junction each nerve action potential is always followed by a muscle action potential, which then sweeps rapidly down the muscle fibre to cause a synchronous contraction of that fibre. The vertebrate neuromuscular junction is thus a kind of relay which simply transmits this impulse.

In vertebrate neuromuscular junctions each myelinated nerve fibre gives off an end bush of terminal non-myelinated branches, of some 1 μm in diameter, which then run in shallow grooves on the surface of the muscle cell. All along this groove, of some 100 μm, a synaptic contact is made between the nerve and the muscle. The surface area of this contact is greatly increased by the presence of numerous folds in the muscle fibre (Fig. 2.8).

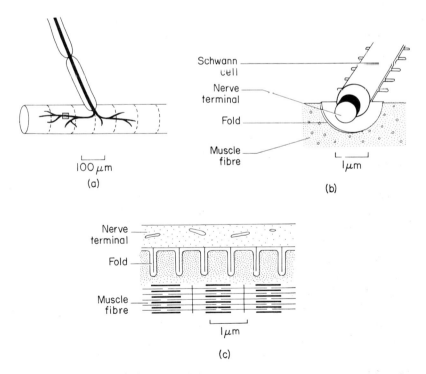

Fig. 2.8. Diagram of neuromuscular junction of the frog: (b) and (c) show two enlarged views of the marked area of the nerve fibre shown in (a). The nerve terminal is the presynaptic site containing vesicles with ACh. The postsynaptic site is the muscle membrane containing receptors which combine with the ACh released (after Katz).

Electrical recording from such synapses shows that there is a delay of some 0·5 milliseconds between the nerve impulse reaching the neuromuscular junction and the muscle action potential being set up. This delay is due to the release presynaptically of a chemical, acetylcholine (ACh), which then takes some time to diffuse across the gap to affect the postsynaptic membrane. To establish this it has been shown (a) that ACh is released from the presynaptic site, (b) that its action is at the postsynaptic site, and (c) that sufficient ACh is released to initiate a muscle action potential.

The presynaptic terminals contain ACh, which is thought to be stored in vesicles. On the arrival of a nerve impulse the nerve terminal is depolarised, Ca^{2+} enters and somehow causes the fusion of these vesicles with the nerve membrane. The vesicles then rupture to discharge their contained ACh into the cleft of the neuromuscular junction. Upon reaching the postsynaptic site the ACh molecules cause a depolarisation of the muscle membrane—the end-plate potential—due to a combina-

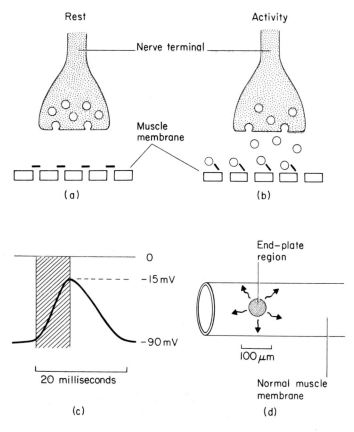

Fig. 2.9. Diagram to show release of ACh, its action on the muscle membrane and subsequent development of an end-plate potential (a, b, c). The end-plate potential set up (d) at the neuromuscular junction then initiates an action potential in the surrounding muscle membrane.

End-plate potentials normally have action potentials superimposed on them unless the latter are blocked, as here. The shaded area in (c) represents the time during which the ion channels remain open during an end-plate potential; the prolongation of the potential is due to the electrical properties of the membrane (a and b partly after Paul).

tion with specific receptor molecules on that membrane. Subsequently, the ACh is hydrolysed to choline by the enzyme cholinesterase found on the postsynaptic membrane. This choline is reabsorbed by the nerve terminals and is used to reform acetylcholine via synthetic pathways in the vesicles.

Vesicles are made in the cell body of the presynaptic axons and travel down the axon to reach the neuromuscular junction; they appear to be refilled several times before degradation.

Single vesicles are being released randomly every second or so at each neuromuscular junction, the packets (or quanta) of ACh so released giving small depolarisations of the muscle cell—the miniature end-plate potentials (MEPPs). Upon the arrival of an action potential at the junction there is a large increase in the number of quanta released, giving the end-plate potential. At present it is thought that the spontaneous release of quanta keep the muscle cells 'normal' by a mechanism which is not understood.

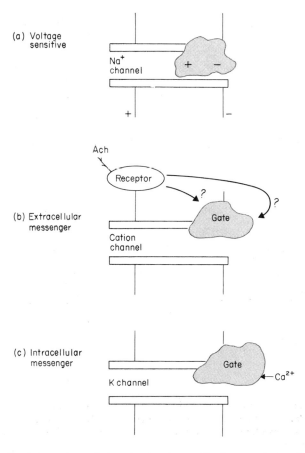

Fig. 2.10. Gated channels can be imagined as shown. The gates can be acted on by (a) voltage changes, (b) extracellular messages via receptors or (c) intracellular messages such as cAMP, cGMP or Ca^{2+}. Anatomically, the first pictures obtained show that channels are composed of 7 proteins packed as in Fig. 2.7b.

End-plate potential

The end-plate potential, in normal muscle, is only set up in the region of the neuromuscular junction (Fig. 2.9). It is thought to be due to the formation of leaky ionic channels, whose protein lining retains fixed negative charges, but which does not discriminate between small cations. So the action of ACh on the postsynaptic membrane is to make it highly leaky to Na^+, K^+ and Ca^{2+}. The potential therefore moves towards a mean value of about $-15\,mV$, determined by the permeability and concentration of these ions. Normally, of course, this end-plate potential acts as a trigger to set up action potentials in the muscle membrane adjacent to the end plate region; these action potentials then sweep away over the muscle cell membrane.

The *fundamental event* in setting up a postsynaptic potential is the opening up of a single conducting channel following the combination of an ACh molecule with a receptor molecule (Fig. 2.10b); this leads to a small leakage of ions with a small depolarisation of the end-plate region. Each end-plate potential consists of the sum of millions of such events. Recently it has been realised that such fundamental events can be analysed by suitable techniques.* These results suggest that each channel opens for 1 millisecond, allows about 40,000 ions to pass through, and produces a depolarisation of $0{\cdot}3\,\mu V$. Thus the presynaptic membrane is characterised by the presence of vesicles whereas the postsynaptic membrane is characterised by these single channel voltages (noise). A miniature end-plate potential contains about 1000 such fundamental events.

* There are three kinds of voltages found at end plates: the EPP (75 mV), the MEPP (1 mV) and 'noise' ($< 1\,\mu V$).

Chapter 3
Muscle

MUSCLE TISSUES

Structure of striated muscle

The common feature of all muscle tissue is its ability to convert the chemical energy of ATP into mechanical work. Striated muscle gets its name from the well-developed cross-striations which arise from the arrangement of contractile proteins within the muscle cells (see later sections). It differs from cardiac and smooth muscle in that it is under 'voluntary' control and is innervated by the somatic nervous system. Each muscle in the body is made up of anything from a few hundred to several hundred thousand individual muscle fibres (Fig. 3.1). In many muscles, the fibres insert directly into tendons attached to bones. However, in other muscles, particularly large ones, the arrangement of individual fibres is more complex. In these, the connective-tissue sheath which surrounds and invades the muscle also provides points of insertion for the fibres. In this way fibres need only be a few centimetres long even in very long muscles.

Each muscle fibre is a multinucleate cell which was formed during embryonic development from the fusion of a large number of mono nuclear myoblasts. In addition to containing the normal constituents of cells, such as mitochondria, ribosomes, etc., muscle cells contain specialised contractile filaments together with an elaborate internal membrane system which functions in the control of contraction. These contractile filaments are contained in myofibrils $1-2\ \mu m$ in diameter running the whole length of the muscle fibre (Fig. 3.1). Each muscle fibre contains several hundred myofibrils. The myofibrils contain around 60 percent of the protein in muscle cells. The localisation and composition of individual myofibrillar proteins are given in Table 3.1 and Figs. 3.1 and 3.2.

Table 3.1. Composition of proteins in the myofibril. The various contractile proteins are organised within the myofibril (as in Fig. 3.2).

Protein	Function	Percentage myofibril by weight
Myosin	Structural/contractile	54–60
Actin	Structural/contractile	20
Tropomyosin	Structural/regulatory	4·5
Troponins	Regulatory	3–5
α-Actinin	Structural	2
β-Actinin	Structural	0·5

The structure of the myofibril

Light microscopy shows that each myofibril has a characteristic banded pattern which aligns with bands on other myofibrils to give the whole fibre its characteristic striated appearance (Fig. 3.1). In the electron

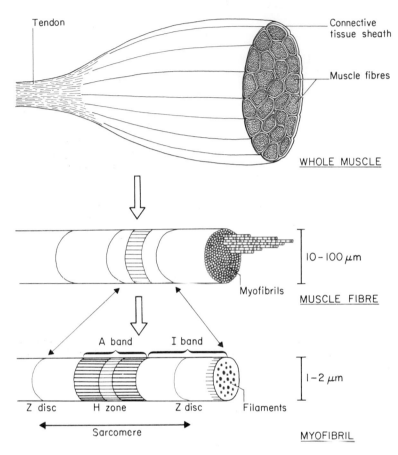

Fig. 3.1. Diagram of the anatomical arrangement of the components of muscle. The I band only contains actin and so gives a simple response to polarised light (*Isotropic*), the A band contains both actin and myosin and gives a complex response (*Anisotropic*).

microscope the myofibril can be seen to consist of a series of interdigitating thick and thin filaments (Fig. 3.2). The thin filaments are arranged in a lattice on either side of a dense structure called the Z line or Z disc. The Z line is characteristic of skeletal muscle (Fig. 3.3).

The thin filaments insert into the Z line and extend through the I band into the A band. The highly refractile material recognised as the A band in the light microscope corresponds to the position of the thick filaments (Figs. 3.1 and 3.2). The H band corresponds to a region of the thick filaments where there is no overlap between the thick and thin filaments (Fig. 3.2). In vertebrates the thick filaments in the H band are arranged in a characteristic hexagonal pattern. Thin filaments are sometimes called I filaments and thick filaments A filaments because of their correspondence to the I and A band respectively.

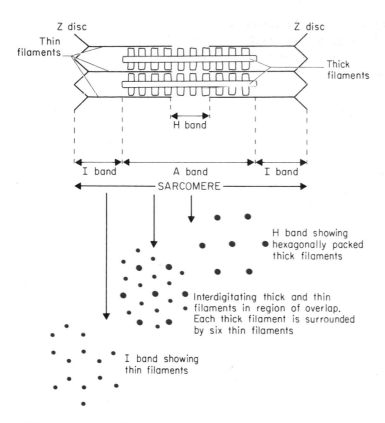

Fig. 3.2. *Top:* Arrangement of thick and thin filaments in each sarcomere gives rise to the banded pattern of striated muscle. *Below:* Section through I band, H-zone and overlap region in A band showing array of filaments.

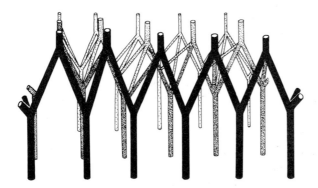

Fig. 3.3. Model showing the basketweave structure of the Z line into which are inserted the thin filaments. The protein α-actinin is located in the Z line region.

Structure and composition of the thin filaments

The thin filament mainly contains actin; each thin filament is composed of two strands of actin (f-actin) arranged in a right-handed helix, together with proteins which control the contraction of the muscle (see subsequent sections). Each actin filament is in turn composed of many actin monomers (G-actin) joined together (Fig. 3.4).

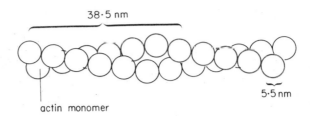

38·5 nm

5·5 nm

actin monomer

Fig. 3.4. The backbone structure of the thin filaments of muscle. Each actin filament (f-actin) is made up of some 360 globular G-actin molecules, each with a molecular weight of about 47,000 and containing a site for interaction with myosin. A tropomyosin filament lies in each groove between the strands of the actin molecules (not shown).

Structure and composition of the thick filaments

The main constituent of the thick filament is myosin, which is a high molecular weight protein (480,000–500,000) large enough to be visible in the electron microscope as a rod-like structure with a globular head at one end (Fig. 3.5). Myosin is interesting in that it is both a structural protein and an enzyme. A great deal of information about the structure of myosin has been obtained from the study of its digestion by proteolytic enzymes such as trypsin and papain. The names given to the various proteolytic subfragments of myosin are illustrated in Fig. 3.5.

The myosin molecules are thought to be packed with their tails located

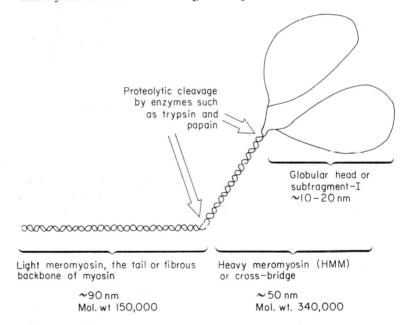

Proteolytic cleavage
by enzymes such
as trypsin and
papain

Globular head or
subfragment–I
~10–20 nm

Light meromyosin, the tail or fibrous
backbone of myosin

~90 nm
Mol. wt 150,000

Heavy meromyosin (HMM)
or cross–bridge

~50 nm
Mol. wt. 340,000

Fig. 3.5. Proteolytic subfragments of myosin. Myosin consists of two polypeptide chains which are wound together in a double-stranded helix in the long tail of the molecule. Each globular head consists of two halves, each of which contains the site for binding to actin and the site which splits ATP.

Each myosin head contains two low molecular weight polypeptide light-chains. Muscles with different contraction speeds differ with respect to the structure of both heavy and light chains. One class of light chain can be phosphorylated *in vivo*, and is involved with controlling contraction in smooth but not skeletal muscles.

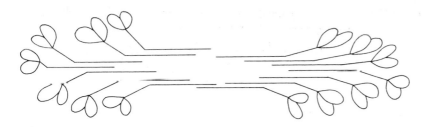

Fig. 3.6. Packing of myosin molecules into a single thick filament. Note the structural polarisation of the myosin heads away from the centre of the filament (see subsequent section on sliding filament theory). There are about 200 cross-bridges per thick filament.

along the core of the filament and their heads (which are thought to form cross-bridges between the myosin and actin filaments) pointed out and orientated away from the midpoint of the filament (Fig. 3.6). The current view is that myosin molecules are arranged in pairs on either side of the myosin filament and that successive pairs are staggered by 143 Å and rotated relative to each other so that the projecting cross-bridges fall on a 429 Å helix (Fig. 3.7).

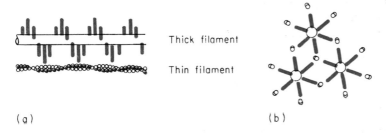

Thick filament

Thin filament

(a)   (b)

Fig. 3.7. (a) Structure of thick filament deduced from X-ray diffraction experiments. (b) Diagram showing relation of thick and thin filaments.

Sliding filament theory of contraction
According to the sliding filament theory of muscle contraction, force is developed by the cross-bridges in the region of overlap between the thick and thin filaments. Cycling of the cross-bridges results in the sliding of the filaments past one another and the shortening of the

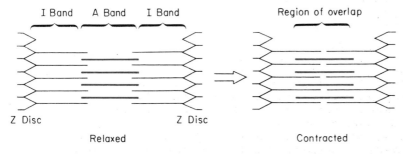

I Band   A Band   I Band   Region of overlap

Z Disc   Z Disc

Relaxed   Contracted

Fig. 3.8. Diagrammatic representation of the sliding filament theory showing sarcomere shortening. Note that the filaments themselves remain the same length.

muscle (Fig. 3.9). Movement of cross-bridges is thought to be asynchronous so as to prevent slippage of the filaments under tension.

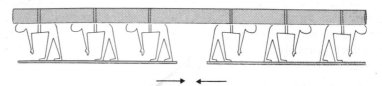

Fig. 3.9. A rough analogy to the sliding-filament theory is to imagine men, bent double with a pole (the thick filament) along their backs, running along a thinner pole (the thin filament). If the thicker pole is fixed then the thin pole will move below them. As the polarity of the cross-bridges at each end of the thick filaments is in opposite directions the filaments are pulled inwards (after A.F. Huxley).

Biochemical and structural basis of cross-bridge action

The globular head region of the myosin cross-bridge contains both an actin binding site and an ATPase site. Energy released from the hydrolysis of ATP results in movement of the cross-bridge and hence a relative sliding of the thick and thin filaments. Although the biochemical mech-

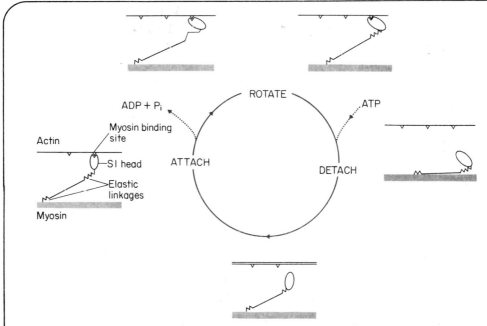

Fig. 3.10. *A model of the cross-bridge cycle.* The model assumes two attached and two unattached states. Note that the energy from the ATP is not used directly in force production, but is used to alter the state of the cross-bridge; force production then occurs as a consequence of the binding of actin with the myosin cross-bridge. In the relaxed muscle, binding between the cross-bridge and actin does not occur, as the Ca^{2+} concentration is too low. For simplicity only one cross-bridge head (of the pair) is shown. In relaxed muscle the cross-bridge is prevented from completing the cycle since the binding sites for myosin on the actin are masked by tropomyosin (see subsequent sections).

anism of the ATPase cross-bridge head (Fig. 3.10) has been worked out, the way in which the energy released from ATP is related to the behaviour of cross-bridges in intact filaments remains unclear.

Muscles are very stiff for small movements, and become much less stiff during and after larger movements; a property which may be due to short range activities of the cross-bridges. This 'thixotropic' property of muscles is probably very useful for accurate control during small movements.

Constant-volume behaviour of muscle

When muscles shorten they also become fatter so as to keep their volume constant. This means that the distance between an actin and myosin molecule increases. Thus the cross-bridge must be capable of interacting with the thin filament over a range of distances when the muscle contracts.

Mechanics of muscle contraction

Two types of mechanical recording are commonly made on muscle: isometric, in which the muscle is held at constant length and tension recorded; and isotonic, in which tension is held constant and the alteration in muscle length is recorded. Two simple types of apparatus for studying the isometric and isotonic contractions of muscle are illustrated (Figs. 3.11 and 3.12 respectively).

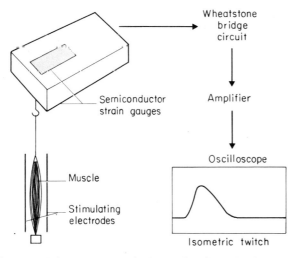

Fig. 3.11. An isometric lever, constructed using semiconductor strain gauges bonded to a suitable piece of steel. Using a Wheatstone-bridge circuit small changes in length of the gauge produce changes in electrical resistance proportional to the tension produced by the contracting muscle.

Isometric contractions

As we have seen in the cell physiology section, skeletal muscle fibres are electrically excitable. Experimentally, muscles can be stimulated by direct stimulation as well as via the motor nerve. The force produced by

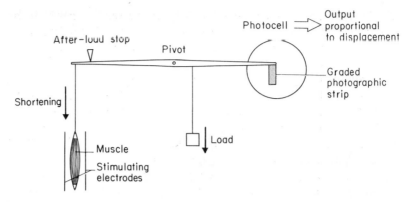

Fig. 3.12. Isotonic lever for recording muscle contractions. When the muscle is relaxed the lever rests against a stop so that the muscle does not support the load. When the muscle is stimulated it initially contracts isometrically until the tension produced equals the applied load on the muscle. At this point the muscle begins to shorten isotonically. The movement of the lever can be recorded using a vane which interrupts a beam of light focused on a photocell which produces an electric current proportional to the displacement of the lever.

a contracting muscle is called tension and has the units Newtons per square metre of cross-sectional area. Maximum tensions exerted by mammalian muscles at physiological temperature are around $4 \times 10^5 \, \text{N}/\text{m}^2$. A single stimulus of sufficient strength to activate all the fibres will produce a rapid rise in tension (a twitch) which rapidly decays.

If a second shock is given to the muscle before the tension response from the first twitch has decayed, the second response is greater than the first one; a third response is similarly greater than either of the first two (Fig. 3.13). At a sufficiently fast rate of stimuli the responses fuse in a smooth contraction—a tetanus. The reason for this behaviour is that a single twitch is a relatively inefficient way of developing tension at the

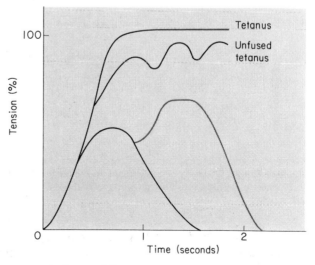

Fig. 3.13. Tension development following a single stimulation (single twitch), two stimuli (red line) or repetitive stimulation (lighter line). On repetitive stimulation at fairly high frequencies twitches fuse to form a tetanus.

end of the muscle, because the elasticity of the tendon needs to be overcome before tension can be developed. With repeated twitches the elastic elements are stretched and a full tension can be developed.

The length–tension curve

As a result of the network of connective tissue surrounding the fibres (see p. 42) resting muscle has a certain amount of elasticity. This connective tissue is mechanically in parallel with the contractile apparatus and is referred to as the parallel elastic component. Other elastic structures, such as the tendons, are in series with the contractile elements in the muscle (series elastic components). The mechanical elements of muscle are illustrated diagrammatically by a simple model in Fig. 3.14a. When a muscle is stretched passively it will exert proportionally more tension the greater the force applied. That is, it does not obey Hooke's law of elasticity. Since the parallel elastic component is in parallel with the contractile component the tension actually recorded when the muscle is stimulated (total active tension) will be a product of the tension due to the contractile and parallel elastic components (Fig. 3.14b). In order to discover how the extra isometric tension varies with length it is necessary to subtract the resting tension from the total active tension to obtain the active increment (Fig. 3.14b). It can be seen from these types of experiments that the active tension increment is at a maximum at lengths close to that of the resting length of the muscle in the body and falls away to zero at either longer or shorter lengths (Fig. 3.14b).

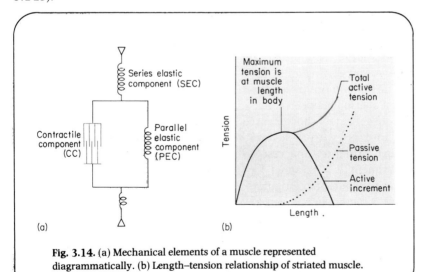

Fig. 3.14. (a) Mechanical elements of a muscle represented diagrammatically. (b) Length–tension relationship of striated muscle.

The length–tension curve and the sliding-filament theory

Muscles develop their maximum tension at the length which they have in the intact body; at lengths greater or less than this they develop less tension (Fig. 3.14b). A critical test of the sliding filament theory is to explain this result in terms of the activity of the cross-bridges which may be activated at each muscle length. To do so it was necessary to make very accurate measurements of the lengths of thick and thin filaments and of the degree of overlap at various sarcomere lengths. Standard filament lengths for frog semitendinosus muscle are shown in Fig. 3.15.

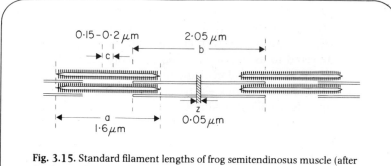

Fig. 3.15. Standard filament lengths of frog semitendinosus muscle (after Gordon *et al.*).

It was also necessary to relate the length–tension diagram to sarcomere length rather than to the length of the whole muscle. A summary of the results obtained in this experiment are given in Fig. 3.16 together with a diagrammatic representation of the appearance of the sarcomeres at particular sarcomere lengths. It can be seen that the length–tension diagram consists of a series of straight lines connected by short curved regions. At the point at which the thick and thin filaments do not overlap the tension development becomes virtually zero as predicted by the sliding filament theory. As the degree of overlap increases the tension rises as more cross-bridges on the thick filament engage with the thin filament producing tension.

The plateau on the length–tension diagram corresponds to the maximum overlap between thick and thin filaments. On the left-hand side of the curve, at still shorter sarcomere lengths, the thin filaments from one side interfere with those from the other side. This interferes with cross-bridge formation and results in a fall in isometric tension. Thus this experiment is accounted for in detail by the sliding filament theory. Recent evidence also suggests that the degree of activation of the muscle fibres is incomplete at very short sarcomere lengths contributing to the fall of tension on the left-hand side of the length–tension curve.

Isotonic contractions

When a muscle is stimulated isotonically it is found that the smaller the load the faster the muscle contracts. The relationship between the velocity of shortening of a muscle and the tension or force it develops (which is equal to the load) is given by the force–velocity curve (Fig. 3.17). With the exception of insect muscles this type of curve can be obtained for all muscles so far examined including cardiac and smooth muscles.

The shape of the force–velocity curve is thought to be related to the effect mechanical constraints have on the splitting of ATP by the cross-

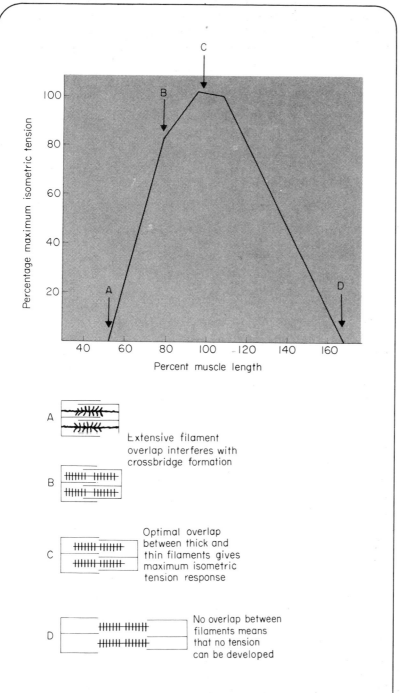

Extensive filament
overlap interferes with
crossbridge formation

Optimal overlap
between thick and
thin filaments gives
maximum isometric
tension response

No overlap between
filaments means
that no tension
can be developed

Fig. 3.16. Length–tension relationship of a single frog fibre with a
diagrammatic representation of the appearance of sarcomeres at four
critical lengths (after Gordon *et al.*).

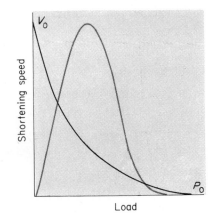

Shortening speed

Load

Fig. 3.17. Relationship between muscle force and velocity. The red line shows how the power produced by the muscle (force × velocity) varies with load.

bridges. In the simple analogy of men running (Fig. 3.9) this kind of curve would be expected; the less the resistance the higher the rate.

The control of muscle contraction

The regulatory proteins of skeletal muscle involved in the control of contraction are called tropomyosin and troponin. Tropomyosin is a long fibrous molecule consisting of two polypeptide chains located in each groove formed by the coils of the actin filaments. The troponins are three globular proteins located on every seventh actin monomer along the thin filament (Fig. 3.18).

If the thin filament (actin) has its regulatory proteins stripped off it will now react with myosin to develop tension, even in the absence of Ca^{2+}. If the regulatory proteins are present on actin then no reaction with myosin will occur until the ionised calcium concentration rises above a certain level. The presence of the regulatory proteins on actin makes it possible to control the system by varying the ionised calcium concentration. This feature has been exploited in the course of evolution to control contraction. At rest, in a relaxed muscle, when the Ca^{2+} concentration is very low (10^{-7} mol/l) tropomyosin masks the myosin combining sites on the actin and inhibits the activity of the cross-bridges. In an activated muscle Ca^{2+} released from the vesicles of the sarcoplasmic reticulum (see subsequent sections) binds to troponin C causing conformational changes in the other components of the troponin complex which results in an uncovering of the myosin combining site of actin allowing full cross-bridge activity. Tropomyosin thus functions to maintain control over all the actin monomers even though only one in seven is in contact with the ATPase inhibitory protein of the troponin complex.

The control of muscle contraction involves a fast and large increase in ATP splitting by the cross-bridges. For example, it has been estimated that in fast muscles the rate of ATP splitting increases more than 100 times in the few milliseconds it takes to develop maximum tension. This is due to the movement of relatively large amounts of Ca^{2+} within the muscle during the excitation–contraction coupling cycle. In resting muscle the Ca^{2+} concentration surrounding the myofibrils is only about 10^{-7} mol/l. Under these conditions cross-bridge cycling is inhibited (see

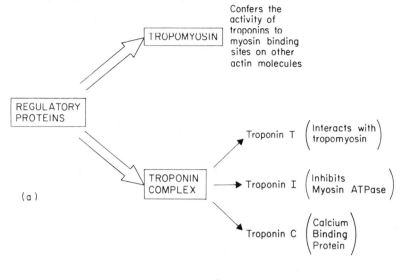

(a)

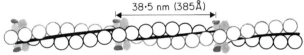

(b)

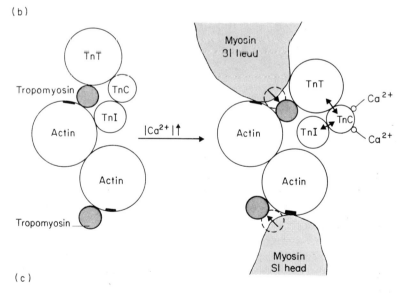

(c)

3.18. (a) Functions of regulatory proteins. (b) Location of regulatory proteins on the thin filament. Tropomyosin is represented as a dark line lying in each groove of the actin filament. The troponin complex consists of equal amounts of the calcium-binding protein troponin C, the inhibitory protein and troponin T, (c) Functions of calcium. With no calcium present the actin combining sites are masked by tropomyosin (left picture); when calcium binds to troponin C, tropomysin moves towards the groove between the actin filaments and so allows the myosin cross-bridge to attach to the actin binding site (right picture). Regulatory proteins such as calmodulin (found in muscle) and troponin C change their tertiary structures when acted on by Ca^{2+} or other triggers and so signal to the next protein in the chain.

previous section). This relatively low Ca²⁺ concentration compared to active muscle is achieved by storing the Ca²⁺ in an elaborate internal membrane within the fibre (sarcotubular system) where it is bound to various calcium-binding proteins. This internal membrane system consists of a network of vesicular elements surrounding the myofibrils called the sarcoplasmic reticulum (SR) (Fig. 3.19), which run close to a system of transverse tubules.

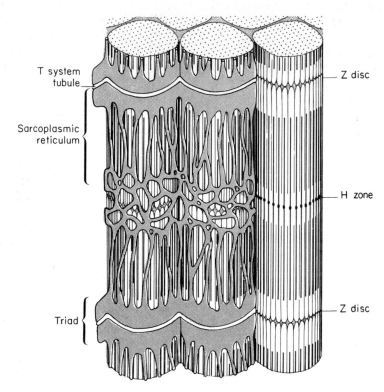

3.19. A reconstruction of the internal membrane system of frog sartorius muscle showing the transverse system and the sarcoplasmic reticulum joined at the triads (after Peachey).

The transverse tubules (T system) are open to, and so part of, the extracellular space; while the sarcotubular system is an intracellular structure. The T system occurs at the Z lines in frog muscle and at the A–1 boundaries in most other striated muscles. These two membrane structures, which are in close contact but not actually joined, occupy about 15 percent of the muscle–cell volume depending on muscle type. The T system, which is an invagination of the surface membrane of the muscle, has been shown to allow the inward-conduction of the action potential from the cell surface to the innermost myofibrils (see section on excitable tissues, p. 31). The effect of the action potential propagated down the transverse tubules is in some way communicated to the SR, causing release of Ca²⁺ into the myofilament space. The released calcium activates the cross-bridges and results in contraction. In activated muscle Ca²⁺ levels rise to around 10^{-5} mol/l; after activation, Ca²⁺ is pumped back into the vesicles of the sarcoplasmic reticulum resulting in the inhibition of cross-bridge activity and a decline in tension (relaxation).

The energetics of muscle contraction

Muscle cells, like all cells, use ATP to fuel their reactions. As in other cells, the ATP available for this purpose is at a concentration of about 4 mmol/kg wet wt. Muscle can, however, increase its work output faster and to a greater extent than other cells and so requires a larger immediately available supply of energy. This is supplied by a 'back up' store of the high-energy phosphate compound phosphoryl creatine (PCr), which holds another 20 mmol/kg of high-energy phosphate. The equilibrium constant of the enzyme generating ATP from PCr is such that ATP is not depleted until the phosphoryl creatine stores are almost exhausted. Normally the stores of phosphoryl creatine are replenished by oxidative metabolism via the ATP produced by the Krebs' cycle; however, during severe exercise, when the muscle is doing work at a faster rate than the blood can supply oxygen and nutrients, the muscle must rely on local glycogen stores and anaerobic glycolysis to meet its energy requirements. The glycolytic machinery of vertebrate muscle cells is well- developed and is able to supply ATP at a high rate with a short time delay (Fig. 3.20). The cell is prevented from passing into a reduced state under these conditions by converting pyruvate to lactic acid, which then diffuses out of the muscle and accumulates in the blood. The disadvantage of anaerobic glycolysis is that it only produces very small amounts of high-energy phosphate (2 ATP) from each molecule of glucose, compared with oxidative phosphorylation (36 ATP). Thus, most muscles contain a mixture of different cell types, some adapted to use oxidative metabolism and highly economic fuels such as lipids, and others which can supply ATP very rapidly but more expensively using anaerobic metabolism. In exhausted muscles there is yet another 'emergency' mechanism for the supply of ATP. This works by combining two ADP molecules to regenerate one ATP and one AMP molecule.

Muscle fibre types

In general, three types of muscle fibre can be distinguished in human skeletal muscles (Table 3.2). They differ not only in their capacity to

Table 3.2. Properties of three types of skeletal muscle fibre.

Property	Slow twitch oxidative	Fast twitch oxidative	Fast twitch glycolytic
Colour	Red	Red	White
Speed	Slow	Fast	Fast
Myosin ATPase activity	Low	High	High
Primary source ATP	Oxidative phosphorylation	Oxidative phosphorylation	Glycolysis
Capillary supply	Many	Many	Few
Fibre diameter	Small	Intermediate	Large
Rate fatigue	Slow	Intermediate	Fast

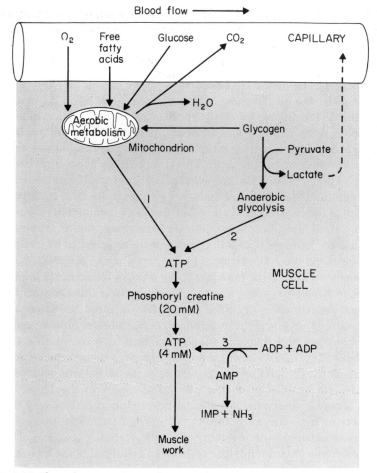

Fig. 3.20. The pathways producing the energy for muscle contraction. The numbers indicate the order in which the various pathways are called upon to supply ATP as the level of muscular effort increases. At low levels of activity the ATP is produced aerobically (Pathway 1); fat is the preferred fuel (transported in the blood as free fatty acid chiefly from adipose tissue). At slightly higher levels of muscular activity other fuels become more important, such as glucose (transported from the liver) and local stores of glycogen. During very vigorous activity the blood cannot supply nutrients and oxygen fast enough, and so the muscle fibres rely on anaerobic glycolysis to supply ATP at a faster rate (Pathway 2). Although some fibres have all these pathways, many are specialised for either aerobic or anaerobic metabolism. When the muscle fibre becomes fatigued then some 'extra' energy may be obtained by Pathway 3 (catalysed by myokinase). This reaction does not proceed unless the AMP formed is removed. This is done by coupling the myokinase reaction to a practically irreversible reaction catalysed by adenylate deaminase.

generate ATP (see previous section) but also in their speeds of contraction and resistance to fatigue. Muscles adapted for continuous activity, such as the postural muscles of the back, have a high proportion of slow red fibres. These fibres are red because they contain the respiratory pigment myoglobin which facilitates the flux of oxygen from the capillary to muscle mitochondria. Since these fibres are of small diameter, surrounded by numerous capillaries and are slow contracting (i.e. use

ATP at a low rate) the blood is able to keep them supplied with all their oxygen and nutrient requirements; hence, these fibres are very resistant to fatigue. In contrast, muscles which are required to have a rapid response and develop tension quickly contain high proportions of anaerobic fast fibres. In these fibres myoglobin is absent and there are few mitochondria and capillaries but a well-developed glycolytic capacity. These fibres are somewhat larger and able to produce ATP at very high rates but quickly fatigue once their glycogen stores are exhausted. There is a third type of fibre which is fast contracting but is red and has a well-developed oxidative potential. This type of fibre has somewhat intermediate properties to the other two types.

Each fibre type is innervated by a different type of motor neurone (Fig. 3.21). A single motor neurone branches to innervate a number of

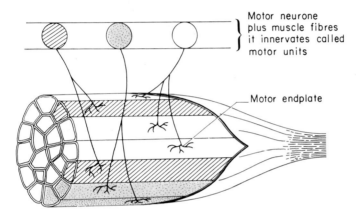

Fig. 3.21. Motor units to skeletal muscle fibres.

different muscle fibres of the same type (called a motor unit). Stimulation of the motor neurone will result in the contraction of all the fibres in that motor unit. The different motor neurones associated with the three fibre types have different thresholds for recruitment.

In general, muscles which are capable of very delicate movements (for example, ocular muscles) have fewer muscle fibres associated with each motor unit in order to give a finer control. Postural muscles are able to produce sustained tension for long periods without fatiguing because there is asynchronous firing of the different motor neurones. Thus, some fibres are contracting while others are relaxing but the total tension exerted by the muscle remains constant. The evolution of three different fibre types with different speeds, metabolic characteristics and thresholds of activation enables skeletal muscle to perform a wide variety of different functions. The properties of a given fibre are not fixed but are determined by the motor nerve. Cross-reinnervation of fast and slow mammalian muscle fibres results in a reciprocal change in their metabolic and contractile characteristics. Recent research has provided evidence that it is the pattern of motor-neurone activity which is important in determining the fibre phenotype.

CARDIAC MUSCLE

In many respects cardiac muscle is similar to skeletal muscle. Both are striated and contain myofibrils packed with regular arrays of thick myosin and thin actin filaments. The molecular basis of contraction and its regulation are thought to be similar in the two types of muscle; however, in contrast to skeletal muscle fibres which are formed by the fusion of many mononucleated myoblasts during embryonic development, each cardiac muscle cell retains its own individuality. Thus, myofibrils are not continuous for long lengths but end at the plasma membranes of individual cells. The junctions between cells are often referred to as intercalated discs (Fig. 3.23).

Cardiac muscle cells are electrically coupled through 'gap-junctions' formed where a region of the outer lamellae of adjacent plasma membranes fuses together (p. 36). Thus, if a contraction is initiated at one point on the muscle then the action potentials are able to spread to adjacent cells.

Since the heart has to work continuously, the muscle cells contain numerous mitochondria and have a very well-developed blood supply.

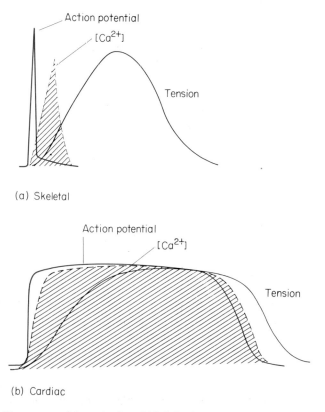

(a) Skeletal

(b) Cardiac

Fig. 3.22. Time course of the activation of (a) skeletal and (b) cardiac muscle. The brief action potential in skeletal muscle acts as a trigger for a correspondingly transient rise in free Ca^{2+}: the tension response only starts towards the end of the Ca^{2+} transient so cannot be controlled by it. In cardiac muscle the action potential, and consequently the Ca^{2+} response is long enough to persist during the contraction, so that they can control tension development. (Not to scale.)

The sarcoplasmic reticulum is only moderately developed and there are no triads as are found in skeletal muscle; however, the T system opens to the extracellular space and is related to the SR via a series of flattened membranous sacs.

A major difference between skeletal and cardiac muscle is that in the case of the latter, the action potential is much longer-lasting and of comparable duration to the mechanical event. Thus the refractory period of cardiac muscle extends into the relaxation period and hence repetitive stimuli do not give rise to a tetanus (see p. 48).

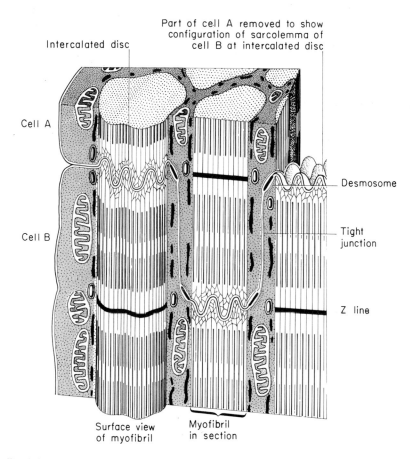

Fig. 3.23. Diagrammatic reconstruction of intercalated disc of two heart cells (A and B) (after BDS).

SMOOTH MUSCLE

Smooth muscle differs from skeletal muscle in that the characteristic cross-striated pattern of skeletal muscle is not present. It is also sometimes called involuntary muscle because its activity arises either spontaneously or through activity of the autonomic system. Examples of smooth muscles are to be found in the vessel artery walls of the vascular system, air passages to the lungs and in various tissues of the intestinal

10 μm

Fig. 3.24. Structure of smooth muscle.

and reproductive systems. Smooth muscle from different sources varies considerably in both its structure and properties, but a number of general features may be distinguished between the different types. The fibres are generally spindle-shaped and considerably smaller than skeletal muscle fibres; characteristically, each has a centrally placed nucleus (Fig. 3.24).

With the electron microscope, contractile filaments of actin and myosin can be observed scattered throughout the cytoplasm; although not arranged in such highly organised filaments as skeletal muscle, cross-bridges can occasionally be seen on smooth muscle thick filaments and the molecular basis of the shortening mechanism in the two types of muscle is thought to be similar. Smooth muscle cells contain only about 10 percent of the actin and myosin of skeletal muscle and have a low actomyosin ATPase activity, consistent with its low speed of contraction and tension. The more random overlap of thick and thin filaments in smooth muscle may explain the broader length–tension characteristics compared to skeletal muscle (Fig. 3.25).

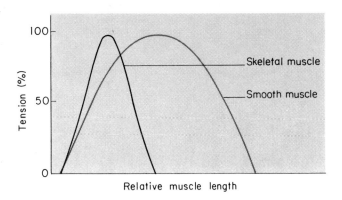

Fig. 3.25. Length–tension diagram of smooth compared to skeletal muscle.

The precise nature of activation of smooth muscle is dependent on its 'origin' but there are a number of broad differences in excitation–contraction coupling compared to skeletal muscle. (1) Smooth muscle has no well-developed sarcoplasmic reticulum although scattered vesicles occur which may function as Ca^{2+} stores. (2) In contrast to skeletal muscle, external stores of Ca^{2+} are important in many types of smooth muscle and calcium pump sites are found distributed across the entire external membrane surface. (3) Recent research also indicates that regulation of contraction at a molecular level may also be different between skeletal and smooth muscle. Although smooth muscle contains tropomyosin and troponins there is evidence that control of contraction is exercised through interaction of Ca^{2+} directly with the myosin cross-bridge. It is thought that one class of myosin light chains may specifically block the actin binding site when Ca^{2+} is not bound to myosin.

It is possible to divide smooth muscle into the following two broad categories of mode of activation.

Multi-unit smooth muscle

Examples of this type of smooth muscle are found in the nictitating membrane of the eye, larger arteries, and in some areas of the intestinal and reproductive systems. Multi-unit smooth muscle is like skeletal muscle in that it shows no inherent activity, but depends for activation on its autonomic nerve supply. Unlike skeletal muscle, this nerve supply is more diffuse and extends over a larger area of the muscle membrane. At these points of innervation are numerous membrane-bound vesicles which are thought to contain the transmitter substances. Following stimulation of the nerve, the vesicles fuse with the nerve membrane releasing transmitter into the extracellular space at several points along the fibre, resulting in the simultaneous activation of a number of the surrounding muscle cells. Interestingly, the total surface of the smooth-muscle membrane contains receptor sites for nerve-transmitter substances. Binding of transmitter results in changes in membrane potential by altering the permeability of smooth-muscle membranes to certain ions. Depending on the nature of the transmitter either excitatory or inhibitory potentials may be produced. Multiple action potentials in the nerve are required to depolarise the muscle cell sufficiently to reach the threshold potential for an action potential. The action potential causes a rise of intracellular Ca^{2+} and hence contraction. There is evidence that the inward current of the rising phase of the action potential is not carried by Na^+ ions as with skeletal muscle but by Ca^{2+}. In support of this idea, the spike is not abolished by incubation in Na^+-free solutions or by tetrodotoxin, a specific sodium channel blocker. In general the larger distances between smooth-muscle membrane and nerve terminals compared to the motor end plate and nerve ending in skeletal muscle results in a much slower development of tension following excitation. Hormones such as noradrenaline may also cause changes in membrane potential and contraction in multi-unit smooth muscle cells.

Single-unit smooth muscle

Examples of this type of smooth muscle are found in the small arteries and veins of the vasculature and in the uterus. This type of smooth muscle is distinguished from multi-unit smooth muscle by its ability to contract spontaneously in the absence of either nerve or hormonal stimulation. Normally a large number of cells contract synchronously hence the term 'single unit'. Action potentials are conducted from cell to cell via low-resistance gap-junctions formed by the fusion of adjacent cell membranes. These action potentials originate from a certain proportion of the cells which show regular slow spontaneous depolarisations known as pacemaker potentials. When the depolarisation reaches the threshold potential and action potential is generated which is transmitted to adjacent muscle cells through the gap-junctions (Fig. 3.26).

The intensity of the mechanical response increases with the frequency of the action potentials (equivalent to tetanus in skeletal muscle). External agents such as certain hormones can either depolarise or hyperpolarise the membrane potential resulting in a modification of muscle activity. Depolarisation raises the membrane potential nearer its threshold resulting in an increased frequency of action potentials. Hyperlarisation of the membrane potential will have the reverse effect.

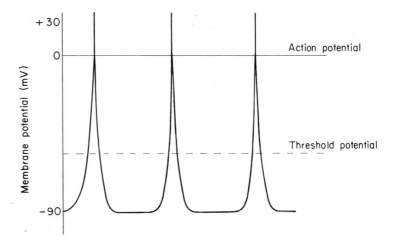

Fig. 3.26. Spontaneous action potentials in pacemaker cells.

Interestingly, noradrenaline can have different effects on smooth muscle from different parts of the body. For example, application of noradrenaline leads to membrane depolarisation and contraction in vascular smooth muscle and hyperpolarisation of membrane potential and relaxation in intestinal smooth muscle. Stretching single unit smooth muscle also causes depolarisation, increased frequency of action potentials and increased tension resisting further stretch. This is a very useful feature in an organ such as the urinary bladder which is required to exert more tension the fuller it becomes.

Part 2
Maintenance Systems

Chapter 4
Blood and
Body Fluids

Each cell in the body carries out, within itself, all the chemical processes necessary for its own maintenance. Therefore all the substances necessary for this must be carried to, and all the waste products taken away from, each cell. The system which does this—the cardiovascular system—transports materials in a closed system of tubes around the body. Exchanges between the cells and this system occur from the capillaries— the smallest blood vessels—which lie within a few thousandths of a millimetre of all body cells.

The main materials which cells require are fats, amino acids, sugar, vitamins, salts, water, oxygen and hormones. These substances are carried to the cells by the blood. The waste products of cells are carbon dioxide and simple soluble compounds of nitrogen such as urea—which are transported to the lungs or kidneys for excretion.

In the higher animals the blood consists of a fluid, the plasma, in which cells of various kinds are suspended, each adapted for special purposes. The blood volume is about 8 percent of the body volume (or $5\frac{1}{2}$ litres in a 70 kg man); of this, about 55 percent is plasma.

CELLS OF THE BLOOD

The cells of the blood do not originate in the blood; they are made in organs outside the blood and merely use it as an avenue while performing their functions, or while in transit from one area to another. There

Table 4.1. The cellular elements of blood

		Size	Life	Function
Erythrocytes				
♂	$5.4 \times 10^6/mm^3$			
♀	$4.8 \times 10^6/mm^3$	$7 \times 2\mu m$	120 days	O_2 transport
Total white blood cells	$9000/mm^3$			
Granulocytes	%			
Neutrophilis	60			
Eosinophils	3	$10\mu m$	1 week	phagocytosis
Basophils	0.5			
Lymphocytes	30	$7–20\mu m$	? 100 days	antibody production
Monocytes	5			killing
Platelets	$300,000/mm^3$	$2 \times .5\mu m$	10 days	coagulation
Packed cell volume*				non specific defence
♂	0.40–0.54			
♀	0.36–0.47			

*The packed cell volume (PCV, haematocrit) is obtained when blood is centrifuged in a tube under standard conditions; it is expressed as the proportion of the column occupied by the packed cells.

are three kinds of such cells: erythrocytes (red cells), leucocytes (white cells) and thrombocytes (platelets—these are really parts of cells) (Table 4.1).

The present view (Fig. 4.1) is that these three cell types are derived

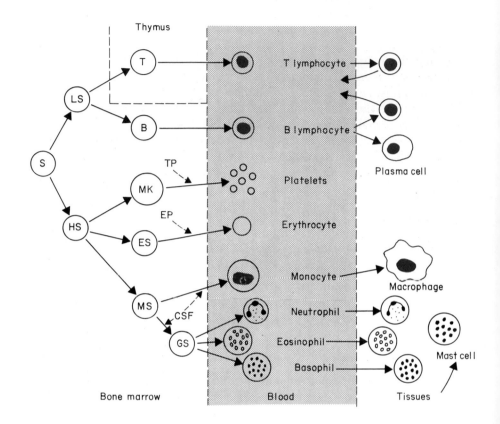

Fig. 4.1. Origin of the blood cells. The stem cells (S) lead via the lymphoid stem cells (LS) to the T and B lymphocytes. The haemopoietic stem cell (HS) leads via the megakaryocyte (MK), the erythroid stem cell (ES), the myeloid stem cell (MS) and granulocyte stem cell (GS) to the other blood cells. Cell production is controlled by the three stimulating factors thrombopoietin (TP), erythropoietin (EF; see Fig. 4.3) and colony stimulating factor (CSF), required for clone formation. The mast cell is similar to the basophil, but has a different origin. The thymus gland, at an early stage of life, is responsible for teaching the T cells to distinguish between 'self' and 'non-self' (after Playfair).

from a common precursor (stem cell) which originates in the bone marrow; an organ scattered (in adults) in the flat bones of the skull, the ribs, the breast bone (sternum) and parts of the hip, and with a total size similar to the liver. The red cells, platelets and most of the white cells are made in the bone marrow; the lymphocytes originate both in the bone marrow and the thymus and multiply in the lymph nodes. The monocytes, made in the bone marrow, subsequently invade the reticulo-endothelial system.

Erythrocytes

The primary function of the erythrocytes is to carry oxygen: a function done by the haemoglobin within them. They are thus container cells continuously circulating in the vascular system, their other properties being adapted to this rigidly defined role. Each cell in man is a small non-nucleated biconcave disc ($7 \cdot 5 \ \mu m \times 2 \ \mu m$) (Fig. 4.2); there is a total of

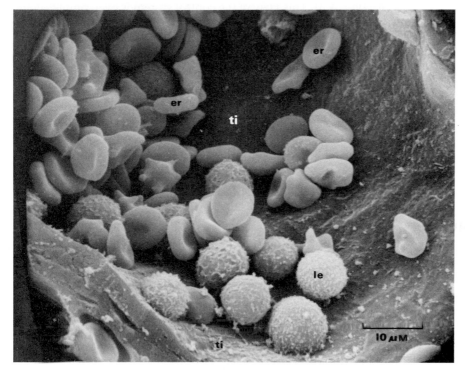

Fig. 4.2. Scanning electron micrograph of biconcave erythrocytes (er) and rounded leucocytes (le) in contact with the tunica intima (ti) of an arteriole. The surface of the erythrocyte is smooth whereas that of the leucocytes is rough due to the presence of numerous projections. (From Freeman & Co, with permission)

about 3×10^{13} cells in the body containing 1 kg of haemoglobin. The average life span of a cell is 120 days. It has been calculated that the biconcave shape is the best for allowing diffusion of oxygen into the cell, but in the small capillaries in which exchange occurs the cells are often very distorted by the narrowness of the vessels.

Erythrocyte production and destruction

Erythrocyte production (erythropoiesis) occurs in bone marrow, starting with the stem cell and progressing through various intermediates to the reticulocyte, which then matures into the erythrocyte in blood. Normal development takes some 7–10 days. At the end of their lives the erythrocytes are destroyed by the scavenger cells of the liver (macrophages or Kupffer cells) and many of the constituent parts recycled.

The rate of production of erythrocytes by the bone marrow is controlled by a feedback system which seems to depend on the level of the oxygen supply to the kidney. If the oxygen supply is decreased—by anaemia or hypoxia—then more erythrocytes are produced; if it is increased—by an increase in erythrocytes or high oxygen tension—then less erythrocytes are produced. The receptors for this effect are unknown. The current view of the process is that the kidney releases erythrogenin, which acts on a plasma globulin, converting it to erythropoietin; this then controls the number of erythrocytes which starts to mature (Fig. 4.3). This view of the control system is over-simplified and not yet well established.

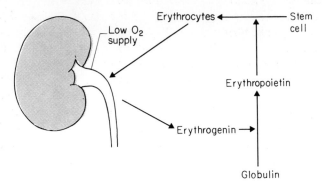

Fig. 4.3. Control of the production of erythrocytes by the bone marrow. A decrease in the O_2 supply to the kidney (for example due to a low number of erythrocytes) causes an increase in the production of erythrogenin, etc. and ultimately more erythrocytes.

Haemolysis of erythrocytes

Erythrocytes put into a hypotonic solution swell, and the degree of swelling is related to the degree of hypotonicity. With extreme swelling the cells burst (haemolysis), and haemoglobin escapes. Curves of erythrocyte fragility are constructed in this way.

The blood pigments

Haemoglobin is the pigment found in erythrocytes and is responsible for their large capacity to carry oxygen. It has a molecular weight of some 65,000 and is made up of four subunits, each consisting of a pigment part (haem) joined to a polypeptide part (globin). The haem part of the molecule is an iron-containing porphyrin derivative (Fig. 4.4), which binds reversibly to O_2. The haem molecule is suspended between two residues of histidine in the polypeptide chain by links to the central Fe^{2+} atom; one of these links is direct, the other can be broken so that O_2 can be inserted when oxyhaemoglobin is formed.

Various different haemoglobins exist. Ultimately these differences are due to differences in the amino-acid sequences of their constituent polypeptide chains; but these differences are often restricted to two of the four polypeptide chains in each haemoglobin molecule. Thus it is convenient to think of the four polypeptides in each molecule of haemoglobin as consisting of two pairs. One pair of polypeptides—the α chains containing 141 amino acids—is constant for nearly all haemoglobins; variations in the other pair give rise to different haemoglobins. In normal adult haemoglobin (haemoglobin A) the second pair of polypeptides are called the β chains. Haemoglobin changes its structure when it takes up oxygen, so that it behaves as a molecular 'lung' rather than a tank holding oxygen (Fig. 4.4a). This is because the 4 subunits of haemoglobin show cooperative effects (Fig. 4.4b). Thus oxygenation leads to changes in the shape of the whole molecule; this 'explains' the sigmoid shape of the oxygen dissociation curve and the buffering power of haemoglobin (p. 195). Haemoglobin follows the biblical parable of the rich and the poor 'For unto every one that hath shall be given, and he shall have abundance: but from him that hath not shall be taken away even that which he hath.' (St Matthew 26, XXIX)

Fetal haemoglobin (haemoglobin F) has γ chains instead of β chains, the changeover to haemoglobin A occurring in the first three months of extra-uterine life (p. 139, Fig. 5.46b); the γ chain contains the same number of amino acids as the β chain but 37 are different. The substance 2,3-diphosphoglycerate (2,3-DPG), normally present in erythrocytes, competes with O_2 for binding to deoxygenated haemoglobin. Fetal haemoglobin binds 2,3-DPG less strongly than adult haemoglobin and so binds more O_2 at any given partial pressure of oxygen. Therefore fetal haemoglobin can receive O_2 from the maternal haemoglobin at the same partial pressure of oxygen.

Abnormal haemoglobins   The amino-acid sequences of the polypeptide chains of haemoglobin are, of course, genetically determined. Amino-acid sequence analysis shows that many haemoglobins exist; these are identified by letter as, for example, haemoglobin A (normal), F (fetal), S, J, C, etc. The differences may be very slight but the consequences very great; the change of one amino acid for another at a critical site leads to a change in the three-dimensional organisation of the haemoglobin with marked consequences on affinity for O_2, physical properties, etc. In haemoglobin S, for example, substitution of valine for glutamic acid gives a haemoglobin which has a low solubility at low oxygen tensions, a circumstance which causes the erythrocytes to become sickle shaped and makes them more liable to haemolysis. Haemoglobin S persists in some 40 percent of the Negro population of Africa because it confers resistance to one type of malaria, although it gives a severe anaemia. Most other abnormal haemoglobins are harmless, presumably because the altered amino acids are present in non-critical parts of the polypeptide chain.

Carboxyhaemoglobin is a combination of haemoglobin with carbon monoxide instead of O_2. As the affinity of haemoglobin for CO is much greater than that for O_2, CO is taken up preferentially so reducing the O_2 carrying capacity of the blood.

Methaemoglobin is the pigment formed when the ferrous iron (Fe^{2+}) is converted to the ferric state (Fe^{3+}). This occurs in the presence of various drugs and oxidising agents (for example, nicotine), but also normally at a slow rate. The normal concentration of methaemoglobin is kept low by an enzyme system (NADH–methaemoglobin reductase) which converts it back to haemoglobin; for the reaction is not reversible otherwise.

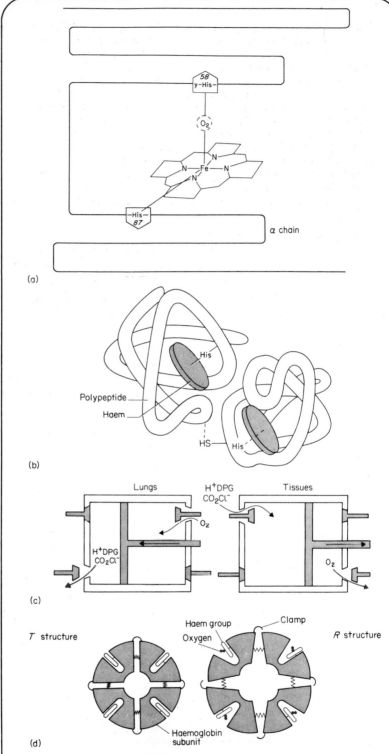

(a)

(b)

(c)

(d)

Fig. 4.4. (a) Outline structure of the α subunit of the haemoglobin molecule showing the 'pocket' into which the oxygen atom must pass in order to bind with the ferrous iron. Oxygenation can only occur between the Fe^{2+} and one of the histidines of the backbone. Each subunit is waxy inside and soapy outside, which makes it soluble in water but impermeable to it. (b) Three-dimensional structure of two subunits showing the flat-disc shape of the haem part of the molecule. The shape of the protein moiety varies with the state of oxygenation.

Fig. 4.4. (c) Reciprocating engine as a model of the cooperative effects of haemoglobin. The piston is driven to the left by the energy liberated in the reaction of haemoglobin with O_2 and to the right by the protons and CO_2. Diphosphoglycerate (DPG) and chloride ions are passengers riding in company with protons and CO_2. (d) Model illustrating haem-haem interactions. In the T (tense) state the subunits are clamped against the pressure of the springs; in the R (relaxed) state all the clamps have sprung open. Uptake of O_2 by the T structure strains the clamps until they all burst open and allow the molecule to relax to the R structure. Loss of O_2 would narrow the haem pockets and allow the T structure to re-form. The probability of O_2 triggering the T→R transition increases as more O_2 is bound and decreases as the proton, CO_2 chloride and DPG concentrations increase (c & d after Perutz).

Synthesis and destruction of haemoglobin

Haemoglobin is manufactured by the developing erythrocytes within the bone marrow. Before losing its nucleus the erythrocyte is much like any other cell in that it can synthesise RNA and lipid in addition to haem and proteins. After maturing to the non-nucleated form metabolism becomes very low, consistent with its 'container' cell function.

Substances essential to erythropoiesis

Protein. Amino acids are required for the globin part of the haemoglobin. When the supply of amino acids is limited, as in starvation, haemoglobin manufacture seems to have priority so that severe anaemia (Hb < 8 g/dl) rarely occurs in starvation.

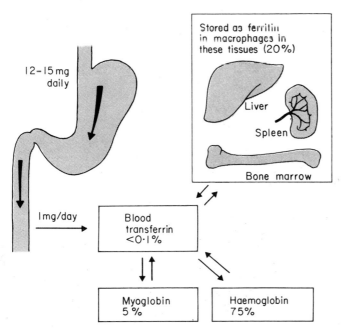

Fig. 4.5. Iron metabolism; iron is taken up by an active process in the small intestine and normally only lost when cells are desquamated from the skin (+ menstruation in the female). The uptake process is 'active' and controlled by bodily needs. Only ferrous (Fe^{2+}) iron is absorbed; the current role of hydrochloric acid in iron uptake is uncertain. There is a total of 3–4 g iron in the body.

Iron. Iron is required for erythrocyte formation because of its essential place in the haemoglobin molecule; if it is lacking an iron-deficiency anaemia (Hb < normal) occurs. Normally very little iron is lost from the body (Fig. 4.5a), so that only about 1 mg/day is required for replacement purposes. A normal diet provides much more than this, thus more may be absorbed if required. It is known that uptake from the gut can occur against a concentration gradient and is controlled by the needs of the body. Iron metabolism therefore works as a closed system in that none is excreted, and so the body must have an efficient system of control of intake for once absorbed it cannot be excreted. The mechanism for control of uptake is, at present, not understood.

Vitamin B_{12} and folic acid

These substances are essential for the normal maturation of erythrocytes; in their absence the maturation of the nucleus of the erythrocyte occurs more slowly than that of the cytoplasm, leading to cells which are larger in size and fewer in number than normal.

Like other vitamins neither substance is made in the body, although B_{12} is stored in large amounts in our liver. In the absorption of B_{12} an intrinsic factor produced by the stomach is essential (Fig. 4.6).

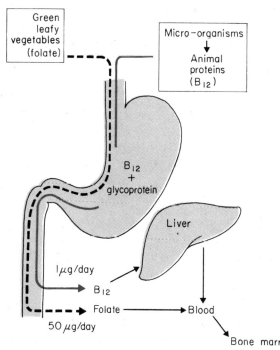

Fig. 4.6. B_{12} and folate metabolism; B_{12} requires an 'intrinsic factor' for absorption to occur; once absorbed it is stored in the liver (1 year's supply). Both are concerned with the metabolism of one-carbon radicals concerned in transmethylation.

Death of erythrocytes, degradation of haemoglobin and metabolism of bile pigments

The precise mechanism of erythrocyte death is not yet clear—ideas ranging from sudden haemolysis to gradual fragmentation have been

suggested— but eventually they are cleared from the circulation by the
mononuclear phagocytic system. These phagocytes split off globin, the
haem molecule is split open and the iron removed for recycling. The
residual pigment is biliverdin which, in man, is converted to bilirubin
and then transported (bound in albumin) to the liver where it is made
water-soluble and excreted via the bile. In the gut the bile pigments are
further converted to stercobilinogen which colours the faeces (Fig. 4.7).

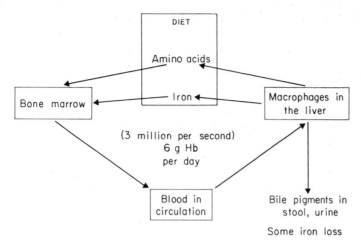

Fig. 4.7. Erythrocyte formation and distribution (after Ganong).

Blood groups

The erythrocytes of all races have an identical appearance. Their
membranes, however, contain a variety of antigens—sometimes called
agglutinogens—which vary in type in different individuals and to some
extent in different races. When these antigens combine with the appropriate antibodies—called agglutinins—the cells become sticky and
clump together like bunches of grapes (agglutination). The different
antigens give rise to the different blood groups. The common blood
groups, A, B, AB and O are due to the presence, in various combinations,
of two antigenic determinates A and B (Fig. 4.8). Thus group A has the
antigen A present on its surface, group B the antigen B on its surface,
group AB has both and O has neither present. These antigenic specific-
ities are due to the carbohydrate chains of glycoproteins and result from
the action of a series of glycosyltransferase enzymes determined by genes
at a number of loci. Type A has the sugar residue acetylgalactosamine
as part of its skeleton, whereas Type B has galactose.

It is normally found that the absence of A or B antigen in a person's
erythrocytes is always associated with the presence of the appropriate
antibody in their serum. Thus the absence of A in the erythrocyte means
that anti- A or α-antibody is present in the plasma. The reasons for this
seem obscure but it can be remembered that for the ABO system there
are always antibodies to A and B present unless their appearance is
inhibited by the presence of A and B antigens in the cell membrane. The
practical importance of this is that cells from one person transfused into
another must be 'matched', so that they are not agglutinated by the

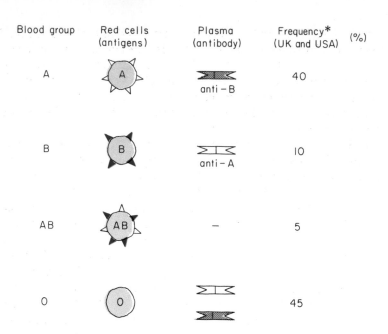

Blood group	Red cells (antigens)	Plasma (antibody)	Frequency* (UK and USA) (%)
A	A	anti – B	40
B	B	anti – A	10
AB	AB	–	5
O	O		45

Fig. 4.8. Diagram of the antigens and antibodies found in the erythrocytes and plasma in the ABO blood-group system. There are several subgroups of A and B. The most important are the A1 and A2; about 20 percent of group A people belong to A2 which has a much weaker antigen than A1. (* Rounded to nearest 5 percent.)

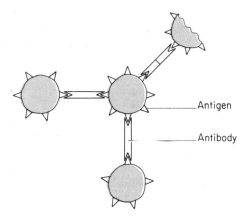

Antigen

Antibody

Fig. 4.9. Agglutination occurs by cross-linking of many red cells with serum antibodies. In transfusion the important reaction to avoid is the clumping of a donor's erythrocytes by a recipient's antibodies.

antibodies present in the recipient's plasma. If agglutination does occur (Fig. 4.9), then haemoglobin is released causing jaundice* in severe cases. The donor's serum has little effect on the recipient's cells because of the large dilution which takes place. The effects of transfusion are tested for directly, by testing the donor's cells on the recipient's serum before transfusion (cross-matching).

These antigens are also found on the surface of other body cells, for

---

* A disease due to an excess of the breakdown products of haemoglobin. It is characterised by a yellowing of the skin.

example liver, lungs, salivary glands, sperm, and in the amniotic fluid.
There are also other antigenic molecules present on the surface of
erythrocytes and other cells, so that probably each of us could be
identified by our cellular surfaces. The most important of these antigens
is the Rh system, discovered when it was shown that a serum prepared
against rhesus (Rh) monkey erythrocytes also agglutinated the erythrocytes of most humans.

The Rh system of antigens is complex as it is really a group of many
antigens. The most important quantitatively is called the D antigen and
this is the one usually tested for. Thus Rh-negative individuals have no
D antigen and normally no D antibody in their plasma. Different races
have slightly different proportions of Rh positive, for example, Caucasians 85 percent, Orientals 99 percent. The Rh factor is important in
transfusion, but it is also of great importance in pregnancy when an
Rh-negative mother has an Rh-positive fetus. Occasionally fetal cells
enter the maternal circulation during childbirth. If the fetal cells are of
a different ABO group than the mother then they are destroyed and no
problem ensues. If, however, the fetal and maternal RBC's are the same
ABO group then the fetal cells survive and induce antibodies in the
mother so that subsequent Rh-positive children will have their red cells
haemolysed by antibodies from the mother, leading to haemolytic disease of the newborn. Usually first children are not affected and (for the
reasons given above) only 5 percent of those who might be affected are,
in fact, so affected. Current practice is to kill the fetal cells which enter
the maternal circulation before they can induce antibodies. This is done
by injecting the mother with Rh-positive antibody within the first 36
hours after childbirth.

The blood groups are inherited according to Mendelian laws. In a
given person the blood group is a complex (phenotype) of the mother's
and father's blood-grouping genes. For example, a person with type B
blood may have inherited a B antigen from each parent or a B antigen
from one parent and an O from the other; an individual of group B may
therefore be B × B or B × O, i.e. may be homozygous or heterozygous.
Studies of blood-group inheritance are important in tracing the migration of populations and in paternity cases.

IMMUNITY: THE LEUCOCYTES AND THE RETICULO-ENDOTHELIAL SYSTEM.

The body, like a city, needs a disposal system which will cope with waste
particles. This waste matter may arise from within the body itself—as
worn-out cells (mainly erythrocytes) and macromolecular complexes—
or by invasion from outside—as dust or bacteria. Disposal works by a
system of scavenger cells which eat up these particles and then either
concentrates them or breaks them down (phagocytosis). If the waste
materials consist of foreign proteins then the body produces antibodies
specific for each protein; these antibodies react with them to produce a
complex, with subsequent precipitation and phagocytosis. The scavenger cells are either free in the blood or tissues and can 'hunt' the particles,

or are fixed in various organs and 'seize' the particles as they pass. The fixed scavenger cells are found in a system of lymphoreticular organs which consist of thin-walled supporting tubes with the phagocyte cells in their walls (Fig. 4.10). Blood and lymph are passed along these tubes, allowing the phagocytes to act on the suspended particles. The lymphoreticular organs contain lymph tissue as well as phagocytes. These organs include lymph nodes, spleen, thymus, bone marrow and the alimentary tract. The proportion of lymphoid tissue to that of structural and phagocyte cells varies in these different tissues. In the liver a similar system of phagocytic cells exists with no lymphoid tissue.

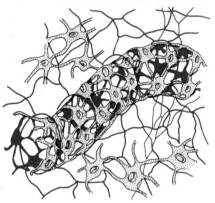

Fig. 4.10. Structure of the wall of a sinusoid showing the fenestrated layer of flattened endothelial cells, with interspersed scavenger cells (red). These cell types are difficult to identify histologically; but the endothelial cells have a basement membrane visible under electron microscopy.

Immunity is concerned with mechanisms which recognise foreign or 'non-self' matter that enters the body. *The immune system* identifies invaders, switches itself on, generates defenders armed specifically to deal with whatever invaders are around, kills them and switches itself off before the body is unduly harmed. The cells involved in this process consist mainly of *killers, helpers* and *suppressors*; with normally more helpers than suppressors. The immune system works as a beautifully constructed biochemical cascade involving both natural and adaptive mechanisms, each in turn composed of cellular and humoral parts (Fig. 4.11).

Almost all cells involved in immunity arise from precursors in the bone marrow; these circulate in the blood, entering and occasionally leaving the tissues as required. The erythrocytes have already been dealt with. The leukocytes, or white (colourless) cells, are nucleated cells lacking haemoglobin; their number in the blood is around 4000–10,000/mm² and varies much more than the erythrocytes (Table 4.1 and Fig. 4.1).

The complement system forms a large part of the natural immune system. This involves a sequential activation and assembly of units which lead to three effects (Fig. 4.12) (1) release of peptides active in inflammation; (2) deposition of an attachment-promotor (C3b, an opsonin, i.e. a substance which makes materials 'tasty' to phagocytes); (3)

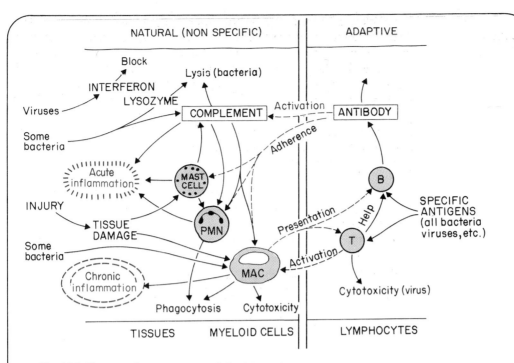

Fig. 4.11. The natural immune system (left side) is rather similar to inflammation (Fig. 4.13), but there is limited ability to destroy some bacteria. Most cells secrete interferon, an antiviral agent. Adaptive immunity is based on the special properties of the T and B lymphocytes. These can respond to and retain a selective memory to thousands of different non-self materials (antigens). This permanent altered pattern of response is an adaptation of the animal to the environment (after Playfair).

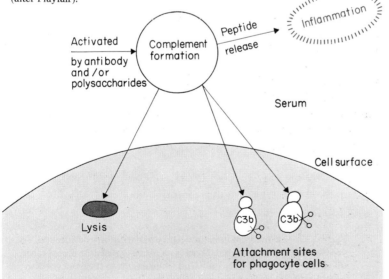

Fig. 4.12. The complement system is formed partly in the fluid phase but largely on cell surfaces. Activation is limited to the immediate vicinity of the injury by the short life of the activators; there are also specific inactivators. Some 15 serum factors are involved in the process (after Playfair).

membrane damage leading to cell lysis. The complement system is activated by a variety of polysaccharides and various components of antibodies.

Inflammation is a system which responds to and repairs all types of damage, including infection (Fig. 4.13). Everyone is aware of the symptoms of inflammation, i.e. redness, warmth, swelling and pain. The mechanisms involved in inflammation are similar to those responsible for natural immunity (see also p. 117).

The reticulo-endothelial system is a system of phagocytic cells, distributed throughout the body, which takes up particles in the blood and tissues. These particles may arise from within the body itself—as worn-out cells, e.g. erythrocytes, macromolecular complexes—or by invasion from outside—as dust or bacteria. The reticulo-endothelial system was originally identified by its rapid uptake of injected dyes etc. The current view is that the cells of the reticulo-endothelial system are not very phagocytic, but are mainly structural. The adaptive immune system is thought to amplify the phagocytosis.

Many cells can ingest foreign material, but only some of these cells can increase their uptake in response to opsonisation by antibody or compliment. Such cells are known as 'professional' phagocytes—the

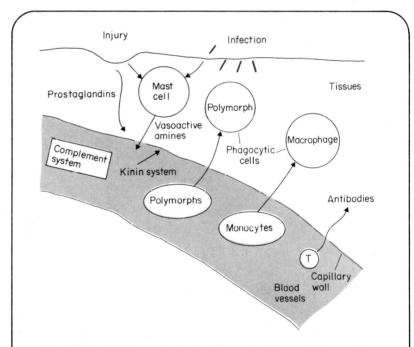

Fig. 4.13. Acute inflammation induced by surface injury or infection. The increase in capillary permeability plays a central role in the process as it allows access of blood cells and serum components to the tissues; it also accounts for the main symptoms of inflammation. The actions of the complement system are as shown in Fig. 4.11 and have been omitted for clarity. The mast cells, basophils and platelets produce vasoactive substances, e.g. histamine, 5-HT etc. which causes the increased permeability (see also Fig. 5.27) (after Playfair).

polymorphs, monocytes and macrophages of the myeloid series. These cells all operate in the same way (Fig. 4.14), by a sequence of attachment, ingestion and digestion. The differences between the cells are that polymorphs are short-lived (hours or days) and often die in the process of phagocytosis; while macrophages, which lack some of the more destructive lysosomal enzymes, usually survive to phagocytose again. Macrophages can also secrete some of their enzymes, e.g. lysozyme. Some of the cells can obtain energy from glycolysis, so that they can live

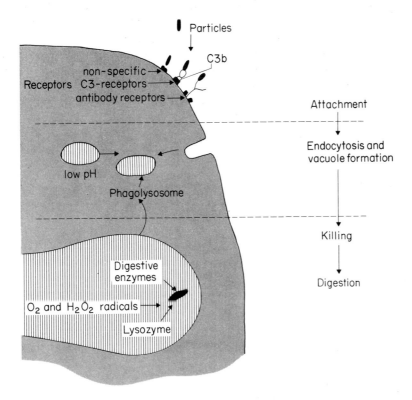

Fig. 4.14. The processes of phagocytosis occurs from above downwards by (1) attachment of the particles to the cell surface; (2) internalisation of the receptor-particle complex (an active step) and then (3) its destruction in the secondary lysosomes. This process occurs in polymorphs, monocytes and macrophages (compare Fig. 1.19) (after Playfair).

in tissues that are deficient in oxygen, inflamed, oedematous or inadequately perfused by blood.

The lymphocyte is the cell of adaptive immunity (Fig. 4.15); it has several unique features: (1) a restricted number of receptors permitting each cell to respond to an individual antigen—this is the basis of specificity; (2) the ability to produce a population of cells from one cell (a clone) and a long lifespan—this is the basis of memory; (3) cells recirculate from the tissues back into the bloodstream—this ensures that the specific memory which follows a local response is distributed all over the body (Fig. 4.16).

There are two main lymphocyte populations, the T (thymus-dependent) and the B (Bursa or bone marrow dependent). The B lympho-

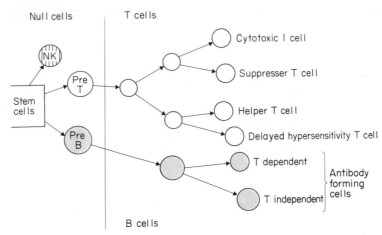

Fig. 4.15. An outline of the functions of lymphocytes, the cell of the adaptive immune system. The 'null' cells resemble lymphocytes, but lack the characteristic surface features. The T and B cells are derived from these null cells. The T cells are the 'officers' and the B cells the 'soldiers'. NK = natural killer cells.

Currently there are thought to be four types of T lymphocyte and two types of B lymphocyte (shown in red). The cytotoxic cell is essential in viral immunity; the suppressor cell inhibits B and T responses; the helper and delayed sensitivity cells attract and activate T cells and other cells respectively. A small amount of antibody formation does not require the T cells, hence the T independent group of B cells (after Playfair).

cytes make antibody; the T lymphocytes 'help' B to make antibody, act as killer cells etc. About 10 percent of the 'lymphocyte like' cells found in the blood and the tissues do not have the surface markers found on the B and T lymphocytes; these are called 'null cells'; this group consists of early T and B cells, monocytes and the 'natural killer' cells.

The platelets

The platelets are produced from stem cells via megakaryocytes; these are large (200 μm) cells lying outside the sinusoids of the bone marrow. The megakaryocytes push processes into the blood through the walls of the sinusoids; these processes break off to form the tiny platelets. The number of circulating platelets (about 300,000/mm³) is probably controlled by a hormone, thrombopoietin. The platelets are rich in enzymes and contain large amounts of ATP, 5HT and phospholipids. Their life span is about 10 days. Their main function is in the clotting mechanism. Normally it is thought that minute clots are continuously formed and destroyed. This is probably the main way platelets die, but 'old age' may also occur.

PLASMA

The plasma is a clear, straw-coloured fluid, except after a fatty meal when it is milky in appearance due to the presence of fatty globules. It is rather like raw egg-white diluted with a salt solution; its main constituents are protein (6 percent) and sodium chloride (0·9 percent). The high water content (90 percent) it contains dissolves the other sub-

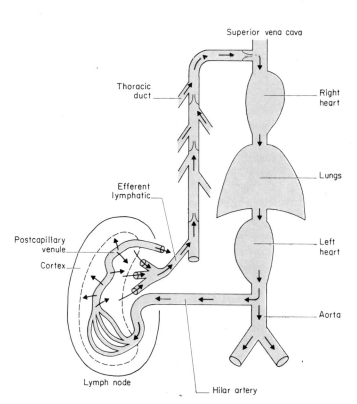

Fig. 4.16. Circulation of lymphocytes. Both B- and T-lymphocytes travel to the lymph nodes in the hilar artery and subsequently migrate into the lymphoid tissue via the endothelium of the post capillary venules. There they mingle with lymphocytes of the cortex (B areas) and paracortex (T areas). Simultaneously equivalent numbers of lymphocytes leave in the efferent lymphatics so that a balanced flow through the node is maintained. The circulation is completed by the ultimate return of the efferent lymphocytes to the bloodstream via the thoracic duct, which is the main entrance of lymphocytes to the vascular system (from Passmore & Robson).

stances and allows the blood to pass easily through the capillaries. Plasma left in a test tube clots, unless substances are added to prevent it. When whole blood is allowed to clot the fluid remaining is serum, which has a similar composition to plasma without the clotting agents (fibrinogen and various factors).

Salts

The main salt in plasma is sodium chloride (NaCl), with smaller amounts of $NaHCO_3$, KCl, $CaCl_2$, MgCl, etc. In addition to their specific functions mentioned elsewhere, such as action potential mechanisms, sodium pumps, etc., these salts act as a protein solvent, allowing the plasma proteins to be carried in solution. The concentration of sodium chloride and the other salts in plasma is regulated within narrow limits by intake and excretory mechanisms.

Blood is well buffered at a normal pH of 7·4. This is achieved mainly by the interaction of CO_2 with the sodium bicarbonate in the plasma and erythrocytes.

Plasma proteins

Traditionally the plasma proteins were fractionated into three groups: albumin, globulin and fibrinogen (Fig. 4.17); the first two by salt precipitation and the last because it could be readily identified in the clotting mechanism. Modern methods of separation have demonstrated the presence of a multitude of different proteins (Table 4.2). Quantitatively the most important is albumin, with the various globulin fractions ($\alpha 1$, $\alpha 2$, β and immuno-) contributing much smaller amounts. Most are made in the hepatocytes of the liver; but the antibodies—immunoglobulins—are produced by the cells of the lymphoid system. About 10 percent of the plasma albumin is replaced per day. The plasma proteins have various functions, summarised here and considered elsewhere: (1) they provide an osmotic pressure of about 25 mmHg across the capillary wall, for they do not readily cross into the interstitial fluid, and so—with the blood pressure—determine fluid distribution between blood and the rest of the extracellular compartment; (2) they provide about 15 percent of the total buffering capacity of the blood, due to the weak ionisation of

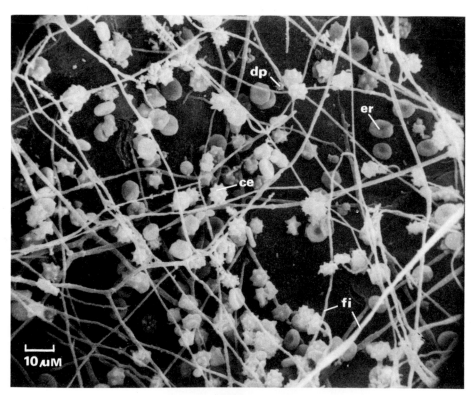

Fig. 4.17. Scanning electron micrograph of a portion of a blood clot showing fibrin threads (fi) forming a meshwork around platelets and erythrocytes (er). The erythrocytes are aggregated into columns called rouleaux formation. Platelets (dp) and leucocytes (ce) also shown. (From Freeman & Co., with permission.)

Table 4.2. Properties of some of the plasma proteins (from Passmore & Robson, 1976)

Name of protein	Plasma concentration (normal values) (g/l)	Molecular weight	Half-life (days)	Comments
Albumin	40	69,000	20	Important in maintaining plasma volume and distribution of extracellular fluid. Has several transport functions; minor role as a buffer.
α_1-globulins				
α_1-antitrypsin	3	45,000	6	Antiproteinase
α_1-acid glycoprotein	0·8	45,000	—	Function unknown
HDL (high density lipoproteins)	3·5	200,000	—	Lipid transport
α_2-globulins				
Ceruloplasmin	0·4	160,000	—	Copper transport
Haptoglobins	1·2	100,000	—	Several phenotypes. Glycoproteins that bind haemoglobin, and which thus may help to conserve iron.
α_2-macroglobulin	3·0	800,000	—	Antiproteinase; also possible transport function
VLDL (very low density lipoproteins)	1·5	10,000,000	—	Lipid transport
β-globulins				
Transferrin	2·5	90,000	11	Iron transport
Hemopexin	1.0	80,000	—	Binds haemoglobin
C_3 (β_{1C}-globulin)	1·2	185,000	—	Third component of complement
C_4 (β_{1E}-globulin)	0·4	240,000	—	Fourth component of complement
Plasminogen	0·7	140,000	—	Fibrinolysis
Fibrinogen	4.0	350,000	3	Fibrin formation
LDL (low density lipoproteins)	4·0	2,300,000	—	Lipid transport
Immunoglobulins				
IgA	2·5	170,000	6	Antibodies, each group having somewhat different functions. They may mainly move as γ-globulins on zone electrophoresis, but some migrate as β-globulins or as α_2-globulins (Fig. 4.15b)
IgD	0·02	160,000	3	
IgE	Trace	190,000	2	
IgG	10	155,000	24	
IgM	1·0	950,000	5	

their constituent COOH and NH_2 groups; (3) their specific functions include the clotting mechanism of the fibrinogen fraction, the carrier mechanism of the binding proteins which provides a reservoir of hormones (e.g. thyroid) in the blood and prevents their rapid loss through the kidneys, the carriage of antibodies to the tissues as the globulins and the carriage of metals, ions, fatty acids, amino acids, etc. by albumin.

Haemostasis

If a small blood vessel is cut a repair mechanism is activated which seals off the cut; this process involves the conversion of fibrinogen to fibrin in the blood which forms a clot at the site of the injury. The clot then forms a scaffold on which new tissue is built. The series of events—haemostasis—which leads to this final result involves several stages: (1) constriction of the blood vessel to narrow the lumen; (2) formation of the plug of platelets; and (3) the conversion of this plug into a clot of fibrin (Fig. 4.18).

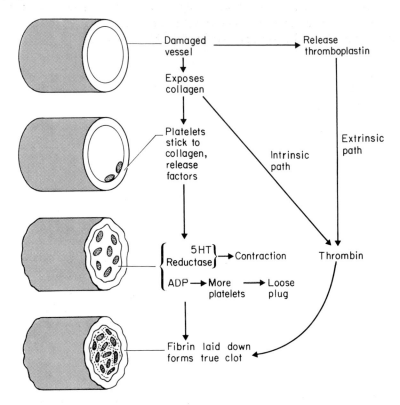

Fig. 4.18. Steps in the formation of a clot in a damaged vessel.

The trigger for haemostasis is an injury to the endothelium of the blood vessel, exposing the underlying layer of collagen. Platelets in the blood stick to this exposed collagen and secrete granules containing 5-HT,* adrenaline, ADP and several clotting factors. The 5-HT and adrenaline cause the blood vessel to constrict, reducing the blood escape, particularly if the cut is 'clean' rather than irregular. The ADP released by the first few platelets make others sticky so that they form a loose plug of aggregated platelets in the vessel. This loose plug is bound together and converted into a clot by fibrin. Fibrin is a long chain protein molecule composed of many subunits; it is rather like a long necklace made up of beads. In its manufacture each 'bead' is derived from a fibrinogen molecule by splitting off a few polypeptides; the residues are then joined together to form fibrin (see Fig. 4.20). The mixture of platelets and fibrin molecules is initially a loose network of strands, but is soon stabilised by covalent cross-linkages to form a tight, strong clot.

* 5-Hydroxytryptamine or serotonin.

The conversion of fibrinogen to fibrin is catalysed by thrombin which in turn is formed in the plasma from its circulating precursor, pro-thrombin (Fig. 4.19). This latter conversion depends on the presence of the activated form of factors X and V, lipids and Ca^{2+}. The formation of the active form of factor X can occur either by an *intrinsic* pathway, which probably starts with activation of factor XII by collagen and proceeds by a pathway involving lipids released by platelets, or an *extrinsic* pathway which depends on the release of a substance—throm-boplastin—from the damaged walls of the blood vessel.

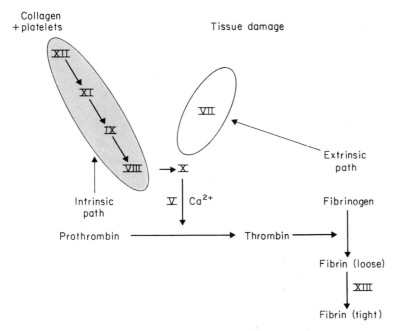

Fig. 4.19. Proposed pathways and reactions involved in the activation of fibrinogen to fibrin (after Ganong).

Anti-clotting mechanisms

Blood does not normally clot in undamaged blood vessels of the body. This is partly due to mechanisms which tend to stop clotting and partly to mechanisms which break up any clots which do form. The former mechanisms include: (1) the removal of activated factors by the liver; (2) the presence in circulating blood of heparin, an anti-coagulant which stops the activation of one of the factors (IX) in the intrinsic pathway and which, with a plasma cofactor, inhibits the action of thrombin; and (3) factors which reduce the adhesiveness of platelets, so that their 'stickiness' at any time is determined by a balance between those which increase and those which decrease it. Once a clot is formed spread is limited because it releases anti-clotting factors.

Clots are removed by a fibrinolytic system, which degrades fibrin and fibrinogen to products which themselves inhibit thrombin (Fig. 4.20). The active component of this system is fibrinolysin (plasmin) which is formed from profibrinolysin (plasminogen) by the action of thrombin

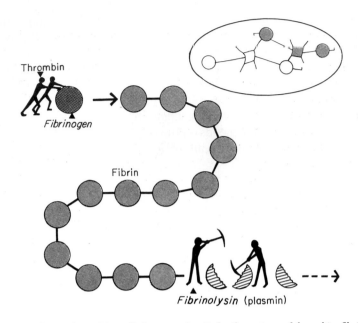

Fig. 4.20. Fibrin is formed from fibrinogen subunits by the actions of thrombin; fibrin is broken down into its degradation products by fibrinolysin (plasmin). The amount of fibrin present at any one time depends on the balance between these processes. The fibrinogen and fibrin monomers are more complex than shown; they consist of single units (pink, see inset) which combine to form a long double strand as shown, linked by co-valent bonds. These strands are then woven together to form a sheet.

and other substances. Fibrin is therefore made and broken down by systems which depend on the activation of thrombin; normally a balance exists with little clot formation. During injury the reactions seem to be biased to favour clot formation. The coagulation mechanism therefore represents a delicate balance between many factors, the specific functions of which are still poorly understood. The current view is that these sequences of events, platelet aggregation, coagulation and then fibrinolysis, occur continuously at a multitude of sites throughout the body where small damage has occurred due to daily traumas.

In medical practice clotting in blood vessels does sometimes happen, for reasons which are at present poorly understood. Such thrombosis tends to occur in blood vessels damaged by arteriosclerotic plaques (e.g. coronary vessels) and in vessels with a sluggish blood flow (perhaps due to a build up of activated clotting factors). Parts of these thrombi may break off (emboli) and lodge further down the vascular tree. Abnormalities of clotting also occur, due to a variety of causes often genetic in origin. For example, congenital absence of factor VIII is classical haemophilia, a sex-linked disorder, and of factor IX is Christmas disease.

Blood may be prevented from clotting in test tubes, etc. by removal of Ca^{2+}, thus preventing the formation of thrombin from prothrombin (this explains non-clotting of menstrual blood). Common methods are to add oxalates, which forms an insoluble salt with Ca^{2+}, or to add citrate—or other chelating agents— which bind Ca^{2+}.

Lymph

The lymphatic system acts both as an overflow mechanism which drains excess tissue fluid and protein back into the veins, and as a system for circulating 'competent' lymphocytes for immunological purposes. Most lymph returns via the thoracic duct (Figs. 4.10 and 4.16). To perform these functions the lymph vessels must have a high permeability to protein and must remain patent when the pressure around them increases. The current view is that the high permeability is due to the absence of basement membrane and the presence of numerous 'flaps' between the cells of its wall; the patency is due to the anchoring of the lymph vessel walls by numerous filaments to the surrounding structures. The amount of protein in lymph is variable, from about 80 percent of that in plasma in the liver to 10 percent in the legs; the concentration is sufficient to cause lymph to clot when left standing. This surprisingly large amount of protein reflects the relatively high permeability of capillary walls to proteins and other large particles. After a fatty meal lymph is milky due to the presence of large numbers of fatty globules absorbed from the intestine.

Chapter 5
The Cardiovascular System

CIRCULATION OF THE BLOOD

The function of the vascular system is to convey nutrient and other materials to and from the various parts of the organism; the blood serving as the intermediary between the environment and the cells. To fulfil this function the blood must be kept in a state of continuous circulation. The heart provides the energy for this circulation and the blood vessels are the channels through which it takes place. The force of cardiac contraction drives the blood to the tissues through thick-walled vessels, the arteries, and back to the heart by a system of thinner-walled vessels, the veins. In the tissues the blood is driven through a fine meshwork of vessels, the capillaries, whose walls consist of a single layer of cells which allows interchange of materials between the blood and the tissue fluid (Fig. 5.1).

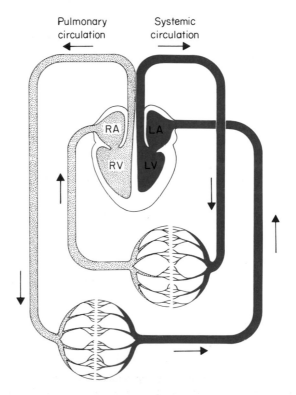

Fig. 5.1. Diagram of the mammalian circulation. The capillary beds can be considered as two 'trees' connected together by their fine branches (= capillaries) (partly after Vander *et al.*).

Outline of function

The heart becomes filled with venous blood during its relaxation, or diastole, and forces the blood into the arteries during its contraction, or systole. The energy for moving the blood comes from the muscular walls of the chambers of the heart; the direction of flow of the blood is largely determined by the presence of valves at the entry to and exit from the

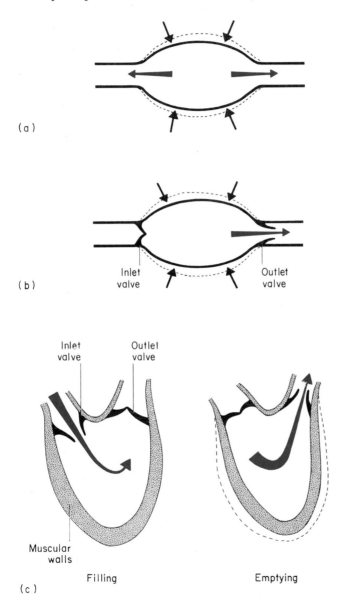

Fig. 5.2. Principle by which the ventricles impart motion to the blood. Compressing a rubber bulb (a) causes the contained fluid to move out of it along both tubes. If valves are positioned at each end of the chamber (b) then fluid is moved in one way only. In the ventricle (c) the walls are muscular and the chamber is bent back on itself so that the inlet and outlet valves are quite close to each other.

ventricles (Fig. 5.2). Each ventricle is preceded by another chamber, the atrium, which 'receives' the blood entering the heart. Under resting conditions the atrium probably does not take much part in moving the blood onwards.

The valves of the heart are fibrous flaps of tissue covered with endothelium. Those guarding the atrioventricular opening have strands of fibrous tissue connecting them to small muscles within the ventricles; these papillary muscles strengthen the valves during contraction and help the ventricle to empty later in the cycle. The valves on the exit from the ventricles (aortic and pulmonary) are rather more filamentous than the atrioventricular valves and float open during ejection of blood from the ventricle (Fig. 5.3).

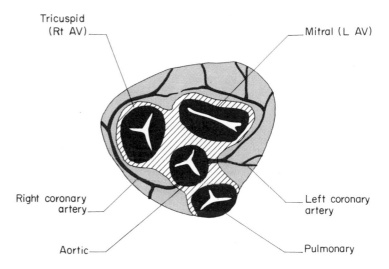

Fig. 5.3. View of the fibrous framework of the human heart from above showing the valves. These valves are surrounded by fibrous tissue rings which form the fibrous skeleton of the heart. To this skeleton are attached the muscles of the two ventricles and the two atria, the aorta and pulmonary artery.

Physiological properties of the heart

Heart muscle cells have properties somewhere in between those of skeletal and smooth muscle. They are small, striated and branching cells with a single nucleus (Fig. 5.4). Each cell is connected to its neighbour by intercalated discs, within which are 'gap-junctions' (p. 36). These junctions allow the passage of electrical currents and small molecules between adjacent cells. A piece of heart muscle therefore behaves as a single unit (syncytium) to electrical stimulation, rather than a group of isolated units as does skeletal muscle. Certain heart cells are specialised to conduct action potentials faster than normal; they also contract with less force than normal heart cells; these cells make up the conducting tissues of the heart. As in other muscle cells the visible contraction of the cell is preceded by an electrical event—the action potential—which depolarises the cell and triggers the mechanical events.

The properties of heart muscle will be discussed in terms of the current

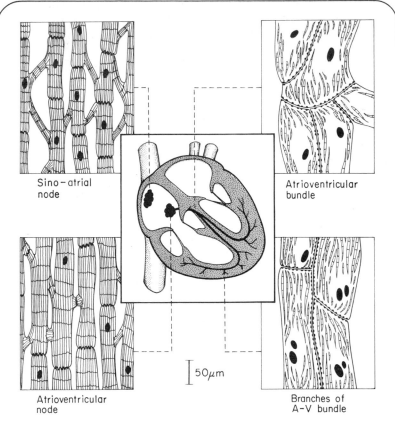

Fig. 5.4. Diagram of cardiac muscle cells in various parts of the heart. Each cell is short and has several branching processes. The processes of adjacent cells are joined end to end at structures known as intercalated discs. The size of these discs and the desmosomes are slightly exaggerated (after Rhodin, Missier & Reid).

views of the heart, first by considering the cardiac action potential and then the contraction.

The cardiac action potential
The cardiac action potential is basically rather similar to that in nerve and muscle (p. 29) but lasts for a much longer time; its precise shape and duration also depends on which part of the heart it comes from. It may be divided into four components: depolarisation, plateau phase, repolarisation and pacemaker potential. In general, the properties of the cardiac action potential are similar to that in nerve—in particular there are permeability differences between heart and nerve which 'explain' the differences in shape of the action potential (Fig. 5.5).

Depolarisation (really the upstroke/overshoot) in heart, as elsewhere, is due to a large increase in sodium permeability, which causes the membrane potential to be reversed producing the overshoot.

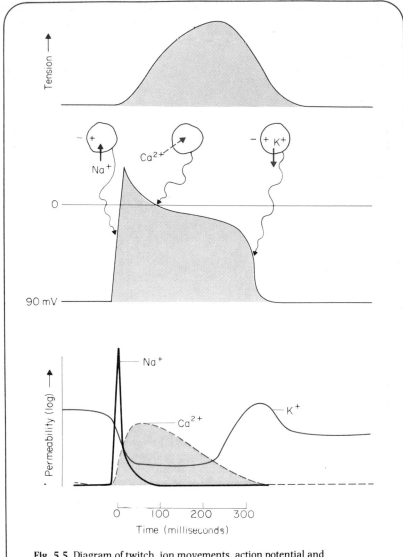

Fig. 5.5. Diagram of twitch, ion movements, action potential and permeability changes in a ventricular muscle cell (partly after Noble). See text for details.

The plateau phase is much longer than that in nerve because: (1) the potassium permeability drops at the onset of the plateau and rises slowly towards its resting value during the plateau; (2) there is a rise in the calcium permeability at the start of the plateau with some decline during the plateau; (3) the chloride permeability of heart is low, so that there is little short circuiting of the potential during the plateau. The combined effect of these changes is to produce a long plateau with a small gain of Na^+ and Ca^{2+} and a small loss of K^+. This means that the cardiac action potential is not too 'wasteful' in ion changes, and the gain in Ca^{2+} is available for coupling to the actomyosin mechanism. The cell remains refractory during the action potential and so may only be stimulated (by a large stimulus) after its end; the refractory period is thus some hundreds of milliseconds long rather than the few milliseconds of nerve or skeletal muscle.

Repolarisation is associated with a rise in potassium permeability and a return of the calcium permeability to normal.

Pacemaker activity. Atrial and ventricular muscle cells have high resting potentials (− 80 mV or more) and show no spontaneous activity; they are normally excited by impulses spreading from adjacent cells. Usually SA and AV node cells have potentials which never reach more than − 60 mV, whereas Purkinje fibres have potentials of − 70 to − 90 mV. Cells in the sino-atrial (SA) node, the atrioventricular (AV) node and the Purkinje fibres do not have a steady resting potential but show pacemaker activity (Fig. 5.6). This means that the potential is at a maximum negative value

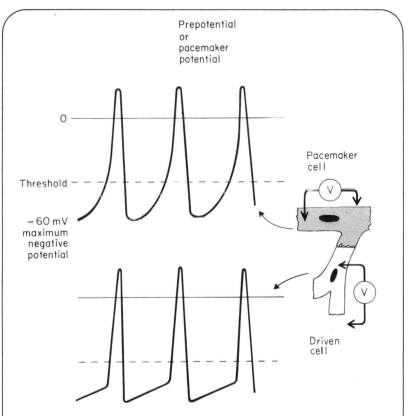

Fig. 5.6. Diagram showing electrical recordings from adjacent cells in the SA node: the pacemaker cell (above); and an adjacent cell driven by the pacemaker cell (below). Note in the pacemaker cell the gradual transition from the pacemaker potential to the action-potential upstroke at the threshold potential but the abrupt change of slope in driven cells. The membrane potential of pacemaker cells is usually low, as here, due to a high sodium permeability. Perhaps only one cell in the SA node acts as a pacemaker at any one time and a different cell acts as pacemaker at different times.

just after an action potential, but then drifts slowly towards the threshold potential; an action potential being set up when it is reached. The current explanation for this behaviour is that (a) there is a decline in the potassium permeability during the pacemaker potential and (b) the sodium permeability of pacemaker cells is higher than in other parts of the heart; the decreasing potassium permeability in the presence of the high sodium permeability leads to the decreasing potential across the cell membrane. Normally the slope of the pacemaker potential in the SA node is steeper than that of the AV node which is steeper than that of the Purkinje fibres; therefore if these parts could be separated the SA node would beat at the fastest rate and the Purkinje fibres at the slowest rate. In an intact heart the SA node, therefore, usually determines the heart rate, and drives all the others at its rate.

These modern observations on pacemaker potentials 'explain' the old observation that isolated heart is spontaneously active whereas skeletal muscle is not.

The transmitter chemical ACh, released by the vagus, increases the potassium permeability of special channels in the membranes of heart cells and so decreases the slope of the pacemaker potential; the threshold for firing an action potential is reached later and so the heart slows down. Noradrenaline, released by the sympathetic nerve endings, or adrenaline, circulating in the blood have several effects on the heart; mediated by β receptors and cyclic nucleotides. One of these is to increase calcium entry, which steepens the prepotential and so speeds up the heart rate (Fig. 5.7).

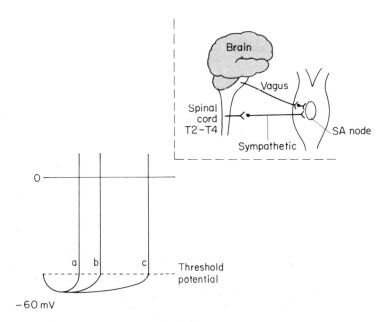

Fig. 5.7. Upstroke of the action potential in the pacemaker region: with no nerve stimulation (b); with sympathetic stimulation (a); and with vagal stimulation (c). The sympathetic effects are mediated by the β-receptors—one of the two classes of receptors excited by adrenergic molecules. The inset shows the cardiac innervation.

Conduction of the action potential over the heart

The SA node makes contact with adjacent atrial cells and so causes them to be depolarised by conduction through the gap-junctions of the intercalated discs. These atrial cells in turn depolarise their neighbours so starting action potentials (see p. 93). In this way a wave of electrical activity spreads throughout the right atrium and, as the atria are connected together, also spreads over the left atrium. This wave has been likened to the ripples of a pool following a stone being thrown into it (Fig. 5.8). Many experiments have confirmed that the atria have no specialised conducting tissue but simply spread electrical activity from one cell to the next.

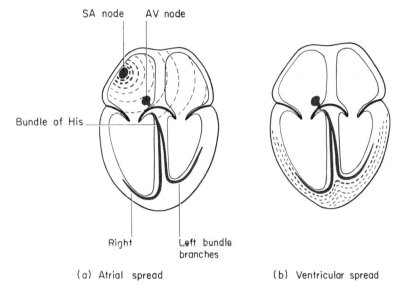

(a) Atrial spread   (b) Ventricular spread

Fig. 5.8. Spread of electrical activity throughout the human heart. The conduction velocity in atrial and ventricular muscle is about 1 m/s, in the AV node $\frac{1}{4}$ m/s and, in the bundle of His, 5 m/s, i.e. in the ratio of 1, $\frac{1}{4}$ and 5. Ventricles from small animals (for example, a mouse) do not have Purkinje fibres, presumably because conduction time is fast enough without them.

The atrioventricular (AV) node lies at the base of the right atrium near the wall between the ventricles. It connects with the Purkinje fibres which lie in the bundle of His in the interventricular septum; this connection is the only electrical link between the atria and the ventricles. Purkinje fibres conduct action potentials at higher velocities than normal ventricular cells and have only very feeble contractile properties.

The wave of electrical activity reaches the AV node, passes slowly across it and then rapidly down the conducting bundle to reach the ventricular muscle. At the ventricular muscle the wave of activity again passes slowly from cell to cell. This arrangement ensures that: (a) there is a delay between atrial and ventricular excitation (say 0·1 second), allowing time for blood to leave the atria; and (b) excitation reaches most ventricular cells simultaneously, ensuring a single coordinated contraction.

Contraction of heart muscle cells

In heart muscle the coupling between electrical and mechanical activity is triggered by the Ca^{2+} which enters the cells during the depolarisation phase of the action potential; the prolonged length of the action potential causes the prolonged contraction. This is clearly a 'good thing' in that blood has a high inertia and so a prolonged force must be applied to it to cause it to be ejected from the heart; short sharp contractions, like those seen in skeletal muscle, would be ineffective. Although Ca^{2+}, which enters heart cells during the action potential, is involved the detailed way in which electrical events are coupled to mechanical events in the heart is still not clear. It has recently been suggested that variation in the concentration of the cyclic nucleotides may explain the variable coupling which occurs between action potentials and their associated mechanical twitches.

Skeletal muscle has a short refractory period because the associated action potential only lasts a few milliseconds. Skeletal muscle can therefore be tetanised (see p. 48), whereas heart muscle cannot. Clearly this is a valuable property of heart muscle; to have a continuous circulation the heart must contract and relax alternately; a tetanised heart muscle would stop the circulation and kill the animal.

Events of the cardiac cycle

Our heart beats in a more or less regular fashion some 3000 million times throughout our lives. It is convenient to discuss the electrical and mechanical events which occur during a single cardiac cycle as representative of all of them.

Electrical events

Each cell in the heart maintains a resting potential of some -60 to -90 mV at rest which changes to $+20$ to $+40$ mV during activity. This change in potential, occurring across the millions of individual cells in the heart, causes currents to flow throughout the body which appear at the body surface as small potentials. The electrical activity of the heart may thus be examined either by recording directly from heart cells with intracellular electrodes, or by recording from the rest of the body or the body surface with extracellular electrodes. The electrocardiograph (ECG) is an instrument for recording these electrical potentials from the skin surface (Fig. 5.9). The letters *P*, *Q*, *R*, *S* and *T* (first used by Einthoven) refer to the electrical waves set up by atrial depolarisation (*P*), ventricular depolarisation (*QRS*) and ventricular repolarisation (*T*). The atrial repolarisation wave is normally hidden by the *QRS* complex. The physiological points shown by the ECG are:

1 The *PR* interval, of about 0·1 second, measures the conduction time of the impulse from the atria to the ventricles. It thus measures the time required for the following events: (a) passage of the atrial depolarisation wave to the vicinity of the AV node; (b) the delay imposed by the AV node; and (c) the rapid spread of activity in the AV bundle and its branches. The main delay is actually that in the AV node.

2 The duration of the *QRS* complex, about 0·12 second, measures the time of spread of the impulse over the ventricles. In high-fidelity recordings a notch can often be seen in this wave due to excitation reaching

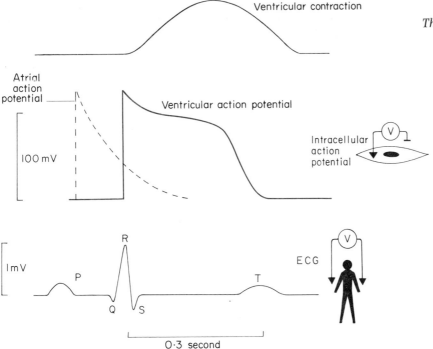

Fig. 5.9. Comparison of electrical and mechanical events in the heart. Note the time relations of the ECG and the intracellular action potentials. The voltage of the ECG is only about 1 percent of that across the membrane of a single heart cell.

one ventricle before it reaches the other.

3   The *RT interval*, of about 0·3 second, measures the mean duration of the action potential in the ventricles.

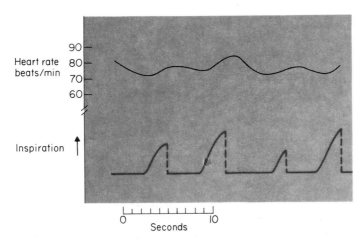

Fig. 5.10. Sinus arrhythmia. The subject took alternate normal and deep breaths. The normal breath changes the heart rate by about 5 percent; the deeper breath by more. The heart rate response starts about 2 seconds after the start of inspiration. The heart rate record is in beats/min. The respiratory record measures the cumulative amount of air taken into the lungs up to the moment of measurement; expiration is not measured (record from Dr S. Jennett).

4   Inspection of a series of *R* waves (Fig. 5.10) gives information about the regularity of the heart beat. The commonest variation in heart rate described in normal—particularly young—subjects is *sinus arrhythmia* (Fig. 5.10a), which is a variation of the heart rate with the phases of respiration; during inspiration the heart rate speeds up, during expiration it slows down. Two explanations are advanced for this phenomenon; either that the cardiovascular centre is affected by irradiation of impulses from the respiratory centre or that it is affected by impulses from stretch receptors originating in the lungs. Other variations in heart rate are due to the blood pressure regulation mechanism (with a periodicity of around 10 s) and to temperature regulation (a periodicity of several minutes).

Clinically the ECG is used to detect alterations in impulse propagation, cell damage, etc., which occurs in various heart defects.

Mechanical events

The events occurring during a single heart cycle can be demonstrated by measuring the pressures in the various chambers of the heart, together with volume measurement and measurement of the ECG and heart sounds.

Students should examine Fig. 5.11 and accompanying text and work their way through a single cardiac cycle, understanding the various changes which occur. Remember that blood, like all substances, flows from a region of higher pressure to one of lower pressure. (In this account the same numbers are used on the main figure, the inset diagrams and in the text.) This account describes the left heart, events of the right heart are similar but the pressures are lower.

Diastole. (1) In early diastole the ventricle is at its lowest volume (65 ml). The pressure in the pulmonary vein is slightly greater than that in the left atrium which in turn is slightly greater than that in the left ventricle; blood therefore flows into the ventricle down this pressure gradient and distends the ventricle. By mid-diastole the ventricle is almost filled although no atrial contraction has yet occurred. In late diastole, electrical activity, followed by contraction of the atria, occurs; this drives a small, extra volume of blood into the ventricle (particularly at fast heart rates) and also causes a back wave in the venous system.

Isometric* ventricular contraction. (2) Electrical activity spreads to the ventricle, followed by ventricular contraction. The pressure in the ventricles starts to rise quickly, exceeds that in the atria and so the AV valves close (= the first heart sound). The ventricles are now closed cavities and so the pressure rises rapidly. When it exceeds that in the aorta then:

Ventricular ejection (3) of blood from the left ventricle occurs. The ventricular pressure remains just greater than that in the aorta (the pressure difference is not large as the filamentous aortic valves offer little resist-

* Isometric: the muscle fibres remain the 'same length'.

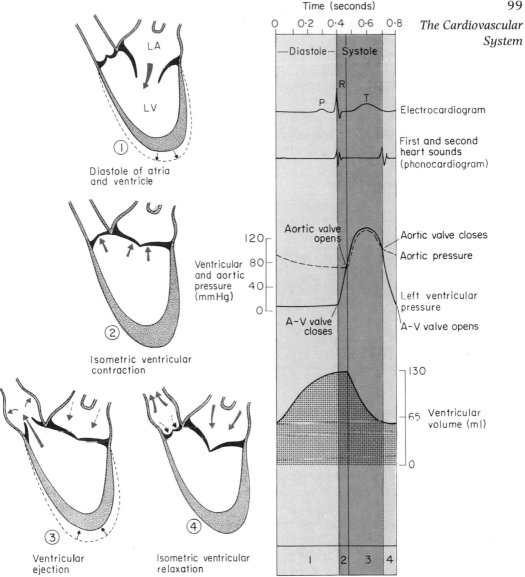

Fig. 5.11. Diagram of electrical and mechanical events in the heart (left side) during a single cycle at a heart rate of 75 contractions/min. The insets show the changes diagrammatically. The small effect of atrial contraction has been omitted.

ance to blood flow) causing the ventricles to eject blood rapidly. The aorta receives blood so fast that: (a) its pressure goes on increasing; and (b) it expands, thus 'storing up' some blood in it. Gradually the rate of ejection of blood falls; the ventricle is repolarised and stops contracting. The ventricular pressure now falls below that in the aorta, the pulmonary and aortic valves snap shut (=the second heart sound) and so the phase of:

Isometric ventricular relaxation (4) occurs; in this the ventricle is once more a closed cavity as both inlet and outlet valves are again closed, so the pressure drops from about 100 mmHg towards zero. When it is less than the atrial pressure the AV valve opens and the cycle restarts.

Additional points

1   The ventricle does not empty completely on contraction; the difference between the end-diastolic and the end-systolic volumes is the stroke volume.

2   The rate of filling of the ventricles is exponential not linear and so filling is largely complete by mid-diastole. This means that the marked shortening of diastole which occurs with increase in heart rate (Fig.

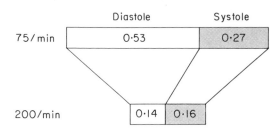

Fig. 5.12. Relative durations of ventricular diastole and systole at normal and fast heart rates. With increasing heart rate diastole and systole both shorten—to minimum values of 26 and 60 percent of the resting durations, respectively. The durations of the ventricular action potential is almost the same as that of systole; the refractory period is slightly shorter. Times are in seconds.

5.12) does not greatly decrease the amount of blood entering the heart per beat. The rate of emptying of the ventricle also has an exponential time course and so is not much affected by the moderate shortening of systole at high heart rates; in addition the rate of rise and fall of tension is greater at fast heart rates due to the presence of catecholamines.

3   *The arterial pulse.* The ventricular contraction not only pushes blood into the aorta but also sets up a pressure wave which travels down the arterial walls. This wave travels much faster (about 5 m/s) than the blood which is ejected into the aorta; so it reaches the wrist in about 0·1 second. The strength of this 'pulse' is related to the pulse pressure* rather than to the actual pressure within the artery and so usually varies with the stroke volume of the heart. If the stroke volume is low, for example, following blood loss or weakening of the heart, then the pulse feels weak; after exercise it is large and can be felt by the individual as a 'thumping' in the body.

4   *Venous pulsations.* There are no valves on the inlet side to the atria (although folds of muscle partly act as such) so that pressure changes in the atria are transmitted back up the veins, and may be observed in the neck. Classically three waves, a, c and v, are described due to atrial systole, the bulging of the AV valve during ventricular contraction and the rise in atrial pressure before the AV valve opens during diastole. The venous pressure waves can readily be seen in the neck of someone lying flat (Fig. 5.13) and can often be noticed in 'close-ups' of someone singing, particularly when sustaining a long note. (The reason is that in this

---

* Difference between systolic and diastolic pressures.

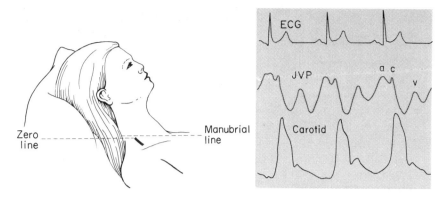

Fig. 5.13. Observation of pulsations in the neck. Either examine someone else or yourself in a mirror (to increase venous pressure hold your breath and blow). Venous pulsations can be seen but not felt, carotid artery pulsation can be seen and felt. The effects of respiration, intrathoracic pressure, etc. can be readily examined. The inset shows the relationship between the ECG, jugular venous pressure (JVP) and the carotid pulsation. The central venous pressure can be measured as the height of the jugular pulsation above the zero line (cm blood ÷ 1·36→mmHg) (inset after Ganong).

procedure the intrathoracic pressure rises and so 'dams back' the blood in the neck.)

5   *Heart sounds.* Heart sounds were originally detected by the physician putting his ear to the patient's chest (perhaps using a silk handkerchief), later by listening through a solid wooden stick or trumpet and now via a stethoscope. They are the first and second heart sounds, traditionally called 'lub' and 'dup'; they are caused by the closure of the AV valve and of the aortic and pulmonary valves respectively.

THE VASCULAR SYSTEM

The left heart ejects blood into the aorta, this creates a high pressure there which drives blood round the various parts of the circulation back to the heart (Fig. 5.14); the pressure in the aorta is therefore greatest and that in the right atrium least. The right heart drives blood through the pulmonary circulation; as the resistance to flow is thus much lower than that of the systemic circulation it requires a much smaller pressure to do so. (The pulmonary circulation is dealt with on p. 180.)

Functionally, the systemic* circulation (Fig. 5.15) may be divided into:

1   A distributory system consisting of the aorta and arteries which contain a small amount of blood, held at high pressure.

2   A variable resistance system consisting of the arterioles, where most of the arterial pressure is dissipated.

3   An exchange system consisting of myriads of capillaries with a vast surface area, where the interchange of substances with the extracellular fluid of the tissues occurs.

---

* That supplying blood to the body from the left ventricle.

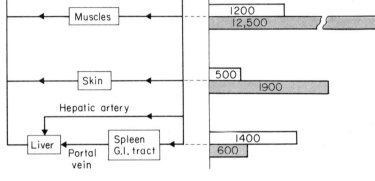

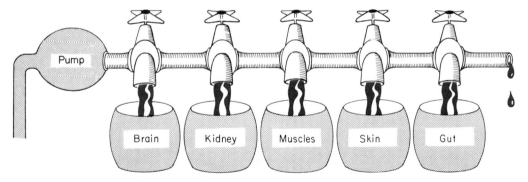

Fig. 5.14. Schematic design of the circulation showing the main functional groupings of blood flow, with typical flows at rest and during severe exercise. The total blood flow in a 70 kg man is about 6 l/min at rest and 18 l/min during exercise (some blood flow occurs through channels not shown).

Fig. 5.15. Cartoon of the circulation. The upper pipe represents the aorta and arteries (high-pressure reservoir), the taps represent the arterioles and the tubs the organs supplied. By varying the size of the arterioles the blood flow to an organ is controlled; opening all the taps fully causes circulatory collapse (after *The Observer* 1978).

4 A collecting and reservoir system of veins, venae cavae and right atrium which contains most of the blood, at low pressure.

Arteries

The arteries have two main functions; first they act as low-resistance elastic pipes taking blood from the heart to the arterioles in the various tissues; secondly they store blood during diastole. Blood is ejected in spurts from the heart at each contraction, is partly stored in the aorta and large vessels and runs away through the arterioles at a continuous rate. This 'smoothing out' function of the arterial system is due to the combination of the capacity effect of the elastic vessels and of the resistance to flow offered by the arteriolar resistance. You may examine this as follows: take a very hot bath, thus dilating the skin arterioles (see later), you will find that a pulsation appears in your nail-beds; this is because the arteriolar resistance of the skin has been decreased thus transmitting pulsatile pressure changes to the capillaries.

Structure

Arteries, arterioles and veins have three coats, an inner smooth layer, a middle layer which contains a variable amount of elastic and smooth muscle cells and an outer layer of connective tissue (Fig. 5.16). The

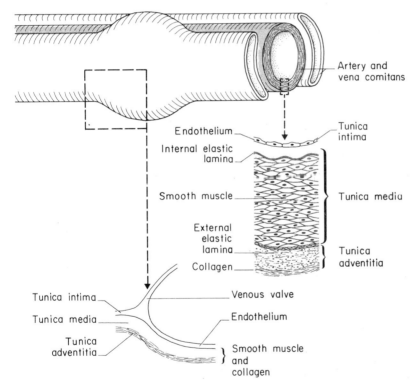

Fig. 5.16. The structure of a small muscular artery and its accompanying veins. The tunica adventitia of the veins and artery are continuous with each other so that pulsation within the artery compresses the veins and so forces the venous blood towards the heart (from Passmore & Robson).

proportions of each component vary with the function of the vessel (Fig. 5.17). The smooth layer reduces friction to blood flow; the elastic tissue gives the vessel a storage function, the muscle regulates the vessel size; and the connective tissue outer coat prevents over-distension as well as anchoring the vessel to surrounding tissue.

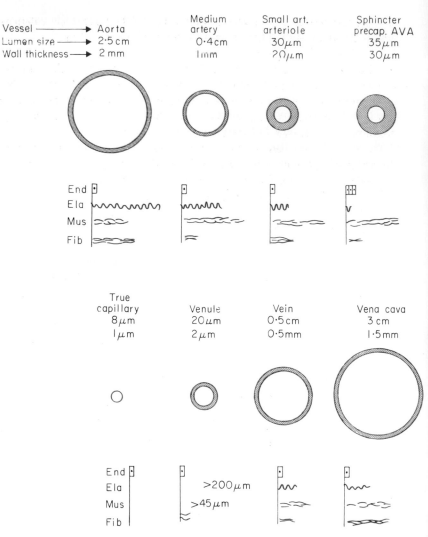

Fig. 5.17. The size, thickness of the wall and relative amount of the main constituent tissues of different blood vessels. End = endothelial lining cells; Ela = elastic fibres; Mus = smooth muscle cells; Fib — collagen fibres (after Burton 1954).

Blood flow in arteries

The blood flow through arteries and the large veins near the heart is pulsatile; the fact that this occurs in vessels whose walls have elastic and viscous properties leads to a complex relationship between pressure and flow, with some odd phenomena. In this section we consider first the much simpler relationships which hold if pressure and flow are considered as non-pulsatile, and later mention some of the more complex findings in pulsatile systems.

Various physical relationships need to be considered before dealing with this:

Pressure, wall properties and radius. Laplace showed that the pressure within a vessel required to keep it open depended both on the properties and on the radius of the vessel walls (Fig. 5.18). Four consequences which are applicable to the vascular system follow from this:

1 A direct consequence of the equation is that very small vessels require much thinner walls to balance internal pressure than larger vessels do; this is why capillaries can be just one cell thick while veins (at a lower pressure) have to be relatively thick. A somewhat similar situation occurs in trying to blow up a balloon; at small sizes large pressures are required, at larger sizes little pressure is required.

A general consequence of Laplace's equation is that it adjusts the wall thickness of a vessel according to the radius and pressure within it. Thin-walled veins used to replace diseased arteries, soon come to have the same wall thickness as the artery they replaced.

2 *Critical closing pressure.* In very small vessels the wall tension may still be too large for the internal fluid pressure; if this is so, the vessel closes. This is the critical closing pressure (Fig. 5.20).

3 *Efficiency of cardiac pumping.* If the heart expands, then any particular degree of contraction produced by the muscular walls ($=T$) will produce a smaller rise in internal pressure (for $T=\frac{1}{2}RP$). Dilated hearts are therefore less efficient than normal sized ones.

4 *Curved vessels.* Any curvature reduces the effect of internal pressure trying to distort a vessel wall; so when a vessel is curved (for example arch of the aorta) the internal pressure has a greater effect on the inner (convex) side rather than the outer (concave) side. To oppose this pressure the wall is thicker on the inside of the curve than on the outside.

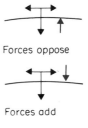

Forces oppose

Forces add

Arch of the aorta

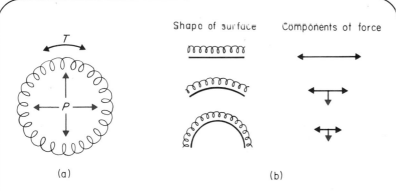

Shape of surface   Components of force

(a)   (b)

Fig. 5.18. (a) Relationship between the pressure (P) required to distend an elastic sphere or tube and the tension in the walls (T). T is a measure of the elastic tension in the wall (shown as a spring) and so must be overcome to keep the vessel open; the force required to do so depends on the radius of curvature (R) and the pressure in the vessel (strictly it should be the pressure difference between inside and outside).

$$\text{For a sphere } P=\frac{2T}{R} \qquad \text{For a tube } P=\frac{T}{R}$$

(b) An approximate reason for this behaviour is as follows: a flat spring only exerts a force in its own plane, a curved spring also exerts a force at right angles to the surface; as the curvature increases the inwards component increases and must be overcome to keep the tube open.

Laminar and turbulent flow. Fluid can either flow along a tube in a laminar or turbulent way. Laminar flow can be considered as a series of concentric layers sliding past each other (Fig. 5.19b), the outer ones nearly stationary, the inner ones moving faster; the energy required to overcome the friction to move one layer past the next is at a minimum and no noise is produced. In turbulent flow (Fig. 5.19a) the molecules hit each other and the vessel walls randomly; the energy required rises disproportionately to any increase in flow and the lateral pressure on the vessel wall increases; vibrations are set up which may be audible. Laminar and turbulent flows can be examined with any domestic tap; with the tap turned on slightly laminar flow occurs, with it turned on further a noisy, turbulent flow develops.

Turbulent flow hardly ever occurs in a normal circulation largely because in the situations where the velocity is high flow is also pulsatile (Fig. 5.19c).

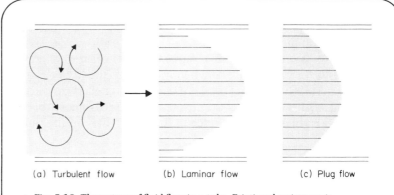

(a) Turbulent flow   (b) Laminar flow   (c) Plug flow

Fig. 5.19. The nature of fluid flow in a tube. Frictional resistance is produced by molecules jostling each other and the walls of the tube. In (a) the molecules move randomly and produce maximum resistance to flow; in (b) the molecules 'march in columns' and produce a marked velocity profile—this occurs if the flow is steady and even; in (c) the molecules also move in columns but the velocity profile is much flatter (plug flow)—this occurs when, as in the vascular system, the flow is pulsatile.

The likelihood of turbulence occurring increases as the velocity increases, but it occurs at much lower velocities with steady (b) than pulsatile (c) flows and so hardly ever happens in a normal circulation.

$$\text{probability of turbulence} \uparrow \text{ as } \left(\frac{\text{radius of tube} \times \text{velocity of flow}}{\text{viscosity}} \right) \uparrow$$

The relationship between pressure, resistance and flow. The vascular system consists of many narrow pipes with fluid flowing along them; so you need to know something about the relationships involved. The factors which determine the resistance to flow are:

1   The diameter of the lumen.
2   The length of the tube.
3   The stickiness, i.e. viscosity, of the fluid.

Common sense would suggest that it is more difficult to drive a sticky fluid, such as treacle, along a long narrow tube than it is to drive water along a short wide tube. 'Difficulty' is measured as a resistance to flow and is obtained from:

$$\text{resistance} \propto \left(\frac{\text{length} \times \text{viscosity}}{\text{radius}^4} \right)$$

The force required to drive fluid through a tube is measured as the pressure gradient between the ends of the tube; the relationship linking all these factors is similar to Ohm's law (Fig. 5.20), i.e.

$$\text{driving force} = \text{flow} \times \text{resistance to flow}$$
$$\text{or flow} = \frac{\text{driving force}}{\text{resistance to flow}}$$

therefore

$$\text{blood flow} \propto \frac{\text{pressure difference} \times \text{radius}^4}{\text{viscosity} \times \text{length}}$$

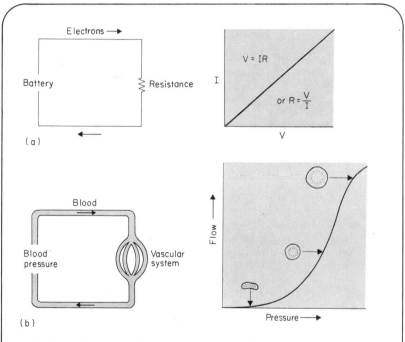

Fig. 5.20. Ohm's law: (a) A voltage (the driving force) drives a current (electrons — a substance) through a connecting pathway which resists the passage of the current. In a perfect system doubling the voltage (driving force) doubles the current flow, so that the 'resistance' behaves as a constant and can be calculated as shown. In many biological systems Ohm's law is used to calculate a 'resistance' to the flow of some substance. For example, (b) pressure–flow relationship in a blood vessel. Blood pressure drives blood through a capillary bed. If the flow for various pressures is measured it is found that a curve is obtained, not a straight line. A 'resistance' can be calculated from the slope of this line at any place; this resistance is not constant however, for it varies with the degree of swelling of the vessel.

This is known as Poiseuille's equation, it can be considered as the Ohm's law of fluid flow.

Normally the length of the blood vessels and viscosity of blood are fairly constant and so need not concern us. The radius of blood vessels does vary, and as flow is proportional to the fourth power of the radius quite small changes in radius lead to large changes in flow. Thus a 16 percent increase in the size of a vessel doubles the blood flow; doubling the radius increases the flow 16 times (or reduces the resistance to 6 percent). As we shall see, the main way in which the distribution of blood within the body is controlled is by variation in the size of blood vessels, particularly the arterioles.

In rigid tubes the flow increases linearly with pressure (Fig. 5.20) so that a 'resistance' can be calculated which is constant; in elastic tubes—such as blood vessels—increasing pressure distends the tube and so a non-linear pressure—flow relationship is obtained, this means that resistance to flow varies with the applied pressure.

Pressure—volume relationships; the stiffness of the vascular system. The 'stiffness' of different parts of the vascular system varies, arteries are very stiff, veins less so and the atria least. This means that arteries are difficult to distend whereas veins and the atria are easily distensible (Fig. 5.21).

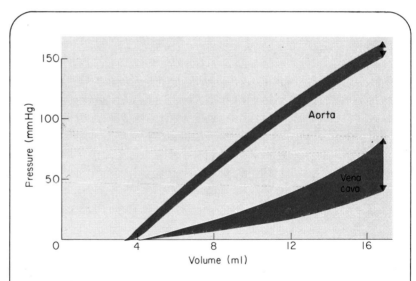

Fig. 5.21. Effects of filling short equal lengths of human aorta and vena cava with fluid. The vena cava can hold a large volume at low pressure, the aorta develops a considerable pressure if the same volume is forced into it. This property of veins is largely due to a change in their shape; at low volumes they are elliptical, at high volumes they become circular. Increase in activity of the smooth muscle in the vessel walls produces large changes in the compliance of the veins but little change in that of arteries (shown as stippled red). (Usually the word compliance is used to describe these properties of vessels; it is the reciprocal of stiffness; the aorta has a high stiffness or a low compliance.)

Blood flow is pulsatile not steady. This means that each 'particle' of blood near the heart is accelerated during systole and then slowed down during diastole; it also of course rubs against its fellows. Force needs to be applied to blood to overcome these two resistive components, inertial and frictional. With pulsatile flow the 'resistance' is replaced by 'impedance'; this contains two components, reactance and resistance. One consequence of this has already been mentioned (Fig. 5.19), i.e. the flattening of the velocity profile; others will be mentioned here.

Any one corpuscle in the aorta moves forward in a jerk with each heart beat, followed by standstill or even some backwards movement during diastole; progressively along the aorta, this pulsatile pattern of flow becomes damped by the storage function of the aorta until, in the smaller arteries, it is more nearly continuous (Fig. 5.22). Measurement of the pulse wave at various parts of the arterial system shows that the maximum instantaneous pressure increases as the wave travels away from the heart (Fig. 5.22), though of course the mean pressure decreases. The explanation for this phenomenon is that when the wave reaches a branch in an artery some reflection occurs from the junction, this adds to the original wave so producing the increase.

Arterial pressure pulse. The pressure in the large arteries varies in a roughly triangular or saw-toothed way (Figs. 5.22 and 5.23). The

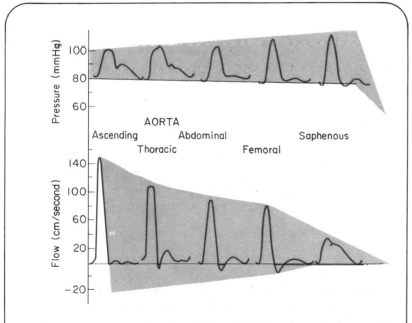

Fig. 5.22. Pressure and flow pulses in the dog arteries at increasing distances from the heart. Pulse pressure oscillation increases while flow oscillation decreases markedly. It will be noticed that the peak of flow slightly precedes the peak of pressure, this is because of the reactance of the system (after McDonald).

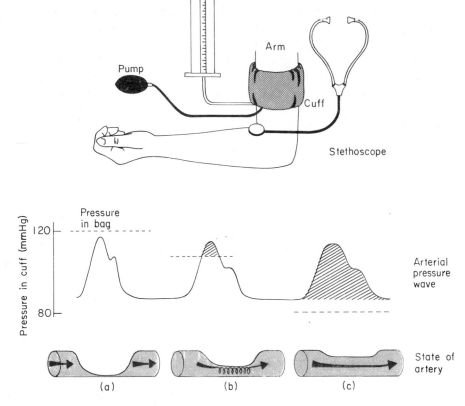

Fig. 5.23. Sphygmomanometer method of measuring blood pressure. A cuff is applied to the upper arm, this cuff contains a rubber bag which is connected to a mercury manometer and can be inflated with air (inset). The bag lies over the brachial artery so that when inflated it compresses it. If inflated above the systolic pressure (a) it completely stops blood flow so no pulse can be felt at the wrist or heard at the elbow; if inflated just below systolic pressure a little blood (at high velocity) pushes through at each systole, this flow is turbulent and so can be heard by the stethoscope (b). As the pressure is decreased further then the period of blood flow and hence turbulence lasts for a longer period each cycle until, when the pressure is below the diastolic (c), there is never any period of closure of the vessel during the cycle and therefore no turbulence occurs. Diastolic pressure is taken either as the pressure at which the turbulent note changes just prior to silence or at the point at which no turbulence occurs.

descending part of the wave has a notch—the dicrotic notch—corresponding to the closure of the aortic valve. The peak pressure is the systolic pressure, the lowest the diastolic pressure and the difference the pulse pressure; the mean pressure may be found by averaging over the cycle but is roughly the diastolic + one-third of the pulse pressure.

Measurement of arterial pressure
Arterial pressure may be measured directly by inserting a hollow needle into an artery and connecting this to a manometer. This procedure is frequently used in experimental animals but infrequently in man. Normally, in clinical practice, arterial pressure is measured by an external means. The usual method used is to apply a pressure to the outside of an

artery by means of an inflatable rubber bag held in a cloth cuff. This procedure interferes with the blood flow through the artery, causing the production of characteristic sounds (Korotkoff). The technique is: the bag pressure is inflated above that of the arterial pressure so that all blood flow stops; the pressure is then released slowly; the systolic pressure is taken as that pressure at which sounds first appear and the diastolic pressure as that pressure at which they become muffled. The sounds are due to partial occlusion of the vessels with the production of a noisy, turbulent flow (Fig. 5.23). The method depends for accuracy on the bag pressure being transmitted to the artery wall; if the limb is very fat then this may not happen adequately giving consistently high readings.

Veins

The systemic veins, like the arteries, have two main functions: first they act as a low-resistance collecting system to carry blood from the body back to the right side of the heart; secondly they act as a low-pressure storage system (i.e. are capacity vessels), holding over 50 percent of all the blood at any one time at a pressure below say 18 mmHg. They often lie alongside the arteries going to an organ, sometimes sharing a connective tissue coat.

Although their structure also consists of three coats they have much thinner and more distensible walls than arteries, with much less elastic and muscle tissue in them (Fig. 5.17). They are therefore less stiff (i.e. more compliant) than arteries (Fig. 5.21). At intervals on the long veins, particularly in the limbs, the endothelial lining protrudes into the lumen to form cusps or valves similar in structure to the semilunar valves in the heart (Fig. 5.16). This arrangement is functionally very useful in that veins, having thin walls, are easily compressed by surrounding muscle; during muscular activity the contraction and relaxation of the tissues around the veins acts as a *muscle pump* helping blood to return to the heart. The close proximity of arteries to veins also produces a similar effect (see Fig. 5.16) at each pulse wave.

The presence of valves in the veins and the way in which they work was used by Harvey in 1628 as part of his evidence for a circulatory system. The experiments he performed (Fig. 5.24) are easily repeated by students and you should do so.

Veins are supplied with autonomic sympathetic nerves which innervate the smooth muscles in their walls. Increased activity in the sympathetic system leads to an increased stiffening of the walls, so that the vessels become less easily distended and hence often narrower.

Venous blood flow and pressure

Flow in veins is less phasic and more continuous than in the corresponding arteries. It does, however, vary with the heart beat and with pressure changes in the thorax. The average velocity of blood increases as it passes from the venules back to the heart, due to the decrease in the cross-sectional area of the vessels.

The pressure in the venous system must be lower than that in the capillaries and yet still high enough to fill the heart and so maintain the

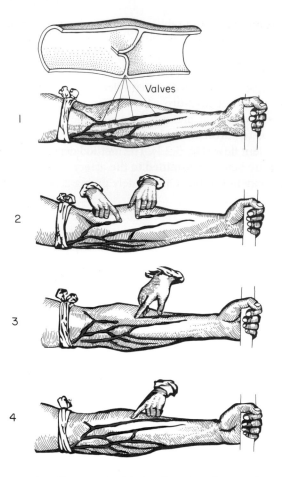

Fig. 5.24. Venous valves. Harvey noticed during dissections that a probe could easily be pushed up a vein past the venous valves towards the heart but could not be pushed backwards down a vein; particularly into side branches. He illustrated their function by (1) which shows the presence of the valves (as 'knots'), (2) how a section of vein could be emptied towards the heart, (3) how the valves prevented blood from flowing backwards from the heart, (4) how the valves allowed blood to flow from the periphery. He used the rapidity with which veins filled from below as part of his evidence that the circulation of the blood was very fast and so must move in a circle. The inset shows a cut-away, enlarged view of a venous valve.

cardiac output. The mean pressure at mid-atrium is remarkably constant at a few mmHg (< 5) in health.*

The transmural, or distending, pressure is the pressure difference between the inside and outside of a vein. Because veins are thin-walled floppy structures, their shape is very dependent on this pressure; if high the veins are circular in cross-section, if low they collapse and are then difficult to feel or see below the skin though they go on carrying blood normally. This different appearance with change in transmural pressure is the basis of estimating the 'central venous pressure' by observation of neck veins (see Fig. 5.13).

---

* It does, however, change quite markedly with posture (see p. 135).

Changes in pressure in the right atrium and in the thorax are transmitted back to the veins in the neck and may be examined as already described (see Fig. 5.13). The normal atrial waves are readily seen, and increases in intrathoracic pressure (for example, blowing up a balloon, blowing a trumpet, etc.) soon distend the neck veins. Normal inspiration produces a reduced intrapleural and increased intra-abdominal pressure; this increases the venous return to the heart giving an increase in the cardiac output and blood pressure above the expiratory values.

The mechanism for maintaining the central venous pressure constant depends mainly on altering the size of the venous reservoir; this is adjusted by variation in the degree of tension in the smooth muscle of the walls of the veins. Such changes in tension alter the compliance of the vein walls and so alter the relationship between volume and pressure in the system (Fig. 5.21). Changes in cardiac function caused either by direct action or reflexly also contribute to the constancy of the venous pressure.

Arterioles and capillaries

The capillary network and associated vessels are the most complex and least understood part of the circulation and that part where the primary function of the circulation, i.e. exchanges of gases, nutrients, metabolites and heat, takes place. To perform this function the blood in the capillary bed is slowed and spread out in a vast thin layer such that no tissue cell is much more than $10\,\mu m$ away from a capillary (except in lens and cartilage).

Structure and function

Arterioles are small muscular tubes with their walls and lumen about $20\,\mu m$ in size. The capillary bed consists of true capillaries which are long thin vessels (wall $1\,\mu m$, lumen $8\,\mu m$, i.e. just sufficient to pass an erythrocyte). There are two types of bypass vessels which connect arterioles directly to venules (Fig. 5.25): (1) the commonest of these is the thoroughfare vessel or metarteriole which occurs in all tissues. These are vessels whose walls are discontinuously coated with muscle cells. The capillaries start as side branches from the thoroughfare vessels; they have muscle cells around their origins which act as pre-capillary sphincters to control the blood flow through them. (2) Some epithelia also contain short A–V shunts. The small venules have no muscle cells, the larger ones do.

The muscle of the arterioles, metarterioles and precapillary sphincters shows histological differences related to their function and innervation. The arteriolar muscle is richly innervated, each cell is insulated from the others and the cells are not much influenced by local metabolites. Metarteriolar and precapillary muscle is poorly innervated, has many connections between the cells and is largely controlled by local metabolites.

Most arterioles are supplied by sympathetic nerves which release noradrenaline, but there are two exceptions to this general rule: (1) some sympathetic nerves to skeletal muscle release acetylcholine, which causes an active vasodilation of the arterioles, perhaps at the start of

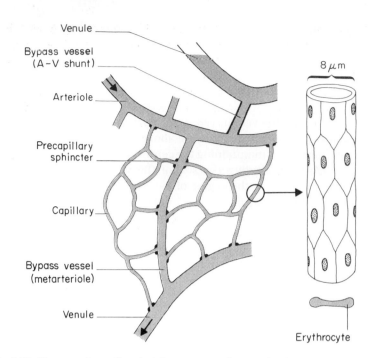

Fig. 5.25. Diagram of a capillary bed showing arterioles, venules, bypass vessels and their capillaries. Note the criss-cross layout of the vessels with no apparent ordering. The direct A–V shunts have only been found in the skin of the hands and feet and in the stomach wall. Capillaries are small and have as high a surface/volume ratio as it is possible to have and still allow red blood cells through them.

exercise; muscle blood vessels are also dilated by circulating adrenaline acting on β receptors (p. 94); (2) the blood vessels of certain areas of the genital tract are supplied with parasympathetic nerves which release acetylcholine and cause a dilation. Certain glands which show an increased blood supply following parasympathetic nerve stimulation probably do so via an increased metabolism rather than by direct arteriolar innervation.

The direct route for blood through a tissue is from arteriole, to bypass vessel, to venule; the capillaries are on a side branch controlled by the precapillary sphincter. The metarteriole, capillaries and venules form the *microcirculation.* All tissues (except the lens and cartilage) have such a microcirculation; but the capillaries in spleen and liver form sinusoids (p. 76) rather than true capillaries. Most organs have so many capillaries that they may be regarded as sponges filled with blood; when a muscle or organ contracts it expresses some of this blood into the general circulation and this action materially influences the circulation.

The detailed arrangement of the capillaries varies with the tissue concerned. In skin they appear as loops coming up towards the surface and then retreating back into the skin (these can readily be seen in the skin beside your nail-beds under a low-powered microscope). In muscle they are arranged longitudinally among the muscle fibres. In membranes such as the pleura or peritoneum (tissues derived from the mesothelium) they appear as flat networks. If any of these capillary networks is observed for a length of time it will be noticed to be 'unstable' in that one capillary can be seen to transmit blood at one time and then close, while another opens and so on (under the microscope you actually see the erythrocytes in the capillaries rather than the vessels themselves). This intermittent arrangement of blood flow means that much of the exchange and equilibrium between tissue fluids and blood may take place when the blood flow is actually stopped.

Red cells pile up behind the larger white cells in capillaries, like cars behind lorries on narrow roads. The gaps between one such string and the next are visible as white lines in your own retina when looking at a blue sky.

Exchanges between tissue fluids and capillaries
This has been dealt with on p. 14 and will only be summarised here. Exchanges of gas and other constituents occur very rapidly between the capillaries and tissue fluids; this is largely because of the very large surface area of blood exposed to the tissues rather than any high permeability possessed by the capillary wall. The route by which such exchanges take place appears uncertain, some exchange may be via vesicles in the lining cells as well as through 'pores' between the cells (p. 17). The mean pressures in the capillaries at heart level almost balance the oncotic pressure of the plasma proteins. As the oncotic pressure is about 25 mmHg and the interstitial pressure a few mmHg negative to the atmosphere, the mean capillary pressure is therefore about 20 mmHg. As already discussed (p. 19) this capillary pressure is correct for the centre of the capillary, but proximal (i.e. upstream) to this point it is higher and distal (i.e. downstream) it is lower giving a 'circulation' of fluid in the tissue space.

Control of the capillary circulation
The blood flow through the microcirculation needs to filfil two partially disparate functions: (1) tissue exchanges and (2) control of the peripheral resistance.
1 The primary function is to exchange substances with tissue cells to meet their current metabolic needs; for this exchange to be effective the critical factor is the total surface area available for exchange rather than the volume of blood in the capillaries. As we have seen this is achieved by having the capillaries as small as is consistent with allowing erythrocytes to pass through them.
2 The total cardiac output flows through the microcirculation so that the size of its vessels (mainly arterioles) determines the total peripheral resistance of the vascular system and, ultimately, the blood pressure (p. 102). For example, if all the arterioles in the body were fully opened up then the peripheral resistance would decrease markedly and the blood pressure drop catastrophically.

These conflicting functions require some degree of independent control over the arteriolar resistance and the microcirculation. The present view (Fig. 5.26) is:
a The nervous system controls the peripheral resistance by altering the tone of smooth muscle in the walls of the arterioles. Over the whole

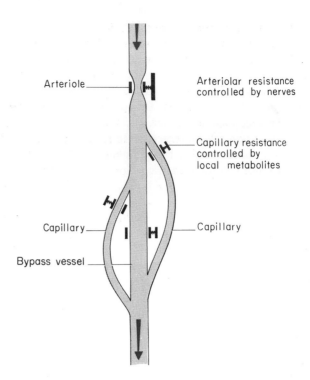

Arteriole

Arteriolar resistance
controlled by nerves

Capillary resistance
controlled by
local metabolites

Capillary

Capillary

Bypass vessel

Fig. 5.26. Diagram of control of arteriolar and microcirculation. The overall blood flow is controlled by the nerve supply to the arteriole, the distribution of blood between the capillary and thoroughfare vessels is determined by metabolites acting on the precapillary sphincters and muscle of the thoroughfare vessel.

body this controls the overall arteriolar resistance; in any one vascular bed this controls the blood flow to it.

b   Local metabolites in the tissues control the precapillary sphincters and hence the blood flow through the large capillary exchange area; increase in metabolites diverts more blood to the capillaries thus washing away these metabolites. There is still some argument about the nature of these metabolites; they do include a raised $P\text{co}_2$, a decreased pH, an increase in K^+ ions, adenosine and perhaps lactic acid; decreased $P\text{o}_2$ also dilates these sphincters. A temperature rise may also be involved. Students sometimes wonder how metabolites released in the capillary bed can act on sphincters 'upstream' of the point of release; the explanation is that metabolites released by cells do not only act on their 'own' sphincters, but also increase the local extracellular concentration of metabolites and these then dilate adjacent sphincters. Because capillaries criss-cross a tissue (Fig. 5.25) this means that increased activity leads to a general increase in blood flow in the area.

These two systems interact to some extent; a metabolic rise first produces a local increase in capillary flow and then opens up the local arterioles. If a severe arteriolar constriction is imposed on an area by central control then the resulting accumulation of metabolites may modify and so relieve the ischaemia caused.

1   Heat applied locally causes an increased blood flow, and cold a decreased blood flow. Generalised heating, sufficient to raise the core temperature, causes a generalised rise in skin blood flow by a reduction in sympathetic 'tone'. These effects act on the arterioles.

2   Exercise leads to a 20- to 40-fold increase in blood flow to the muscles concerned. The increase in blood supply occurs because of a direct effect of 'metabolites' on the local sphincters in the muscle (both arteriolar and precapillary); this is called *active hyperaemia*. This is clearly an efficient arrangement; increased work produces increased concentration of metabolites, these adjust the local blood flow to a rate sufficient to remove the metabolites from the muscle. Muscle work and blood flow are related to each other, the link is the local metabolite concentration acting on the arterioles and precapillary sphincters in the tissue.

3   If the blood supply to a tissue is cut off for a time, release is followed by a large rise in the blood flow over the next few minutes. This increase may reach 30 times the resting value; it is probably caused by accumulation of the metabolites mentioned above. This is called *reactive hyperaemia*.

The response of skin vessels to trauma

The white reaction. When a pointed instrument is drawn lightly over the skin, a white mark appears on the part touched by the instrument; this response is said to be due to contraction of the cells lining the vessels which give the colour to the skin, i.e. the capillaries (Lewis). It still occurs if the circulation is stopped or the capillary pressure is raised to 70 mmHg (Lewis). It develops in about 15 seconds and lasts a few minutes; it is not to be confused with the immediate effect of touching the skin, which is merely a passive emptying of the vessels.

Triple response. When the skin is stroked more firmly, a series of events ensue called the triple response. (a) A *red reaction* soon occurs (10

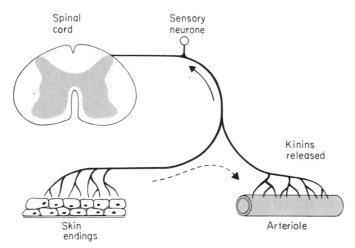

Fig. 5.27. Antidromic axon reflex of the flare reaction. Stimuli from skin receptors travel up the sensory axon to its junction with a sensory neurone from the arteriole, they then travel backwards to the arteriole and dilate it (see also p. 78 and Fig. 4.13).

seconds), limited to the area of contact, followed by (b) an irregular, diffuse *flare* which affects the surrounding area and finally by (c) a swelling, or *wheal*, which is due to local accumulation of fluid. This sequence of events is probably due to the local release of a substance (H substance, which probably includes histamine); it is the normal response of tissues to injury, the skin is simply a convenient place to observe it. The red reaction is probably due to opening up of the precapillary sphincters with an increase in capillary blood flow (so it is very localised); the flare is due to an axon reflex which causes adjacent arterioles to dilate, thus—because they supply a bigger area—leading to a diffuse irregular blotch (Fig. 5.27); the oedema is due to increased vessel permeability allowing protein and white cells to leave the circulation. The whole response can be thought of as a mechanism for bringing maximum amounts of antibodies and white blood cells to a site of injury.

Peripheral circulation as a whole

In the preceding sections we have discussed the various parts of the vascular system separately; here we wish to discuss the way in which these parts work together. To do so we shall consider the blood supply to a single organ or part of the body (for example, the adrenals), a situation in which all the blood entering through the artery leaves through the vein, without any cross-circulation.* Fig. 5.28 shows several features of such a circulation:

1   The pressure in the aorta and large arteries (in mmHg or kPa), created by the contraction of the heart, drives the blood through the capillary system back to the veins (flow occurring from an area of high to one of low pressure). The main pressure drop occurs in the arterioles—due to their high resistance—little drop occurring in the conducting vessels.

2   The volume flow through the system, usually called blood flow, (litre per minute [l/min] or cubic metres per second [m³/s]) is constant, as fluid is neither added nor subtracted by the tissue.* In the kidney, of course, the venous outflow is less than the arterial inflow by the amount of urine produced.

3   Resistance, as we saw on p. 107 by analogy with Ohm's law, is the ratio of the driving force applied to a fluid divided by the flow (i.e. $[P_1-P_2]/F$). As the flow in the system is constant, then the resistance is higher where the pressure drop is greatest, i.e. in the arterioles. The energy imparted to the blood by the heart is dissipated as heat to overcome the resistance of the vascular system. This frictional loss occurs due to the jostling of the molecules of blood against the vessel walls and with each other (see Fig. 5.19).

4   The surface/volume ratio of the vascular tree is lowest in the supply and drainage vessels to reduce friction, and highest in the capillaries where exchanges with the tissue cells take place; structure is thus well adapted to function. In man the surface/volume ratio of the capillary bed is about 500 times greater than that of the aorta or the vena cava.

5   The velocity of blood flow is the reciprocal of the cross-sectional area.

---

* A small fraction also leaves through the lymphatics.

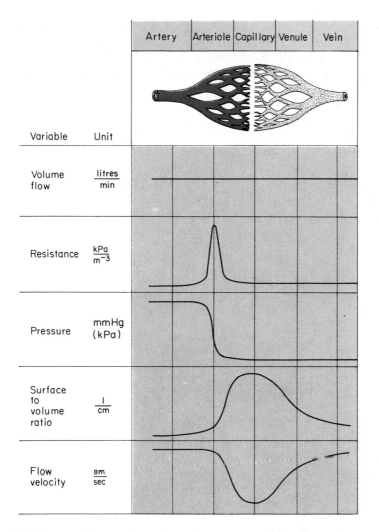

Fig. 5.28. Schematic diagram of a single circulating system and the relation of its physical variables (after Passmore & Robson).

As the cross-sectional area of the aorta is less than that of the capillary bed then the blood velocity is highest in the aorta and lowest in the capillaries. In man the peak aortic velocity may exceed 100 cm/s (about 4 km/h) (Fig. 5.23) while that in the capillaries is 0·07 cm/s. (This means that an erythrocyte moves its own length in 10 ms.)

Methods of measuring blood flow

These are two methods of measuring cardiac output in man; these are the direct Fick method (p. 15) and the indicator dilution method.

The indicator dilution technique (Fig. 5.29) can be readily understood as follows: imagine you and a friend are standing by a stream separated by a few metres; your friend, upstream of you, throws a bucket of red dye into the stream. You start a stop-watch as he does so and then time the appearance of the dye when it reaches you; you also collect samples of the dyed stream at intervals as it passes. You would expect the dye to

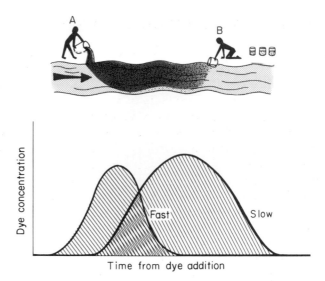

Fig. 5.29. Diagram illustrating the principle of the dye dilution technique. (A) throws dye into a stream and (B) collects samples of the dyed stream at various timed intervals. The curves show possible results obtained with fast- and slow-running streams. The mean concentrations of the dye (obtained from the area under each curve) divided into the original amount of dye used gives the flow over the time interval of the curve,

volume flow = amount/concentration

This volume flow divided by the time under the curve gives the volume flow per unit time.

take some time to reach you and to be diluted when it did so. If the stream starts to run faster, then the dye will reach you sooner and be more dilute (for more water will pass your friend as he empties the dye into the stream). With an experiment such as this (Fig. 5.29) you could work out the volume flow of the stream. This example makes it clear that for this method to work adequate mixing of the dye in the stream is necessary, partly mixed dye would give false results.

In clinical practice dye is injected into a vein and blood collected or monitored in an artery; it is supposed that adequate mixing has occurred in between. In the body the dye recirculates so that the curves do not return to the baseline. A good adaptation of this technique is to inject cold solution instead of dye and measure change in temperature. This may be done repeatedly, for the amount of cold solution does not permanently change the blood temperature.

A modification of this principle is commonly used to measure blood flow in organs or tissue (especially muscle). In this technique an injection of a radioactive substance (for example, Na^+) is made into the muscle and then the rate of removal of the substance measured; the faster the blood flow the faster the removal. This may be understood by considering the effect of throwing dye into a fast or slowly moving stream, the fast stream soon carries away all the dye, the slow stream takes longer.

The function of central control is to maintain a high pressure in the arteries with an appropriate cardiac output, this high pressure then allows blood to flow to the various tissues when the arteriolar 'taps' are turned on by local metabolic controls (see Fig. 5.15). Such problems and their solutions are by no means confined to biology, every electricity and water supply company face similar ones; the electricity company has to maintain the supply voltage at a fixed high level (say ± 5 percent). This then allows the users to draw off current for their various appliances simply by turning a switch. If the mains voltage swings greatly then our radios, televisions and lamps would also vary in their performance; if we wanted a constant performance from these devices we would then need complex local regulatory devices.

Four main systems are involved in the regulation of the vascular system: (1) regulation of cardiac output; (2) control of the blood pressure; (3) control of the local circulation; and (4) control of the blood and extracellular fluid volume. The last two points are dealt with on p. 115 and on p. 418.

Interactions between cardiac output, peripheral resistance and blood pressure

The cardiac output is fixed by the current metabolic demands of the body (e.g. see Fig. 16.4) and so varies from moment to moment through-

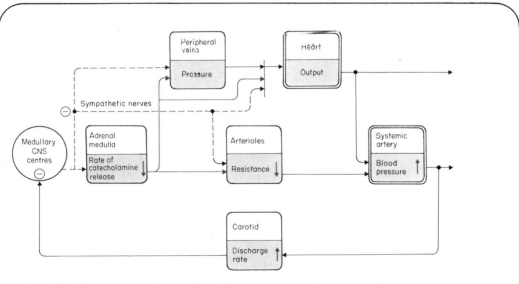

Fig. 5.30. The control of cardiac output and blood pressure. Cardiac output is the primary parameter regulated; blood pressure is then regulated by adjustments of the peripheral resistance (the corrections following a rise in BP are shown by the red arrows). Metabolism determines cardiac output, but the link is unclear. For simplicity the parasympathetic supply to the heart has been omitted.

In exercise the effectiveness of this system is probably increased by 'feedforward' regulation (p. 10); i.e. the CNS signals to the CVS that the muscles are being stimulated.

out life. The blood pressure is then kept fairly constant by varying the peripheral resistance (Fig. 5.30). So the blood pressure is dependent both on the cardiac output and on the peripheral resistance. For convenience we now discuss those factors separately (see p. 116 for details about the peripheral resistance).

The arterial blood pressure exists because the heart forces blood into the elastic aorta, from which it has some difficulty in escaping through the peripheral resistance. The pressure which is produced in any given circumstance is therefore dependent on both the cardiac output and on the peripheral resistance; control over the blood pressure can thus be exercised by actions on either of these. If you have difficulty with this concept then think about the following.

Imagine a situation in which the heart continued to pump but all the arterioles were either closed or fully open. If closed the blood could not leave the large arteries, the blood pressure would rise markedly and the heart fail; if open the blood would rush out of the large arteries and the pressure would fall towards zero.

Control of blood pressure
The relationship between cardiac output, peripheral resistance and blood pressure can be quantified by using the derivation of Ohm's law—Poiseuille's law—which we used before (p. 107). You will remember that:

$$\text{driving force} = \text{flow} \times \text{resistance.}$$

The driving force here is the pressure difference between the arterial and venous sides of the heart, which is approximately equal to the arterial pressure (for the venous pressure is very low); the flow is the cardiac output and the resistance is the total arteriolar resistance (the peripheral resistance). So

$$\text{arterial pressure} = \text{cardiac output} \times \text{peripheral resistance.}$$

From this it is clear that the arterial blood pressure could be varied by changing the peripheral resistance—as argued above—or by changing the cardiac output. This relationship is sometimes used to calculate the total peripheral resistance of the circulatory system; with an arterial blood pressure of 100 mmHg and a cardiac output of 5 litres/minute (l/min) this gives a resistance of

$$\frac{100\,\text{mmHg}}{5\,\text{l/min}} = 20\,\text{mmHg l}^{-1}\,\text{min}^{-1}.$$

Of the three interdependent variables of blood pressure, cardiac output and peripheral resistance the blood pressure is the least variable in any given individual over short periods of time, and also shows a long-term stability. As with many other physiological variables it shows a change throughout the 24 hours, with a minimum during the night (Fig. 17.1). The 'normal' values for the population show a fairly wide frequency distribution with (in Western man at least) an increase with age. Typical daytime values for a middle-aged man are a systolic pressure of 130 mmHg and a diastolic pressure of 80 mmHg (i.e. a blood pressure of

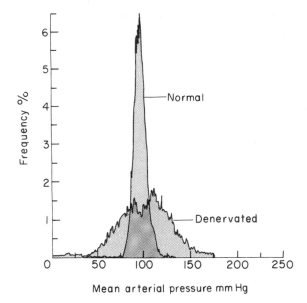

Fig. 5.31. Variation in mean blood pressure over 24 hours in a normal dog and one several weeks after denervation of the baroreceptors. The blood pressure in the denervated dog is much more volatile than in a normal dog, but the mean pressure is the same (after Guyton).

130/80); 80 percent of the normal population are within a range of ± 10 mmHg of these mean values.

Over short periods of time, variations in blood pressure are greatly reduced by the various baroreceptor mechanisms. (Fig. 5.31 and see next section). The mechanisms responsible for longer term control of blood pressure are less clear. Fig. 5.32a shows one scheme for such control. This implicates a variety of control systems with time courses of minutes to days. The longest-term mechanisms are based primarily on processes which affect the blood volume in relation to the capacity of the vessels.

Sensory receptors in the heart and aorta

If we wish to have an automatic heating or cooling system in a house then we need to arrange to measure the house temperature and then use this, via some kind of switch, to control the heating or cooling unit so as to keep the house temperature constant. The vascular system is controlled in the same way; there are sensory receptors to measure the performance of the heart and blood vessels, a cardiovascular 'centre' which makes decisions about the result and a motor system which controls the heart and blood vessels.

The sensory input system from the heart and blood vessels relies on measuring the mechanical events in the heart and blood vessels, rather than detecting the electrical events in the heart directly; thus the mechanical consequences of the heart's action are measured. All the receptors which do this measurement are stretch receptors in the walls of the blood vessels (tunica media and adventitia) and in the heart itself. The quantity which they measure depends on the properties of the vessel or chamber they are situated in; if they are in a very stiff structure (i.e. low

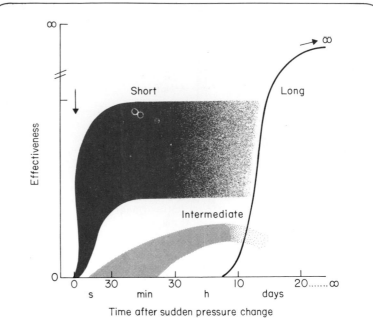

Fig. 5.32a. The effectiveness and time courses of various mechanisms controlling blood pressure. The short-term mechanisms (baroreceptors, chemoreceptors, CNS ischaemic response) act in seconds; the intermediate mechanisms (stress relaxation, renin-angiotensin, vasoconstriction and capillary filtration) act in minutes; the long-term mechanisms (renal-blood volume pressure control) starts in hours and persists indefinitely. Both the short and intermediate responses soon adapt to new pressures (not shown). The effectiveness is measured as feedback gain p. 13 (after Guyton).

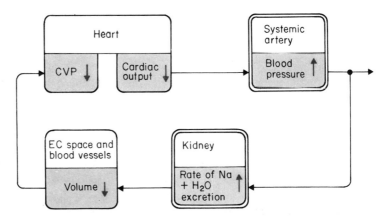

Fig. 5.32b. Blood pressure regulation by the Renal Volume Control system. We show the response to a rise in BP; the time course of the response occurs over days.

compliance), such as the aortic arch, then they effectively measure pressure, if they are in a very floppy structure (i.e. highly compliant), such as the vena cava or atrium, they effectively measure volume.

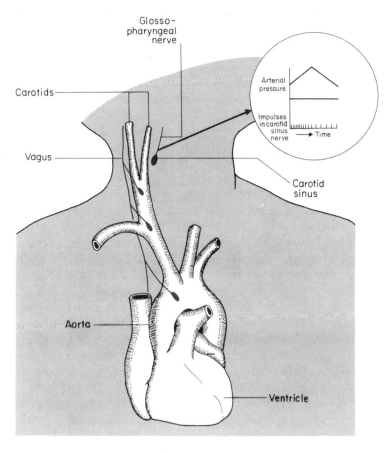

Fig. 5.33. The baroreceptor areas in the right side of the human neck and atria with an inset showing the response of a single carotid sinus nerve to the changing pressure of one cardiac cycle.

High-pressure receptors. The baroreceptors, or high-pressure receptors, in mammals are in the aortic arch, pulmonary arch, carotid arteries and other main arteries near the heart (Fig. 5.33). In man the carotid sinus is the main receptor site; this is situated at the bifurcation of the external and internal carotid arteries in the neck. These receptors are thus situated between the head and the heart on the direct route to the brain so they are well sited to monitor the pressure of the blood supply to the head, of obvious importance in animals which can adopt either an upright or a horizontal position. These receptors are stimulated by stretch of the arterial walls, which in turn is caused by a rise in intra-arterial pressure; as the pressure rises the rate of firing of any one receptor increases, as it falls the rate of firing decreases. A single receptor will only respond to pressures within a rather narrow range, so that many receptors are required to cover the whole pressure range which normally exists. In addition to responding to the absolute pressure, these receptors are also affected by the rate of change of pressure and so respond differently to a pulsatile pressure with the same mean value as a steady pressure.

The impulses travel to the 'centres' in the upper medulla concerned with the nervous control of the blood vessels and the heart. The motor impulses from this centre travel via the parasympathetic nerves to the heart and the sympathetic nerves to the heart and blood vessels. Sympathetic control is differential as shown, for example, by the brain vessels which take little part in these adjustments. Stimulation of the carotid sinus by raised blood pressure leads to a vasodilation of the blood vessels, a decrease in the peripheral resistance and hence a decrease in blood pressure (p. 121).

Low-pressure receptors. It is now clear that low-pressure receptors in the atria and great vessels feed impulses into the cardiac centre in the medulla to alter its activity. An increase in the volume of the right atrium and great veins causes increased firing in these receptor nerves and a reflex increase in heart rate and contractility and peripheral vasodilation (Bainbridge reflex). Thus both the input and the output from the heart are monitored.

Other inputs. Inputs from the respiratory system and higher centres also influence the medullary centres; thus rage, fear, excitement, etc., can cause large rises in blood pressure. In clinical practice this often means that casual readings of blood pressure are unreliable; the act of making the observation itself influences the result.

The vasomotor nerves form part of the sympathetic autonomic nervous system; they therefore run from the medulla down the spinal cord (lateral columns) to the thoracic and upper lumbar regions, where they synapse with the lateral horn cells. The preganglionic fibres leave in roots T1–L2 and enter the mixed peripheral nerves to supply the smooth muscles of the vascular system. (There is no parasympathetic supply to these muscles.) The cardiac nerves arise in T2–T4, synapse in the three cervical ganglia and reach the heart via the cardiac branches of the sympathetic chain; the parasympathetic nerves reach it via the vagus (Fig. 5.7). Normally there is always some activity in the vasomotor nerves to the blood vessels—leading to their partial narrowing—and in the parasympathetic supply to the heart—leading to slowing of the heart rate below its denervated value.

Regulation of cardiac output

The most important single function of the cardiovascular control systems is to maintain an adequate cardiac output; if it falls below about half of its normal value at rest, death occurs.

The cardiac output, which is the volume of blood ejected by one side of the heart per minute, is:

$$\text{cardiac output litres/minute} = \text{heart rate beats/minute} \times \text{stroke volume litres/beat}$$

At rest an average man's heart beats about 72 times per minute and each ventricle ejects 70 ml of blood per beat. The cardiac output is thus $72 \times 70/1000 \simeq 5$ l/min. Each side of the heart has the same cardiac output when averaged over many beats, but over a few beats the outputs may be different, for example when bulk movement of blood from skin to lung vessels is occurring.

Clearly the cardiac output can be varied by changing either the heart rate or the stroke volume; as we shall see the main control is exercised by changing the heart rate, lesser control is by changing stroke volume. The rate can change by more than three times, the stroke volume by less than twice. These two parameters are discussed separately.

Control of heart rate

The heart rate is determined by the rhythmic discharge of the sino-atrial node, which depends on the slope of the pacemaker potentials (see Fig. 5.6); this in turn is influenced by the sympathetic and parasympathetic supply to the heart and by circulating adrenaline. These influences set the heart rate to a value within the range of 50–200 beats/min (Fig. 5.34). At rest the heart rate is set by the level of parasympathetic 'tone', a fact which can be shown by blocking with atropine (this causes a rise in rate). Factors such as blood temperature, pH, ionic composition and other hormone concentrations have a smaller influence on heart rate.

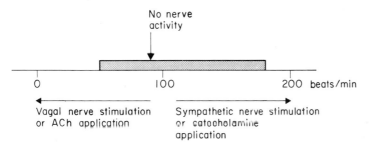

Fig. 5.34. Effect of nervous activity on the human heart rate. Heart rate normally varies between about 50 and 180 beats/min (shaded area) depending on nervous activity. In the absence of all nerves (for example, following a transplant) the heart rate is about 90 beats/min; normally, at rest, there is a background vagal activity slowing the heart.

Control of stroke volume

Harvey thought that the heart emptied completely each time it contracted; we now realise that only about half of the blood is expelled at each contraction. The stroke volume is therefore the difference between the ventricular volumes at the start and end of each contraction, i.e. the difference between the end-diastolic and the end-systolic volumes (Fig. 5.35). Control of the stroke volume is obtained by changing either of these volumes.

The Frank–Starling law of the heart. The relationship between contractility and length of muscle fibres.

The strength of contraction of a heart muscle fibre varies with its length; ultimately depending on the overlap of the actin and myosin filaments and therefore on the number of interacting sites between them. The relationship between the energy of contraction of heart and the

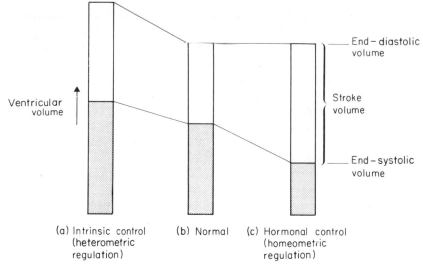

(a) Intrinsic control   (b) Normal   (c) Hormonal control
(heterometric   (homeometric
regulation)   regulation)

Fig. 5.35. Control of stroke volume—the stroke volume is the result of partial emptying of the heart from the end-diastolic to the end-systolic volume (b). If the end-diastolic volume is increased (a) then the resulting stretch of the heart causes a larger stroke volume (intrinsic control). If various hormones act on the heart (c) a greater emptying can occur at the same end-diastolic volume, this is hormonal control.

resting length of the muscle fibres was first measured by Frank and later by Starling; Fig. 5.36 shows this relationship, where the stroke volume is used as a measure of energy of contraction and end-diastolic volume as a measure of length. As stated above, increasing length of the muscle fibres produced by increasing distension of the heart leads (within limits) to increased output.

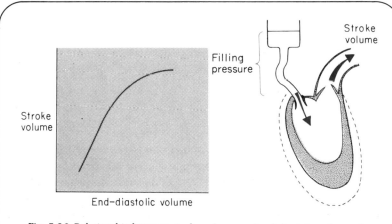

Fig. 5.36. Relationship between stroke volume and end-diastolic volume of heart. The inset shows the heart preparation used to demonstrate this (Starling's heart–lung preparation). Raising the reservoir increases the filling pressure (and hence the end-diastolic volume) and eventually leads to a greater stroke volume.

    The change does not occur instantaneously; usually for several beats the diastolic filling exceeds the systolic emptying so that progressive swelling of the ventricle occurs until the output equals the input.

This mechanism can be regarded as a form of autoregulation, the more the muscle is stretched the greater the subsequent contraction; because the length changes it is sometimes known as heterometric regulation. On p. 105 we argued that a distended heart was less efficient than one of normal size (because of Laplace's Law) so that a heart which increases its stroke volume by distending is less efficient than one which does so by emptying more (see Fig. 5.35). This latter type of response is sometimes known as homeometric regulation (because the muscle fibres at end-diastole are the same length). It can be most readily studied by looking at the family of Frank–Starling curves in a heart perfused with different concentrations of catecholamine (Fig. 5.37); at any one drug concentration the heart works along one of the curves. This means that in the body the stroke volume of the heart can be changed at constant diastolic volume by changing the concentration of catecholamine present. Other variables such as blood pressure, blood pH, plasma [Ca^{2+}], etc. also produce families of curves similar to those of Fig. 5.37.

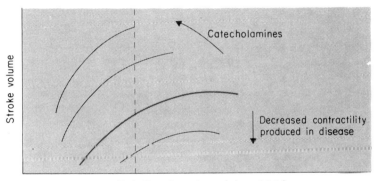

Fig. 5.37. Catecholamine effects on the Starling curves of the ventricle. In intact animals the level of circulating catecholamines moves the length–tension curve of the ventricle upwards and to the left. The dotted vertical line shows that stroke volume can be increased for the same end-diastolic volume by increasing concentrations of catecholamines; this may come from sympathetic nerves or from the adrenal medulla. Clinically the right atrial pressure is used as a measure of end-diastolic volume.

How then is the stroke volume normally controlled in the body? The general rule is that the heart remains the same size during diastole but empties more during systole (Fig. 5.35a), i.e. a homeometric regulation occurs. It is thought, however, that this effect is rather slow in developing so that sudden changes of work are dealt with by some cardiac swelling (i.e. heterometric regulation) until the heart can adjust itself to increased contractility. Thus, a sudden increase in the load on the heart (or heart rate) leads to swelling and increased output, but after 10–20 beats the heart shrinks again but continues with a higher output. This system optimises the efficiency of the heart. Cardiac failure is often associated with a loss of this homeometric response, the heterometric response remaining as the only mechanism.

Regulation of the peripheral resistance
Most of the details of this have been dealt with already, here a summary only is given. Different organs show both a difference in their resting rates of blood flow per gram of tissue and in their ability to change this basic flow (see Fig. 5.14); the kidney and brain have a high, fairly constant blood flow in health, whereas muscle at rest has a low blood flow which can increase some 40 times during exercise. The actual flows are, in all cases, dependent on the degree of contraction or relaxation of the vascular smooth muscle for the blood pressure is fairly constant.

Autoregulation. Vascular smooth muscle, even if denervated, has a resting 'tone'. Normally this tendency to contract is opposed by the production of local metabolites by the tissue, so that the balance between the two determines the local blood flow; this constitutes the basis of the autoregulation shown by many tissues. The nervous supply to the arterioles and microcirculation can be regarded as a central control system superimposed on this local one.

Summary of cardiovascular control
The following features are apparent:
1   There is a well-developed system of local control of blood flow in the tissues; in general this can override central control systems.
2   The two pumps which provide the energy for the system, the right and left hearts, are very sensitive to the venous pressures on their inputs; normally these input venous pressures are kept constant and the outputs of these pumps are accurately matched.
3   The cardiovascular system is regulated by the medullary 'centres' so that a cardiac output appropriate to the bodily demand is maintained.
4   A fairly constant blood pressure is maintained despite large variations in demand by the tissues.

Integration of the cardiovascular system
Students usually learn the function of separate parts of the cardiovascular system but often have difficulty in comprehending how it works as a whole. The problems of doing so are illustrated in Fig. 5.38. The venous reservoir, venous return, cardiac output and arteriolar resistance are all so intimately related that none can be altered without affecting the others. Here we shall concentrate on the relationship between venous return and cardiac output which must, of course, be equal over any length of time.

The central venous pressure is the pressure in the intrathoracic portions of the superior and inferior venae cavae; it is of great importance because the difference between it and the surrounding intrathoracic pressure—the transmural pressure—represents the distending pressure of the heart. The transmural pressure is maintained within quite narrow limits in health regardless of the position of the body in space, the redistribution of blood in dilated capillary beds or the accumulation of blood in distended dependent veins. Its lower limit is zero otherwise filling of the heart would be deficient during the diastolic intervals. Its upper limit

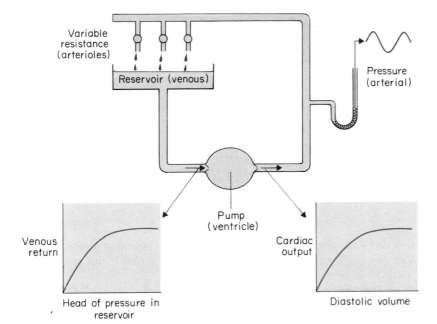

Fig. 5.38. Simple model to illustrate the principles of integration in the cardiovascular system. Because the fluid moves in a circle alteration of one part of the system affects all the other parts. The lower curves show that 'venous return' depends on the pressure difference between the venous reservoir and the heart—the pressure head; the cardiac output varies with the volume of the 'ventricle' (and the rate), which in turn depends on the pressure distending it.

cannot be too high otherwise it would raise the gradient of pressure in both the venous and the lymphatic systems, which would cause accumulation of fluid in the tissues. The true distending pressures of the heart are the transmural pressures in the right and left ventricles during diastole; these are slightly lower than the central venous pressure and difficult to measure.

Although the venous return must equal the cardiac output over any extended period of time, for a few beats at a time the two may be different. Consider the effects of a sudden reduction in venous return; the cardiac output will continue unchanged for a beat or two, thus decreasing the central venous pressure and then decreasing the stroke volume until venous return and cardiac output are equal again. Conversely, if the venous return increases suddenly the cardiac output will continue unchanged for a time, leading to a build-up of blood and an increase in pressure in the venae cavae and heart until the cardiac output increases to match the new venous return. In this way the changes in venous return which occur in life are continually matched to the cardiac output by changes in the central venous pressure.

Exercise as an illustration of integration (Fig. 5.39)

The cardiac output increases greatly (say × 4) and the resistances decrease greatly but neither the blood pressure (BP) nor the central venous pressure (CVP) change much. The velocity of the circulation is greatly

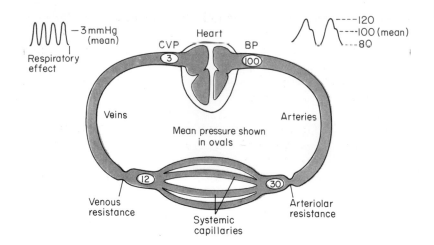

Fig. 5.39. An outline of cardiovascular function. The blood is driven round the circuit by the difference between the arterial blood pressure and the central venous pressure. The resistances are lumped together as arteriolar and venous. The pressures, BP and CVP, are relatively constant because the resistances and blood flows change in opposite directions. The mean pressures are shown in the ovals (after MACMAN).

increased, thus carrying more O_2 and fuel to, and metabolites away from the muscles. On the arterial side, the mean peripheral resistance decreases greatly (although some tissue resistances increase, e.g. in gut) thus delivering more blood to their capillaries. The mean capillary pressure rises. On the venous side the 'effective' venous resistance decreases, thus returning much more blood to the heart. The pressure gradient here is normally the venous pressure − CVP (i.e. 12 − 3 mm Hg). The venous pressure rises due to capillary and arteriolar dilation and because of the squeezing action of the muscles (muscle pump). The resistance of the direct venous path decreases due to more venous capillaries opening up. During exercise the cardiac output increases at the same CVP (and thus end diastolic volume) by the action of catecholamines on the Starling curves of the heart (Fig. 5.37). Students should consider other examples such as blood loss, decrease in cardiac contractility, fluid retention, etc. When doing so try to work out the effects of these conditions on blood pressure, heart rate, cardiac output, right atrial pressure and capillary pressures.

Circulation through special areas

The blood flow through the various parts of the systemic circulation occurs through a number of resistance vessels in parallel (Fig. 5.14), each of which has special features, some of which will now be described.

The pulmonary circulation differs from others in that it is in series with the rest of the circulation; its special features have been described in the respiratory section on p. 180.

Coronary circulation

Three small arteries supply blood to the heart. They all arise from the aorta just past the aortic valve; one from the front and the other two as a single artery from the rear which then divides into two. Each artery divides into numerous branches to supply the muscle, but little communication occurs between branches so that narrowing of one leads to a lack of blood supply to part of the heart muscle. Such narrowing or blockage often leads to death of heart muscle (a 'heart attack'). Venous drainage from the ventricles occurs mainly via the coronary sinus back to the right atrium but partly by direct drainage into the ventricular cavity.

Increased muscular work requires an increased cardiac output and so an increase in the work of the heart; to perform this extra work the heart requires an increase in its supply of free fatty acids, lactate, glucose and oxygen to its cells. Fig. 5.40 shows, however, that the blood supply to

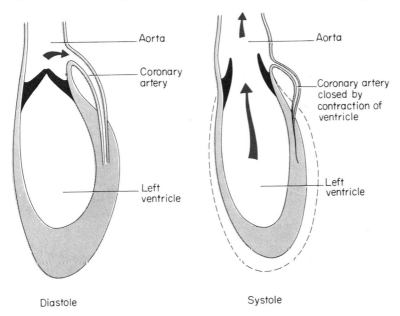

Diastole Systole

Fig. 5.40. The coronary blood flow occurs mainly during diastole and is stopped during the height of systole, due to the compression of the vessels by contraction of the ventricular muscle.

the left ventricle occurs mainly during diastole and least during systole; so that during increased cardiac work (with its increased contraction and shorter diastolic pauses) the conditions for an increase in coronary blood flow are less favourable than at rest.

The oxygen extracted from coronary blood is very high—usually 15 ml/100 ml compared to a mean elsewhere of say 4 ml/100 ml—so that increases in O_2 supply to the heart can only occur by an increased coronary blood flow. This occurs in exercise and hypoxia but the mechanisms for doing so are not well understood; probably nerves play little part and local control via metabolites is the main factor.

At rest, the heart obtains nutrition from glucose, free fatty acids and lactate in roughly equal amounts. During exercise, lactate comes to supply about two-thirds of the total energy. This has two advantages: firstly, the heart is automatically supplied with more energy because of the high circulatory lactate concentration and second, lactate removal helps to stabilise the blood pH.

Cerebral circulation

The brain is an area of high and fairly constant metabolism which is particularly sensitive to hypoxia; its blood supply is correspondingly large (0·75 l/min) and the total flow fairly constant. The blood flow to grey matter is almost 10 times that to white matter per unit volume. The cerebral circulation is unique in that it occurs in an area which is unexpandable, so that changes in the blood volume in the brain can only occur if something else (usually cerebrospinal fluid) is displaced.

The factors affecting the total cerebral blood flow are: (1) those common to all regions, i.e. the perfusion pressure (the arterial–venous pressure difference at the brain level), the degree of vasoconstriction or vasodilation of the cerebral blood vessels and the blood viscosity; and (2) the intracranial pressure (Fig. 5.41).

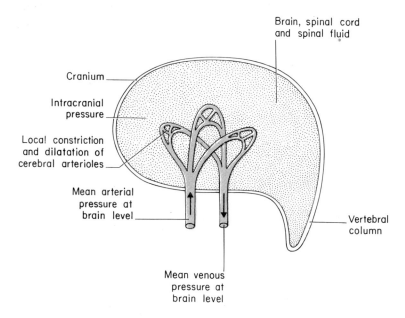

Fig. 5.41. Factors affecting cerebral blood flow (after Ganong).

The total cerebral blood flow shows a considerable degree of auto-regulation over the range of blood pressures of 50–150 mmg, a result very similar to that in the kidney (see Fig. 8.4). The blood supply to local regions of the brain varies markedly with mental activity involving that area, but the mechanisms causing this are not yet known. Changes in PO_2 and PCO_2 do change cerebral blood flow (e.g. as in muscle) but such changes never occur in normal life, probably because other metabolites alter blood flow before any marked change in PO_2 or PCO_2 can occur. Metabolites involved in this control are thought to include a rise in adenosine, K^+ and acidity and a fall in Ca^{2+}. The autonomic nerve supply to the cerebral blood vessels may be involved in control of blood flow via local reflexes. Considerable rises in intracranial pressure reduce blood flow due to compression of the vessels; a reflex causing a rise in arterial pressure has been described in these circumstances (Cushing's reflex) which helps to maintain normal blood flow.

Effects of posture on the circulation

So far we have considered the cardiovascular system as operating in 'gravity-free' conditions, a condition approached by man or animals lying horizontally or when immersed in water. About two-thirds of our lives are, however, spent in the upright position and so this should be considered as 'normal' for man. Gravity has a considerable effect on the pressures in the various parts of the vascular system in upright man, increasing the pressure in vessels below the heart and decreasing the pressure in those above the heart. When standing upright the arterial pressure in a hand held above the head is about 40 mmHg and that at the feet about 180 mmHg.

The main effects (Fig. 5.42) of gravity are indirect:

1   There is increased exudation of fluid from the capillaries, causing an oedema (i.e. swelling) of the tissue spaces with an increase in the tissue pressure. Our feet swell during the day, our faces swell at night in bed

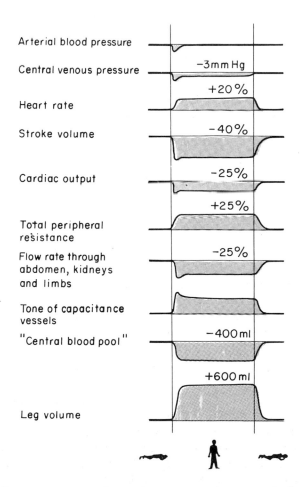

Fig. 5.42. Posture on various cardiovascular parameters. Average results shown; individuals vary greatly.

In the erect posture the venous and arterial pressures at the feet increase by about 80 mmHg (per cm) and the head pressure decreases by 20 mmHg (after Schmidt & Thews)

(due to redistribution of fluid); men should shave after breakfast not before, allowing time for the swelling to decrease.

2 On standing upright some 500 ml of blood is transferred from the chest (i.e. lungs and heart) to the legs. This (equivalent to a brisk haemorrhage) causes a drop in the filling pressure to the heart, with a consequent drop in stroke volume and a drop in cardiac output. In order to minimise this pooling of blood in the feet a reflex vasoconstriction of arterioles and venules occurs (this affects vessels in the lower rather than the upper part of the body); the heart rate also increases and its volume decreases a little. This reflex is relatively slow (seconds) in action; this slowness explains the feelings of faintness and darkening of the visual fields which you may have sometimes experienced on rising suddenly from a horizontal position (explanation: the sudden drop in blood pressure reduces the brain blood flow, producing these effects). This postural reflex disappears quite quickly if not used; in space flights it is gone in a few hours and staying in bed for several days reduces it. It appears that its continual use during life is necessary for its retention. The 'muscle pump' (together with the venous valves, p. 112) helps to reduce pooling of blood in the feet by aiding venous return; this of course only occurs during muscular activity.

These mechanisms do not, however, completely correct the effects of posture, as the cardiac output remains some 30 percent lower when upright than when lying horizontally. As metabolism stays the same, more O_2 is extracted from the venous blood (6 rather than 4 ml per 100 ml) to keep the supply to the tissues constant.

Students are often puzzled as to how blood can 'climb' back up from the feet to the heart. The vascular supply to the legs can be compared to

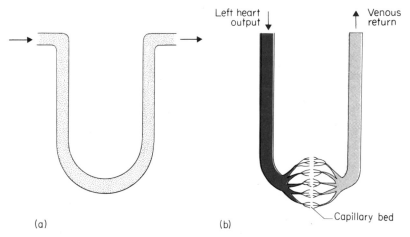

Left heart output

Venous return

(a)   (b)   Capillary bed

Fig. 5.43. Arrangement of venous return from the legs. If fluid enters one side of a U-tube (a) filled with fluid then it displaces fluid out of the other side. (Water returns from a U-tube because it has a high 'tensile strength' and so is difficult to separate.) The arrangement of the vascular supply to the legs and feet is similar (b), except of course there is an arteriolar and capillary bed in the pathway which are relatively distensible, but fluid entering one side must return from the other.

When we stand up from a lying posture, the increased pressure in the vessels of the foot causes them to expand somewhat; this is partly counteracted by increased nervous activity to the foot vessels with a consequent vasoconstriction.

a tube with arterial blood entering one of its sides and venous blood returning from the other (Fig. 5.43a). Common experience tells us that fluid which enters one side returns from the other side. The same happens in the vascular system (Fig. 5.43b) except that there is super-imposed upon the gravity gradient a vascular gradient produced by the arterioles, capillaries, etc. The venous pressure in the feet is always greater than that in the right atrium if adjusted to the same level.

PLACENTAL AND FETAL CIRCULATION

Uterine circulation

The blood flow to the uterus shows cyclical changes during the menstrual cycle; during pregnancy it increases with the increase in size of the uterus and contained fetus. The increase in the mass of the fetus and placenta is much greater (1 cell → 5 kg) than the 20-fold increase in the maternal uterine blood flow which occurs, so that the O_2 extracted from the uterine blood is greater at the end of pregnancy than it is at the start leading to a decrease in the O_2 saturation of the uterine venous blood. Just before parturition the uterine blood flow decreases sharply.

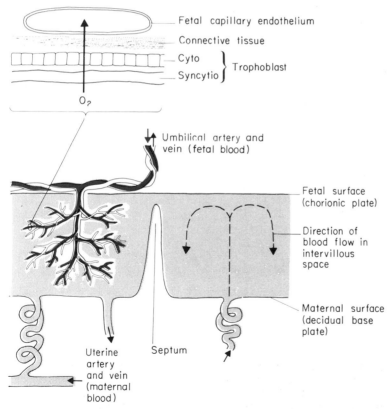

Fetal capillary endothelium
Connective tissue
Cyto
Syncytio } Trophoblast
O_2

Umbilical artery and vein (fetal blood)

Fetal surface (chorionic plate)

Direction of blood flow in intervillous space

Maternal surface (decidual base plate)

Uterine artery and vein (maternal blood) Septum

Fig. 5.44. Diagram of placenta showing two villus units (fetal cotyledons). Each villus is like a 'tree' whose branches consist of an artery and a vein running side by side; at its tip the artery runs into a capillary which drains back to the vein. The villi are closely crowded together and not freely floating in space as shown. Inset shows the four layers between maternal and fetal blood.

Placenta

The fetus is genetically different from the mother yet they must live within the same body for 9 months without rejection problems. This partly explains the complexity of the placenta.

The placenta is the site where O_2 and CO_2 exchange occurs, through which all nutritional materials enter the fetus and where all waste products are discharged; it therefore acts as the fetal lung, gastro-intestinal tract and kidneys. The maternal portion of the placenta acts as a 'lake' into which the villi of the fetal arteries and veins project (Fig. 5.44). Exchanges take place across the four layers which separate the maternal pool of blood from the fetal capillaries; as pregnancy proceeds the overall thickness of these layers decreases, perhaps to facilitate the increased diffusion of materials required as the fetus becomes larger. The fetal and maternal circulations do not, however, normally mix.

Fetal circulation

Blood returning from the placenta (Fig. 5.45) is split into a part which goes directly to the right heart (via the venous duct) and a part which goes there via the liver; this is a sensible arrangement because the placental blood contains both O_2 and nutrients (glucose, fats and amino acids). At the right atrium the larger part of this placental blood is shunted through an oval hole into the left heart and is then sent to the head via the upper aorta; this again is sensible, for the head is thus

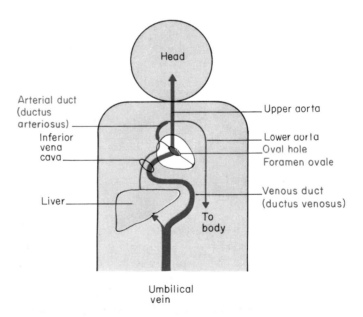

Fig. 5.45. The functional distribution of 'cleaned' blood from the umbilical vein to the fetal organs. The parts and associated names in red are those ducts which close at birth to give the adult-type circulation of blood. The thickness of the arrows shows the relative flow rates. The blood in the umbilical veins has a Po_2 of about 30 mmHg with the Hb about 70 percent saturated with O_2. About two-thirds of the cardiac output goes to the placenta. Note that the two sides of the heart operate in parallel in the fetal circulation. Anatomically, the foramen ovale is situated in the wall between and left and right atria, not as indicated.

supplied with the most highly oxygenated blood. The smaller stream of blood remaining in the right atrium is augmented by that returning from the upper body and enters the right ventricle. It then largely bypasses the lungs (via the arterial duct) to the descending aorta and hence to the lower body and placenta again. This latter arrangement means (a) that the less oxygenated blood goes to the lower body, and (b) that the lungs, which are of course non-functional, receive only a small blood supply.

Fetal Hb receives its O_2 from the maternal Hb, through the rather thick exchange membranes of the placenta. To facilitate this exchange of O_2 the fetal form of Hb has a higher affinity for O_2 than the adult form (Fig. 5.46).

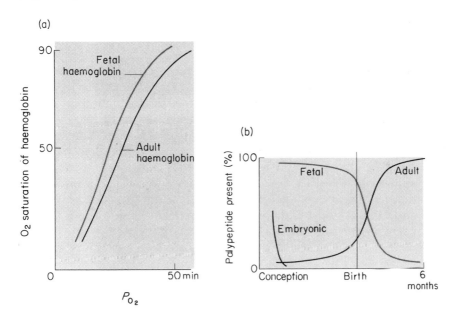

Fig. 5.46. (a) O_2 dissociation curves of adult and fetal haemoglobin. (b) Changes in the second pair of chains during intrauterine life and early childhood. The α chains remain largely constant during this period (see also p. 69).

Changes at birth

At birth there is an immediate need to change from placental to pulmonary respiration (Table 5.1). As soon as the infant ceases to be 'weightless' the respiratory centre receives a flow of new sensory impulses, notably from the skin and muscles. The partial anoxia and mild acidosis during labour may also play a part. These cause the first few respiratory movements (mainly due to the diaphragm) which give very negative intrathoracic pressures -40 to $-60\,cmH_2O$ (4–6 kPa) during inspiratory effort and large positive pressures (say $20–30\,cmH_2O$) during expiratory effort. These overcome the initial resistance to entry of air to the lungs. When the lungs expand the Po_2 rises to normal values and the pulmonary arteriolar resistance falls to about half. Much more blood now flows through the lungs and so the left atrial pressure rises, stopping the left to right shunt through the foramen ovale. The ductus

Table 5.1. Circulatory values in the human fetus and newborn.

	Fetus	Newborn	Adult
Systemic arterial pressure (mmHg)	70/45 (umbilical/arterial)	70/45	120/80
Pulmonary arterial pressure (mmHg)	—	35/15	30/10
Heart rate/minute	140	140	70
Blood flow in pulmonary artery	small	large	
through ductus arteriosus	large to aorta	small from aorta	nil

The cardiac output in the newborn is twice that of the adult if expressed per body mass, it is the same if expressed per surface area (adapted from Passmore & Robson).

arteriosus closes as the Po_2 rises; a mechanism apparently triggered by increased oxidative metabolism in the muscle cells.

The placental blood flow decreases after birth due to contraction of the umbilical vessels; this initially causes a 'transfusion' of about 100 ml of blood into the circulation but also causes the right atrial pressure to drop, thus helping to close the foramen ovale. The ductus venosus starts to close soon afterwards, reducing the bypass of blood past the liver.

Although all these changes start at birth it takes some weeks for complete changeover to be completed. The unstriped muscle in the pulmonary arterioles gradually atrophies and so their walls become thinner. After blood flow through the foramen ovale, ductus arteriosus and venosus stops, all those channels gradually close. Table 5.1 shows normal values in man for various parameters.

Chapter 6
Gastrointestinal System
and Nutrition

GASTROINTESTINAL SYSTEM

General features

The gastrointestinal (GI) system acts together with the cardiovascular system (p. 88) and organic metabolism (p. 357) to deliver an appropriate mixture of fuel molecules to the cells of the body. The various parts of the process are coordinated so that they work together smoothly. Thus the sight or smell of food starts the mouth 'watering', food in the mouth starts secretion in the stomach, digestion in the gut starts the release of insulin etc. Each stage therefore sends information 'down the line' to warn that food is coming, helping to ensure an integrated response. Regular eating habits also programme the GI system to expect food (p. 175).

The GI tract acts mainly to convert food into smaller molecules, and to pass them on to the interior of the body. It does this by a combination of the mechanical processes of fragmentation, mixing and transport and the chemical processes which follow from the secretion of digestive juices. Various enzymes within these digestive juices, or on the surface of the cells lining the GI tract, act to split proteins, fats and carbohydrates into particles small enough to be absorbed (digestion). These products, together with mineral salts, water, and vitamins pass across the epithelial layer into the blood and lymph.

The GI tract is a long muscular tube of variable diameter, running from the mouth to the anus and lined with mucous membrane (Fig. 6.1). The secretions of the various organs, glands and of the mucous membrane itself enters the lumen of this tube and digests the food. The outermost layer of the tube (serosa) is slippery and allows coils of gut to move over each other in the abdomen, the outer two muscle layers with the myenteric plexus in between gives it motility, the submucosa has the nerve plexus (Meissner's) which supplies the mucosal glands and the mucosal layer carries blood vessels and a small muscle which can change the folding of the attached epithelium.

Control of gastrointestinal activity

The activity of the GI tract is controlled by three mechanisms:
1   The intrinsic nerve plexuses (Fig. 6.1). The myenteric plexus mainly affects motility, whereas the submucosal plexus affects both motility and secretion. Both require to be functional for motility and secretion to occur, but they will work in the absence of autonomic inputs.
2   The extrinsic innervation is by the sympathetic and parasympathetic divisions of the autonomic nervous system, with the activity of the latter predominating. The general rule is that activity of the parasympathetic

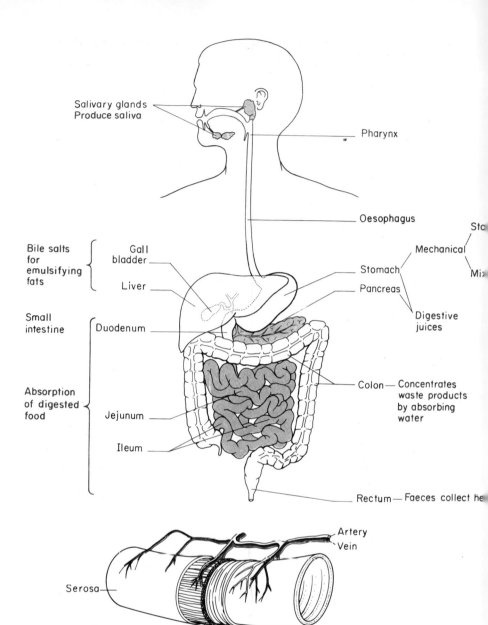

Salivary glands
Produce saliva

Pharynx

Oesophagus

Sto

Bile salts
for
emulsifying
fats

Gall
bladder

Liver

Mechanical

Stomach

Mi

Pancreas

Small
intestine

Duodenum

Digestive
juices

Absorption
of digested
food

Colon — Concentrates
waste products
by absorbing
water

Jejunum

Ileum

Rectum — Faeces collect he

Artery
Vein

Serosa

Longitudinal muscle   Circular muscle

Fig. 6.1. Structure and function of the GI tract. Inset shows the 4 layers of gut wall.
The nerve nets lie between the longitudinal and circular muscle layers, and within the
submucosal layer. Blood vessels receiving digested food are in the mucosa. Function
shown in red text.

system increases smooth muscle activity and gland secretion, while
sympathetic activity decreases secretions and causes sphincters to contract. There is increasing support for the view that there is another
division of the autonomic, the enteric, which can function independently
of the CNS and of the other two divisions of the ANS. A multiplicity of
transmitter substances have been found recently (e.g. 5-hydroxytryp-
tamine, prostaglandin, adenosine triphosphate, and peptides) whose

Table 6.1. Regulatory peptides. Some peptides seem to be released from endocrine cells in the mucosa and also to be stored in neurones (after Polak & Bloom).

Peptide	Main action
Gastrin	Stimulates gastric acid secretion
Secretin	Stimulates secretion of pancreatic bicarbonate·
Cholecystokinin	Contraction of gall bladder
	Secretion of pancreatic enzymes
Gastric inhibitory peptide (GIP)	Stimulates insulin release
	Inhibits gastric acid
Motilin	Stimulates motility
Vasoactive intestinal peptide (VIP)	Vasodilation
	Muscle relaxation
	Secretion
Somatostatin	Antagonises peptide release
Substance P	Sensory nerve transmitter
	Vasodilation
	Muscle contraction
Bombesin	Releases peptides
	Anti-somatostatin
	Trophic agent
Neurotensin	Vasodilation

precise function is still unclear. The activity of both extrinsic and intrinsic nerves is affected by the basal activity of the smooth muscle of the gut itself; this makes interpretation of experimental findings difficult.

3   Control by circulating or locally released hormones These were originally thought of and described as local or gastrointestinal hormones but, because they have been found to exist and function in many areas of the body, are now called regulatory peptides (Table 6.1).

Basic mechanisms of motility. At the entrance to and exit from the GI tract (i.e. mouth, throat, upper oesophagus and the rectum), voluntary muscles control movements. The movements of the rest of the GI tract are brought about by the three smooth muscle layers of its walls. There are four kinds of movement (Fig. 6.2). These have the functions (1) of *transporting* material along the GI tract (by peristalsis), (2) of *mixing* the contents (by segmentation) and (3) *separation* of functionally different regions (by sphincters), this also allows forward transport without reflux.

Basic mechanisms of secretion. Synthesis of digestive juices occurs in the secretory cells of the salivary glands in the mouth, of the glands of the stomach and intestine, and the exocrine part of the pancreas and in the liver. The primary secretion consists of enzymes with sodium, potassium, chloride and an osmolality similar to plasma; this is then altered as it travels through the duct system (see Fig. 6.11).

Glands which produce digestive enzymes have to avoid the problem of digesting themselves. They do so by secreting inactive precursors

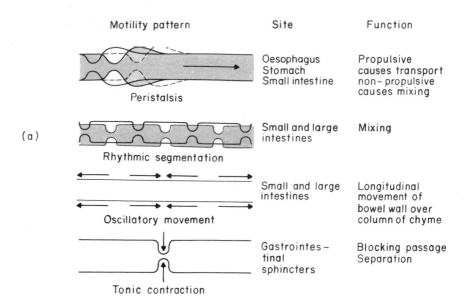

Motility pattern	Site	Function
Peristalsis	Oesophagus Stomach Small intestine	Propulsive causes transport non-propulsive causes mixing
Rhythmic segmentation	Small and large intestines	Mixing
Oscillatory movement	Small and large intestines	Longitudinal movement of bowel wall over column of chyme
Tonic contraction	Gastrointestinal sphincters	Blocking passage Separation

(a)

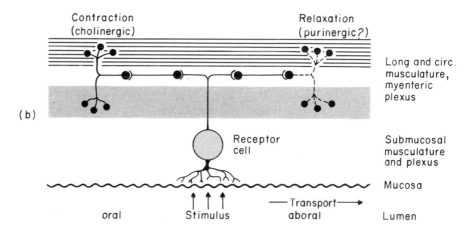

(b)

Fig. 6.2. (a) Motility patterns in the GI tract, with an indication of the sites where they occur and their function. (b) Neurones involved in peristalsis. The stimulus excites stretch receptors, which via interneurones, causes contraction on one side and relaxation on the other so a wave of relaxation is followed by a wave of contraction (after Schmidt & Thews).

which are activated in the lumen of the GI tract, whose walls are protected by mucus etc. Precursors (proenzymes) are made in the ribosomes on the endoplasmic reticulum of secretory cells, and stored in vesicles; these vesicles can be seen in histological sections and are often called zymogen granules. When secretion occurs these vesicles fuse with the surface membrane of the cells, and discharge their contents into the lumen of the gland.

Proteolytic enzymes need (1) to be activated from precursors, (2) to recognise the peptide bond which is to be attacked, (3) to attack this bond and (4) to be inactivated when the process is over. Recent evidence

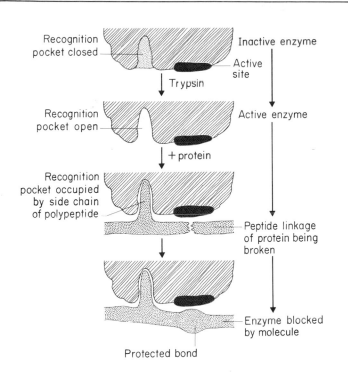

Fig. 6.3. Diagram showing activation, recognition, catalysis and inhibition of proteolytic enzymes.

The substrate is recognised by a pocket which lies next to the active site in the enzyme; this pocket is so tailored that it attracts the side chain of the amino acid preceding the peptide bond to be split. In the inactive precursors of the enzymes this side pocket is closed; trypsin activates these precursors by opening this pocket. Pancreatic trypsin inhibitor mimics a substrate by presenting to the enzyme a side chain which the pocket 'recognises', but which has the peptide bond following that side chain shielded so that the enzyme cannot split it. The inhibitor thus remains firmly attached to the enzyme in the position fleetingly occupied by the normal substrate during catalysis.

This mechanism of activation, recognition, catalysis and inhibition has been described in some detail because it appears to be exploited over and over again for different functions. Examples are the clotting of blood, the dissolution of blood clots, the disposal of complexes of antigen and antibody by compliment and some functions of kinins (after Perutz).

using the techniques of molecular biology has given much new information on this process (Fig. 6.3).

Mouth and oesophagus

Chewing. When food is taken into the mouth it is broken by chewing into smaller particles which can be more easily swallowed. This is mainly an involuntary reflex, but is obviously also a voluntary process. The contact of food on receptors in the palate and tongue causes a reflex opening of the mouth, which in turn results in the normal anti-gravity reflex closing it, thus setting up a rhythmic movement. Chewing continues until ingested food is soft, but this is greatly influenced by habit and external events such as conversation. The amount of chewing probably does not influence the digestion of food very much, apart from its beneficial effect in breaking up the vegetable fibres. The forces involved range from the normal of about 20 N to around 600 N.

Saliva is produced mainly from glands which lie outside the mouth and pass their secretions into the mouth by ducts (Fig. 6.1), and partly from the many mucous glands in the mucosa of the mouth itself. The parotids and the sub-mandibular glands only secrete when stimulated; the sublingual and the salivary glands in the mucosa secrete a thin watery fluid all the time at a rate of about 0·5 ml/min (but it decreases at night). When eating this may increase by 15 times, due mainly to the presence of food in the mouth. The presence of anything (even grit) in the mouth acts as a stimulus to this unconditioned reflex, although certain substances have a greater stimulatory effect than others. Smell also acts as a stimulus to salivary glands. In man, however, the Pavlovian conditioned reflex is not a particularly strong stimulus. The main enzyme in saliva is α-amylase (ptyalin), it also contains mucus, traces of blood group antigens and immunoglobulins and electrolytes. The pH, ion concentrations (and hence the tonicity) of saliva depends on the flow

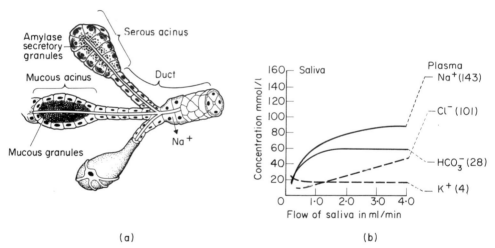

(a) (b)

Fig. 6.4. (a) Structure of the submandibular gland. (b) Composition of parotid saliva at various rates of secretion; as the secretory rate increases, the electrolyte concentrations become nearer that of plasma. The pH is around 6 at low rates and 8 at high rates of secretion (after Passmore & Robson).

   The *sensory innervation* of the receptors in the oral mucosa is by branches of the lingual and glosso-pharyngeal nerves. The motor innervation of the salivary glands is by sympathetic and parasympathetic nerves, but their effects are complex. Stimulation of the parasympathetic causes a copious flow of watery saliva; effects due to both actions on the cells producing the secretion and on those which modify it in the ducts. Parasympathetic stimulation also causes a marked increase in blood flow; an action thought to involve both kinins and VIP. This copious secretion is accompanied by an increase in the oxygen consumption of the gland which shows that saliva is being formed by a metabolically active process in the gland. When the sympathetic supply is stimulated these glands produce a viscid saliva with a high concentration of mucus.

rate (Fig. 6.4). The functions of saliva can be worked out from its composition. They are: (1) amylase starts the digestion of starch which continues in the stomach, (2) the large amount of water dilutes the food and is a solvent for some of the components (e.g. it allows tasting), (3) the mucus makes the food slippery to help in swallowing, (4) saliva keeps the mouth moist, rinses and disinfects it.

Swallowing. Once food has been chewed and mixed thoroughly with saliva, it is formed into a small ball or bolus and can be swallowed. Swallowing involves a complex series of contraction and inhibition of many voluntary and involuntary muscles; it is often divided into three phases, buccal, pharyngeal and oesophageal. The tongue initiates the process by pushing the bolus into the back of the mouth, resulting in stimulation of receptors there. Impulses from these receptors in the mouth pass to a 'swallowing centre' in the medulla which coordinates the train of events. You can convince yourself that swallowing cannot always be initiated voluntarily by trying to swallow your saliva three times in succession; after swallowing twice, no saliva remains in the mouth to stimulate receptors and you cannot initiate a third swallowing movement. Once swallowing has been initiated it cannot be voluntarily stopped.

As the tongue pushes the bolus backwards in the midline the soft palate is raised to shut off the naso-pharynx, respiration is inhibited, the vocal cords are brought together, the larynx raised and the glottis closed to prevent food entering the trachea. This movement of the bolus backwards and downwards folds the epiglottis back over the glottis helping to protect the trachea from the entry of food. Occasionally a small particle of food or fluid may reach the respiratory tract before closure of the glottis and cause coughing by stimulation of the respiratory tract receptors.

The oesophagus is the tube which carries food from the pharynx to the stomach. It has skeletal muscle around its upper third and smooth muscle in its lower two-thirds. Normally the upper end of the oesophagus is kept closed by the tonic activity of the crico-pharyngeus muscle. At the start of swallowing this sphincter opens, the bolus passes into the oesophagus and then the sphincter closes. Once a bolus has entered the oesophagus a *peristaltic wave* of contraction passes along the muscle of the oesophagus from the oral to the gastric end. When the wave reaches the gastro-oesophageal or cardiac sphincter, this relaxes to let the bolus through. The relaxation of the normal high resting tone is mediated by non-adrenergic, non-cholinergic vagal fibres. Normally a peristaltic wave passes along the oesophagus in about 10 seconds. (You can time this for yourself by listening for liquid reaching your empty stomach.)

The passage of a bolus is normally assisted by gravity in man so that food may reach the lower oesophageal sphincter before the wave of peristalsis. It remains there until the peristaltic wave arrives and the sphincter then relaxes. It is possible to swallow when standing on one's head; then movement of food along the oesophagus depends entirely on peristalsis.

If part of the bolus sticks in the oesophagus this distending stimulus results in a secondary wave of peristalsis, which again depends on an intact vagal reflex.

The lower end of the oesophagus normally remains closed and so prevents reflux of gastric contents. Although the muscle fibres involved do not form an anatomical sphincter, the lowest 4 cm of the oesophagus is normally in a state of contraction and acts as a physiological sphincter. As intra-abdominal pressure is always higher than the normally negative intrathoracic pressure (and at times very much higher) it is surprising that stomach contents do not get pushed back into the oesophagus. This is prevented mainly by the physiological sphincter and partly by the other anatomical arrangements (Fig. 6.5).

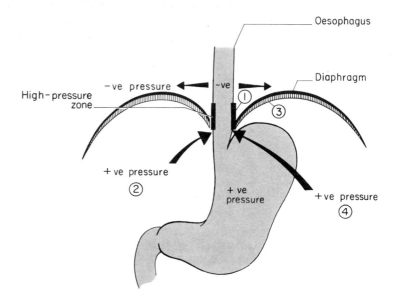

Fig. 6.5. Anatomical arrangements around the cardiac sphincter. (1) The physiological sphincter. (2) Part of the oesophagus lies below the diaphragm; so any increase in intra-abdominal pressure squeezes the stomach and this part of the oesophagus and so helps to keep it closed. (3) The oesophagus is surrounded by the crura of the diaphragm; this may act as a pinch-cock, particularly during a deep inspiration. (4) The oesophagus enters the stomach at an acute angle; filling of the fundus of the stomach helps to prevent reflux of gastric contents by pushing across as a flap-valve. The 'high-pressure zone' causes a high-resistance zone.

Stomach

The stomach is not an important organ for enzymic digestion or for absorption; its main role is as a temporary store which allows food to be swallowed faster than it can be passed into the small intestine. Once in the stomach, the food is mixed with the gastric juices and released in small amounts into the duodenum.

The mechanical activities of the stomach can be divided into two parts: an upper storage hopper and a lower mixing part.

1   The upper part is able to hold considerable quantities of food without an increase in pressure. This is due partly to the basic plasticity of

smooth muscle, partly to a rearrangement of smooth muscle fibres and partly to a vagal reflex (receptive relaxation).

2   The lower part grinds and mixes the food with gastric secretions to form *chyme*, a semi-liquid mixture. This is done by the various rhythmic contractions of the gastric smooth muscle (Fig. 6.2). The mixing process is rather slow, so that the chyme in the middle of the stomach still has a pH of around 5 after 1–2 hours. This means that the action of salivary amylase continues into the stomach.

The longitudinal smooth muscle layer has a basic electrical rhythm (BER) at a rate of about three per minute which sets the rate of peristalsis. This is driven by a pacemaker on the greater curvature of the stomach; the electrical impulses travelling to the lesser curvature and the pyloric region (Fig. 6.6). The force and frequency of gastric contractions depends on both extrinsic and intrinsic nerve activity and on the concentrations of circulating regulatory peptides. After cutting the vagus, function is disturbed for around a month.

Most stomach contractions fade out in the antrum; those that continue there get larger and propel some chyme into the duodenum. During this process the antrum is nearly closed off and the pyloric canal shortens. Gastric emptying is exponential (i.e. proportional to the

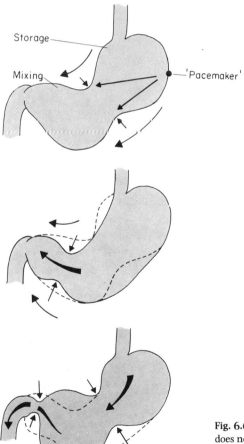

Fig. 6.6. Gastric peristalsis. As contraction does not close the lumen of the stomach completely, some contents escape back into the fundus and cause mixing.

amount in the stomach), and varies with the type and consistency of the food. The order is: carbohydrate (fastest), protein and then fat (which may take 4 hours or more). The chyme leaves the stomach to be 'processed' in the duodenum; so we would expect the duodenum to have some control over the rate at which it receives chyme. A vagal reflex from the distension of the duodenum is partly responsible; it has also long been known that fats and HCl in the duodenum have a similar effect, but the mechanism is not too clear. Probably secretin, gastric inhibitory enzyme (GIP) and cholecystokinin-pancreozymin (CCK-PK), released in the duodenum are responsible for this reflex. Gastrin also inhibits antral motility and delays emptying.

The empty stomach has occasional contractions which sometimes may be heard (fasting contractions); they occur at the basic rate of three per minute.

The gastric secretions consist of mucus, pepsinogens, hydrochloric acid and intrinsic factor, all of which arise from cells of the gastric mucosa

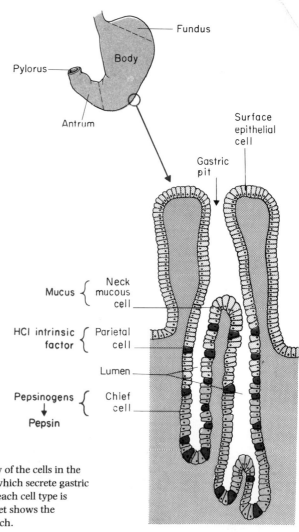

Fig. 6.7. The histology of the cells in the body of the stomach which secrete gastric juice; the function of each cell type is shown in red. The inset shows the anatomy of the stomach.

(Fig. 6.7). Secretory activity of the mucous membrane varies throughout the stomach; the middle part (mainly body) secretes acid and pepsinogen and makes up 75 percent of the total mucosa, the lower part (antrum and pylorus) secretes gastrin; mucus is secreted in all parts.

Hydrochloric acid is secreted by the parietal cells. A characteristic feature of these cells are the membranous processes within the cytoplasm. At rest these form vesicles; during secretion the vesicles fuse and form secretory canaliculi which are continuations of the lumen into the cytoplasm of the cell (Fig. 6.8). The maximum rate of HCl production coincides with the maximum appearance of canaliculi. Staining techniques show that the canaliculi are strongly acid, but the inside of the cells is like other cells (pH 7·2). For these, and other, reasons it is thought that during acid secretion the vesicles place K^+/H^+ (potassium/hydrogen) pumps into the cell membrane; these then secrete HCl into the

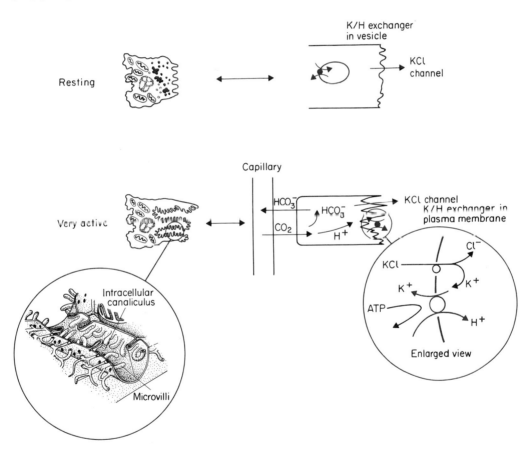

Fig. 6.8. Parietal cell at rest and secreting. On the left is shown an anatomical view; on the right a current view of the mechanism. The vesicles contain K^+/H^+ exchangers which are inserted into the canalicular walls during activity. KCl leaves the cell through a channel and the K^+ recycles into the cell on the K^+/H^+ pump during acid secretion. The KCl channels are always present in the plasma membrane. The H^+ comes from splitting CO_2, for which carbonic anhydrase is necessary. Note the large number of mitochondria to supply energy. The secretory membranes are in red (partly after Schmidt & Thews).

lumen (Fig. 6.7). The exchange of carbon dioxide from the blood with bicarbonate results in gastric venous blood being much more alkaline than the venous blood from other organs. So HCl secretion results in an 'alkaline tide' in the blood.

Production and function of macromolecules. The chief cells of the mucosa produce pepsinogens and store it in zymogen granules. When required these granules fuse with the plasma membrane and the pepsinogen is released into the lumen. The enzyme is activated by splitting off an inhibitor protein; a process started by HCl and progressively accelerated by pepsin. So far seven pepsins have been isolated; those produced only by the fundal cells have a pH optimum appropriate to the acid environment of the stomach (i.e. around 2), those from the body, antrum and duodenum have a pH optimum around 4 and can survive in the duodenum.

Gastrin is secreted by the G cells, which are rounded cells found in the antral area of the stomach. Gastrin is carried in the blood to the parietal cells and is the strongest known stimulus for HCl secretion. The human form consists of 17 amino acids, and exists in either of two forms. All of the activity is present in the terminal pentapeptide; an artificially produced form of this is used clinically. A large number of small and large 'gastrins' have been isolated.

Histamine is produced by histaminocytes, cells found close to parietal cells. It is thought that gastrin acts via histamine on the parietal cells (Fig. 6.9). Blockers of histamine receptors (the 'H$_2$ blockers'), inhibits HCl secretion caused by gastrin and reduces that caused by the vagus.

Mucus is secreted by the neck cells of the gastric glands; it adheres to the surface of the mucosa and plays a role in the protection of gastric mucosa from digestion by gastric juice. It has lubricating properties which allow the chyme to move over the mucosal surface, and the proteins and bicarbonate in it act as local buffers protecting the mucosa from the hydrochloric acid without altering the pH of the bulk of the gastric contents.

The parietal cells also produce an important protein, *intrinsic factor*, which combines with Vitamin B$_{12}$ in the food and aids its absorption by ileal epithelium (p. 158). This is actually the only component of gastric juice essential for life.

Regulation of gastric juice. The secretion of gastric juice before, during and after eating is a controlled process; three overlapping phases are recognised; the cephalic, gastric and intestinal.

1   The *cephalic phase* is part of the 'forward signalling' system mentioned earlier; food in the mouth acts reflexly via the CNS and the vagus to cause gastric secretion.

2   The *gastric phase* lasts for hours; it is controlled by the vagus, local cholinergic reflexes and histamine and gastrin release in the wall of the stomach. The main stimulus for gastrin release is the presence of protein in contact with antral mucosa, but distention of the stomach and the

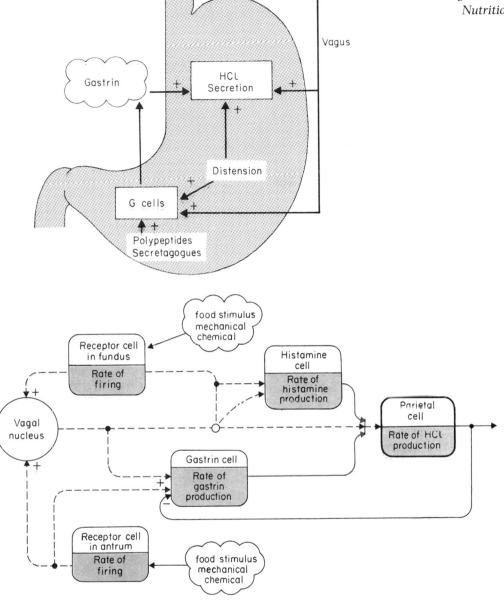

Fig. 6.9. Two diagrams showing the inter-relationships of some of the factors causing HCl secretion. Vagal activity, mechanical and chemical factors causes HCl secretion directly and via gastrin and histamine.

presence of substances like alcohol and other secretogogues also act as stimuli. Distension acts on the parietal cells both by a local and a vago–vagal reflex. These controlling mechanisms for gastric secretion are inter-related and indeed synergistic (Fig. 6.9). The acid secreting system tries to keep the contents of the body of the stomach at a pH of around 1

(Fig. 6.10), however, if the pH of the antrum falls below 3, gastrin (and hence HCl) release is inhibited.

3   In the *intestinal phase* there is firstly an increase and then a decrease in gastric secretion. The initial increase occurs if recently eaten, non-acidified food enters the duodenum; it is due to gastrin release from the duodenum. The later decrease occurs when the duodenal pH drops below 4, and if fatty chyme enters the duodenum. It is not yet clear which hormones are involved (see p. 150).

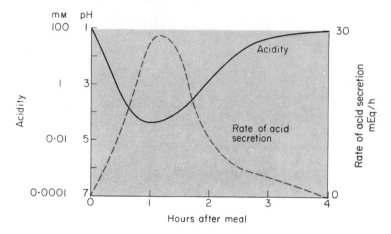

Fig. 6.10. Changes of acidity and the rate of acid secretion in the stomach following a meal. The initial high acidity is decreased when food enters the stomach; this is due largely to the buffering properties of the protein (NH_3, COOH groups). Peptides from the proteins, and distension, then cause increased acid production which restores the acidity. The maximum rate of acid production occurs around 1 hour after the meal is eaten.

These various mechanisms can be regarded as a regulatory system to keep the acidity of the body of the stomach low (100 mmol/l, a pH of 1), but to ensure that the chyme leaving the stomach is much less acid (antrum: 1 mmol/l, a pH of 3; duodenum 0·1 mmol/l, a pH of 4). Among the many other factors which affect gastric secretions, we all know from personal experience the prominent part played by emotion.

Small intestine
In the small intestine the acid chyme from the stomach is mixed with the alkaline secretions from the pancreas, liver and intestinal glands. The enzymes from these secretions are responsible for most of the digestive process. Most absorption also occurs in the small intestine.

Motility. Mixing movements of the small intestine ensure a thorough blending of the chyme from the stomach with the various enzymes of the small intestine. These consist of non-propulsive peristalsis, rhythmic segmentation (Fig. 6.2) and (?) pendular movements; these latter are regional contractions which displace the contents to another part of the gut. The villi of the intestine contract and relax during digestion; this has the dual function of exposing their outsides to new chyme and of moving lymph along their insides (see Fig. 6.16). Towards the end of digestion, propulsive peristalsis drives the contents into the large intestine; a series of such contractions almost sweeping the small intestine clean. Transfer of food from the small to large intestine starts about four hours after ingestion, and ends in around ten hours.

Motility continues after the vagus is cut, so it is thought that it is a property of the local nerve nets (mainly the myenteric plexus). Propulsive peristalsis is always towards the anus, experimentally reversing a segment of gut upsets transport. The villi are partly controlled by the submucous plexus, and partly by *villikinin*, a factor released by the mucosa in the presence of acid chyme.

Pancreatic secretion. You can regard the pancreas as a kind of abdominal salivary gland (Fig. 6.11). It produces about two litres of an alkaline isotonic juice per day. The high bicarbonate concentration raises the pH of the chyme to neutral; a pH favourable to pancreatic enzymes.

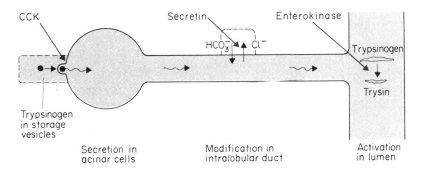

Fig. 6.11. Production, modification and activation of the exocrine secretions of the pancreas. Trypsin shown as an example. Stimulating substances shown in red.

The enzymes are made in the acinar cells as precursors and stored as zymogen granules. Cholecystokinin (CCK) stimulates the release of granules and of electrolyte solution; and also increases the rate of synthesis of new proenzymes. This primary secretion is modified in the duct under the influence of secretin and activated (proteolytic only) in the gut lumen. The ratio of enzyme to bicarbonate is controlled by the CCK and secretin concentrations.

Table 6.2. Pancreatic enzymes.

Enzyme	Substrate	Actions
Trypsin	Proteins and polypeptides	Endopeptidase acting on lysine or arginine
Chymotrypsin	Proteins and polypeptides	Endopeptidase acting on tyrosine and phenylalanine
Carboxypeptidases	Proteins and polypeptides	Exopeptidase cleaving carboxy terminal
Pancreatic lipase	Triglycerides	Frees/monoglycerides and fatty acids
Phospholipases	Lecithin and lysolecithin	Frees fatty acids
α-amylase	Starch	Hydrolysis of 1,4a linkages
Ribonuclease	RNA	Frees/nucleotides
Deoxyribonuclease	DNA	Frees/nucleotides
Collagenase	Connective tissue	Cleaves side chains
Elastase	Elastin	Cleaves side chains

Regulation of pancreatic secretion. Like most alimentary tract glands, the pancreas has a resting level of secretion which increases when food is eaten. Most secretion from the gland occurs when gastric contents enter the duodenum; although some begins during the cephalic phase and continues during the gastric phase of gastric secretion. The characteristics of the secretion depend on the type of chyme. If the chyme is acid then secretion is rich in bicarbonate which neutralises the acid; this response is mediated by *secretin* release from the duodenal mucosa which stimulates the duct cells of the pancreas. If the chyme is rich in amino acids or fat then a pancreatic juice rich in enzymes is released; this response is mediated by cholecystokinin-pancreozymin (CCK-PZ or CCK). For proper action of the system secretin, CCK and the vagus all need to work together.

Liver and biliary system

The liver is the most important metabolic organ in the body. It receives two blood supplies: about 80 percent comes from the intestinal capillary bed, joins with blood from the spleen and enters the liver in the portal vein; this blood is low in oxygen but rich in substances newly absorbed from the gut. The other 20 percent is normal arterial blood in the hepatic artery and so is rich in oxygen.

The hepatocytes are the cells which give the liver its unique properties; they have a complex structure in that they have surfaces facing the blood, the bile canaliculi and each other. The surface of the hepatocytes in contact with blood is increased in area by infolding of the cell membrane. Table 6.3 outlines some of the functions of the liver, these are considered in more detail in other chapters.

Table 6.3. Functions of liver.

Protein metabolism
  synthesis
  urea formation
  interconversion of amino acids
Carbohydrate metabolism
  storage of glycogen
Fat metabolism
  lipid synthesis
Formation of bile
Detoxification and conjugation of steroid hormones
Inactivation of polypeptide hormones
Vitamin storage

Bile secretion takes place into the bile canaliculi; small channels which then join together like the tributaries of a river to form the bile duct leaving the liver (Fig. 6.12a). From this duct the bile can either flow directly into the duodenum, or be diverted to the gall-bladder where it is stored and concentrated (by active sodium pumping, water following). Just before it enters the duodenum the common bile duct is usually joined by one from the pancreas, so that the resultant duct carries both bile and pancreatic juice into the gut. At the point of entry the duct is surrounded by smooth muscle *the sphincter of Oddi*, which is normally closed; this causes bile from the liver to be diverted into the gall-bladder. The secretory mechanism involved in bile production is similar to that of other GI secretions (see p. 143). Table 6.4 lists its composition.

Table 6.4. Average composition of hepatic bile.

	Concentration mmol/l
Inorganic	
sodium	as plasma
potassium	as plasma
chloride	as plasma (approximately)
bicarbonate	27–50
Organic	
Bile salts	36
Phospholipid	9
Cholesterol	2·5
Bile pigment	1·5

Bile has two functions:

1 The bile salts act in the digestion and absorption of fat. The liver produces and secretes bile as shown in Fig. 6.12b. About 75 percent of the bile salts are reabsorbed through the terminal ileum, mostly by an active transport mechanism, and returned to the liver to be secreted unchanged; the rest is altered but mainly also recycled. A particular bile salt circulates in this way about three times during a meal.

Hepatocytes can be stimulated by *choleretics* and bile salts themselves have this property. In this way they regulate their own synthesis from cholesterol in hepatocytes.

2 The bile acts as an excretory pathway, especially for non-water-soluble substances which cannot be excreted by the kidney. Generally these substances (e.g. bilirubin) are combined with glucuronic acid to make them water-soluble so that they can enter the bile.

The bile pigments, mainly bilirubin and biliverdin, are breakdown products of haemoglobin, myoglobin and cytochromes and give bile its yellowish-green colour. They are converted to colourless stercobilinogen in the gut which is in turn oxidised to stercobilin. Some of the stercobilinogen is absorbed, partly being recirculated to the liver and partly excreted in the urine as urobilinogen.

The excretion of cholesterol depends on the formation of aggregates of cholesterol surrounded by bile salts (*micelles*, Fig. 6.13). The water-soluble (charged) part of the bile salt faces outwards, and the steroid (water-insoluble) part inwards thus acting as a kind of membrane round the cholesterol.

Regulation of the secretion and release of bile. The resting rate of bile secretion is doubled during digestion, and contains more bicarbonate. This is mainly due to secretin. During digestion the gallbladder contracts and the sphincter opens, so bile flows from the gallbladder to the small

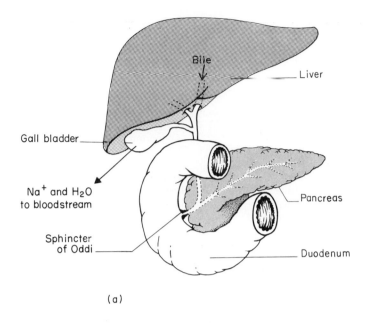

(a)

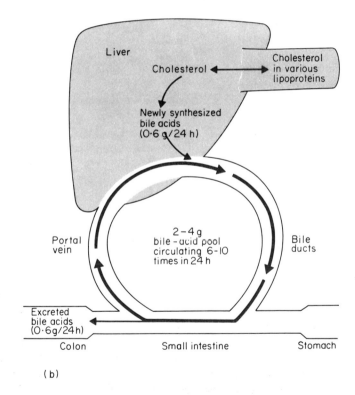

(b)

Fig. 6.12. (a) Storage and concentration of bile in the gall-bladder and the route to the duodenum. (b) The enterohepatic circulation of bile salts.

The primary bile acids, cholic and chenodeoxycholic acid are synthesised from cholesterol by liver cells, conjugated with either glycine or taurine and excreted in bile as the sodium or potassium salts. Reabsorption of the bile salts is mostly by an active process in the terminal ileum. Synthesis by the liver replaces that lost in the faeces every day, and keeps the bile salt pool constant (b after Dietschy).

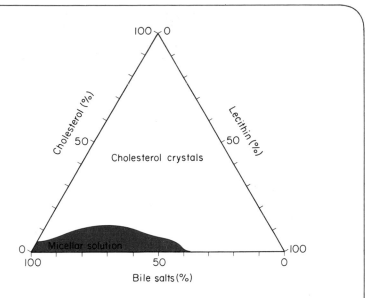

Fig. 6.13. Solubility of cholesterol in the bile as a function of the concentration ratios of bile salts, lecithin and cholesterol. Cholesterol is in micellar solution in the red area, but precipitates when the concentrations of bile salts and lecithin decrease.

intestine. Two factors control this: firstly the increased activity of the vagus which accompanies eating, and second the release of CCK from the duodenal mucosa.

Intestinal secretion occurs from *Brunner's Glands* in the wall of the duodenum. The secretion contains much mucus and bicarbonate, making it both very viscous and alkaline (pH of 8–9). It probably contains no enzymes.

Large intestine and rectum

Motility of the large intestine. The longitudinal muscle of the colon is arranged in bands (taeniae), so that the colon normally has a sacculated appearance. Mixing is by slow non-propulsive peristalsis and rhythmic segmentation; both controlled by the myenteric and submucosal nerve nets. Propulsive peristalsis perhaps occurs two to three times a day (peristaltic rushes), starting in the caecum and sweeping over the entire colon. The *gastrocolic reflex* is an example of this type of contraction triggered by a meal. Food residues remain in the colon for at least 12 hours, and may persist for three days.

Production of faeces. Up to half a litre of contents leave the small intestine daily and enters the large intestine. In the colon this is concentrated by the removal of electrolytes, water and the water-soluble vitamins but the reabsorption is small compared to that in the small intestine.

The colon is normally inhabited by various aerobic (e.g. *E.coli*) and anaerobic bacteria which break down some carbohydrates by fermentation and proteins by putrefaction. Fermentation produces acids (lactic and acetic), alcohol, CO_2 and H_2O. Putrefaction produces poisonous amines (indole, skatole), amines with biological activity (histamine, tyramine) and hydrogen, hydrogen sulphide and methane. Normally the processes of fermentation and putrefaction are in balance, e.g. the acid pH of fermentation inhibits putrefaction. Abnormal diets or infections upset the balance, so that faeces typical of fermentation or putrefaction are passed.

Stercobilinogen is oxidised in the colon to stercobilin and with other breakdown pigments like bilifuscans is responsible for the brown colour of faeces. If bile is prevented from reaching the gut the lack of these bile pigments results in the faeces becoming greyish-white.

Defaecation. The urge to defaecate occurs when the rectum is filled by mass movement of contents from the colon. The reflex involves stretch receptors in the walls of the colon, the sacral cord (anospinal centre), and the parasympathetic nerves to the *internal anal sphincter*. In defaecation this reflex is followed by a voluntary relaxation of the external anal sphincter and contraction of the abdominal muscles.

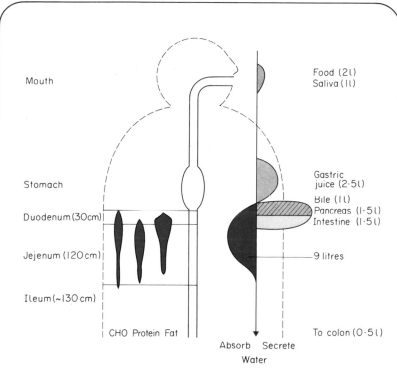

Fig. 6.14. Food and water movements across different parts of the GI tract. The left shows the relative absorption of carbohydrate, proteins and fat in the small intestine; the right shows daily water movements. Full red indicates absorption; dotted red secretion. Numbers for water movement are rounded (partly after Schmidt & Thews).

In digestion, the macromolecular components of food are hydrolysed by the enzymes in the digestive juices to molecules of an absorbable size. Proteins are split to *amino acids*, carbohydrates to *glucose* and other monosaccharides and fats to *glycerol* and *fatty acids*. Absorbed molecules do not have the same characteristics as the original food, so digestion protects the body from foreign proteins. Digestion and absorption occur mainly in the small intestine (Fig. 6.14).

Digestive-absorptive surface. The digestive enzymes of the small intestine are located in a special filamentous layer (glycocalyx) on the surface of the epithelial cells (Fig. 6.15). This surface of the cells is much enlarged by the brush border of microvilli on the surface facing the lumen. The enterocytes of the GI tract only live for a few days and are then shed into

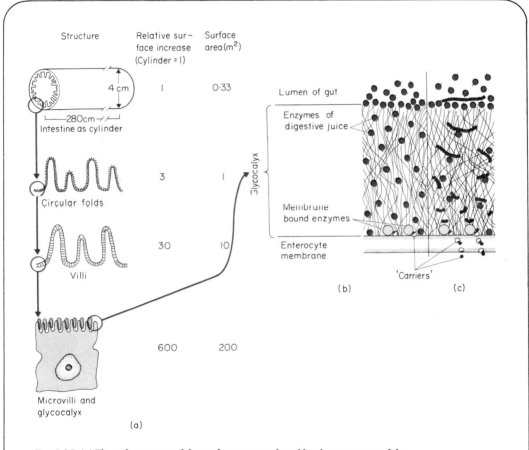

Fig. 6.15. (a) The enlargement of the surface area produced by the structures of the small intestine. (b) The spatial distribution of enzymes and carriers on the surface coat of the microvilli. The enzymes of the pancreatic juice are absorbed onto and then embedded in the glycocalyx. (c) The interaction of these enzymes and carriers with substrate molecules of various sizes. The adsorbed enzymes produce oligomeres, which are then split to monomeres by the enzymes embedded in the cell membrane (after Schmidt & Thews).

the lumen, where they are broken down and their parts reabsorbed (some $\frac{1}{4}$ kg of cells are shed daily). This rapid turnover is probably necessary because these cells live in an environment made very hostile by the digestive enzymes.

Absorption is the process by which substances are taken from the outside of the body (lumen of the GI tract), into the inside (blood, lymph etc.). Most occurs in the small intestine, which in man has a length of

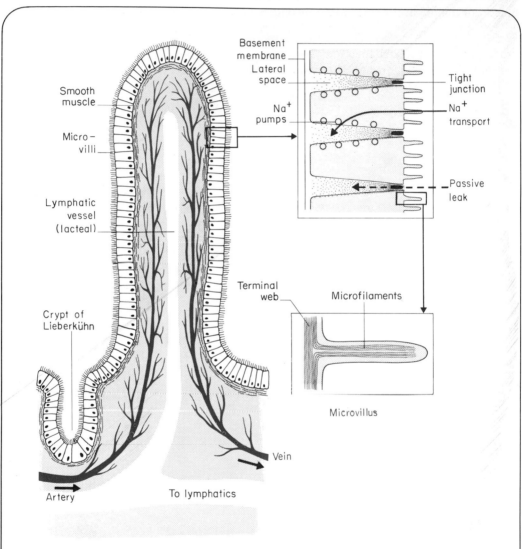

Fig. 6.16. Intestinal villus supplied with a large lymph vessel in addition to a capillary network. Note the smooth muscle fibres from the muscularis mucosae layer which cause shortening of villi and so aid movement within the lacteals. Inserts show details of cells and a single microvillus; each microvillus has a central cone of microfilaments which intermingle with the meshwork of microtubules and filaments called the 'terminal web'. In some species actin has been identified in the microfilaments and the addition of ATP causes them to contract.

about three metres and a surface area of about 200 square metres (Fig. 6.15). The GI tract receives about 10 percent of the cardiac output during eating, of which half goes to the mucosa. When digesting food the mucosal blood flow almost trebles, but the O_2 tension at the tip of the villi is very low.

The mechanisms of absorption in the GI tract are similar to that occurring across membranes elsewhere in the body (Fig. 6.16), and consists of active transport, passive diffusion down concentration gradients, pinocytosis and solvent drag (fluid sweeping dissolved substances with it) etc. Absorption here, as elsewhere, is the net result of undirectional movements in both directions.

Protein digestion and absorption (Figs. 6.16, 6.17 and Table 6.5)
Pepsin, released in the stomach, attacks many proteins, but only at a few points within the molecule. Pepsin digestion, therefore, does not proceed further than the polypeptide stage. In the duodenum, the enzymes trypsin and chymotrypsin similarly only attack some bonds within the protein molecules whereas carboxypeptidase splits off end peptides. The former group are called *endopeptidases*, the latter *exopeptidases*. Further hydrolysis of these peptides takes place by enzymes associated with the brush border and in the cytoplasm of the mucosal epithelium (Figs. 6.16, 6.17 and Table 6.5).

Peptides are *absorbed* into the enterocytes by specific carriers; the sodium gradient between the lumen of the gut and the cytoplasm of the enterocyte supplying the energy for the process. The peptides absorbed are single amino acids, dipeptides or tripeptides. The amino acids leave the cells by a variety of ways at the basal and lateral surfaces, and enter the blood. Glycine is the commonest of the amino acids; although carried specifically on one of the four amino acid carriers present on the gut cells it really isn't too fussy about its carrier and can enter cells on any of them.

Newborn infants, but not adults, absorb proteins intact; this allows antibodies to be transferred in milk from mother to child to provide some protection against infections in the first few days of independent life.

Digestion and absorption of carbohydrates (Fig. 6.18 and Table 6.6)
Fig. 6.18 shows the main events in *carbohydrate digestion*. The starch in food consists of long chains of glucose molecules, with some branched chains. The straight chains are split by salivary and pancreatic amylases to form two, three and six membered polysaccharides and glucose. The enzymes of the brush border split the branched chains, and reduce the other molecules to monosaccharides. Glucose and other monosaccharides are *absorbed* into the enterocytes by a carrier system which depends on the energy in the sodium gradient. Once in the cells these sugars diffuse down their concentration gradients into the blood. No polysaccharides or disaccharides are absorbed.

Digestion and absorption of fats
Fat is digested and absorbed in the duodenum (Fig. 6.14). The pancreatic lipases which *digest* fats are water-soluble and so the fat must be con-

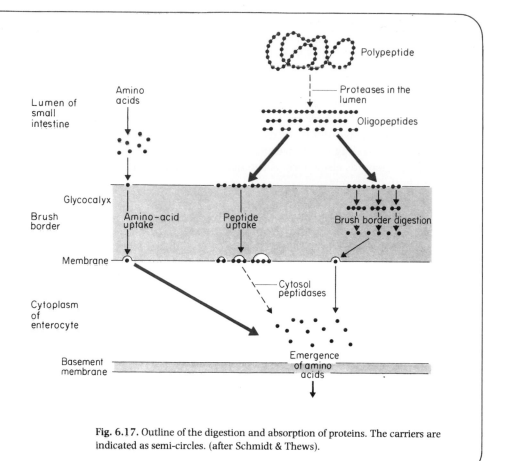

Fig. 6.17. Outline of the digestion and absorption of proteins. The carriers are indicated as semi-circles. (after Schmidt & Thews).

verted to a form with which the lipase can make contact. The mixing activity of the stomach produces droplets of fat, but these would coalesce if it were not for the presence of emulsifying agents in the duodenum. Bile salts together with cholesterol, lecithin and fatty acids keep the emulsion fairly stable with particles some 1 μm in diameter. This process allows the water-soluble lipase to contact fat over a large surface area. Hydrolysis of triglyceride results mainly in the formation of 2-monoglycerides and fatty acids with a little glycerol (Fig. 6.19). These products are not water-soluble but are solubilised by the formation of *micelles*. These polymolecular aggregates are very much smaller than the previous fat droplets, being about 5 nm in diameter, and form a translucent solution. They diffuse between the villi and disgorge their contents into the enterocytes. Although the micelles diffuse more slowly through the unstirred layer than individual molecules, the concentration of fat products presented to the epithelium is 1000 times greater in this way. The fat fragments are processed in the endoplasmic reticulum to triglycerides, converted to lipoproteins in the Golgi apparatus and released into the lymph as chylomicrons and very-low-density-lipoproteins (VLDL).

Table 6.5. The main events in protein digestion.

Site of production	Enzyme (proenzyme)	Mechanism of action	Substrate	End products
Gastric glands, chief cells	Pepsin (pepsinogen)	Endopeptidases; split mainly peptide bonds between NH_2 groups of other amino acids; pH optimum 1.5–3.5	Proteins	Polypeptides
Exocrine pancreas	Trypsin (trypsinogen)	Endopeptidase; splits mainly between COOH groups of lysine or arginine and NH_2 groups of other amino acids; pH optimum 7.5–8.5	Proteins	Polypeptides
	Chymotrypsin (chymotrypsinogen)	Endopeptidase; splits mainly between COOH groups of aromatic amino acids and NH_2 groups of other amino acids except glutamic acid and asparagine; pH optimum 7.5–8.5	Proteins, polypeptides	Polypeptides, oligopeptides
	Carboxypeptidase A, B (procarboxypeptidase)	Exopeptidases; Type A splits off aromatic, nonpolar amino acids at the C-terminal end; Type B splits off basic amino acids in the same position	Polypeptides, oligopeptides	C-terminal amino acids and peptide residues
Duodenal mucosa(?)	Enterokinase	Endopeptidase; splits between isoleucine (pos 7) and lysine (pos. 6)	Trypsinogen	Trypsin and hexapeptides
Brush border of enterocytes (membrane-bound)	Tripeptidase	Exopeptidase; splits off N-terminal or C-terminal amino acids	Proteins, poly and oligopeptides	N- or C-terminal amino acids and poly- or oligopeptides
	Aminopolypeptidase	Exopeptidase	Tri-, dipeptides	Amino acids
	Aminopeptidase	Exopeptidase; splits amino bonds	Tri-, dipeptidase	Amino acids
	Many dipeptidases, some specific			

These then enter the blood via the thoracic duct. After a fatty meal the blood contains so much fat that it is milky. Short- and medium-chain fatty acids diffuse directly through the enterocyte into the blood.

Under normal conditions almost all the fat we ingest is absorbed and only in abnormal conditions do we find fat in the faeces.

The fat-soluble *vitamins* (A, D, E, K) are absorbed with fat; the water-soluble ones mainly by diffusion.

Water and electrolytes (mainly sodium) move across the walls of the GI tract in both directions to keep the contents isotonic with the plasma;

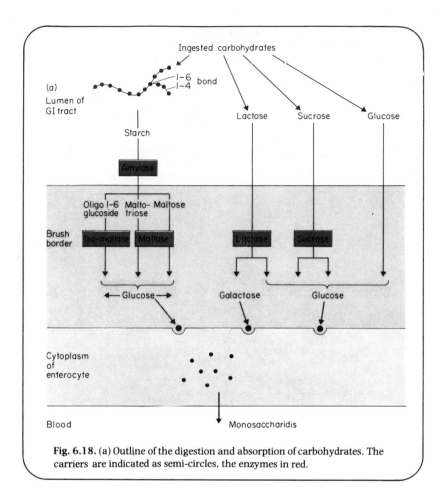

Fig. 6.18. (a) Outline of the digestion and absorption of carbohydrates. The carriers are indicated as semi-circles, the enzymes in red.

most exchanges occur in the small and large intestine and to a lesser extent the stomach. Electrolyte and water reabsorption occurs in the upper small intestine (by active sodium transport) (see Fig. 6.14), leaving about half a litre to go on to the colon. A counter-current circulation in the villi may aid water absorption.

NUTRITION

During life animals use energy to stay distinct from their environment; on death their organisation disintegrates and their materials return to their surroundings. The energy required to maintain the integrity of animals comes from breaking down large molecules originally created by plants using photosynthesis. Foods consist of materials which supply energy plus vitamins, salts, trace elements, crude fibre, water and various spices. The total food intake is regulated by the sensations of *hunger* and *thirst*, but eating habits are rarely rational for they are surrounded by tradition and ritual. Our ancestors probably had more difficulty in getting enough food, than having too much; this may explain why the mechanisms for preventing excess eating are less effective than those for starting eating.

Table 6.6. The main events in carbohydrate digestion.

Site of production	Enzyme (proenzyme)	Mechanism of action	Substrate	End products
Salivary glands	α-amylase (ptyalin)	Endoenzyme, α-amylase) splits α-1,4-bonds (amylose fraction of starch); pH optimum 6,7	Starches	Oligosaccharides and amylopectin (1,6-linked chains)
Exocrine pancreas	Pancreatic amylase	Endoenzyme, α-amylase; cf. ptyalin; pH optimum 7.1	Starches	Oligosaccharides
Brush border of enterocytes (membrane-bound)	Amylase	Glucoamylase	Starches, oligosaccharides	Maltose and glucose
	Oligo-α-1,6-glucosidase	splits α-1,6-bonds (amylopectin fraction of starch)	Glycogen, amylopectin	Oligosaccharides, maltose and glucose
	Many disaccharidases, some specific:			
	Sucrase	β-fructosidase	Cane sugar (saccharose, sucrose)	Fructose and glucose
	Maltase	α-glucosidase; splits α-1,4-bonds	Maltose	Glucose
	Isomaltase	Corresponds to oligo-α-1,6-glucoasidase		
	Lactase	β-galactosidase	Lactose	Galactose and glucose

Composition and function of foods

Foodstuffs are used for two purposes; firstly to provide energy (*functional metabolism*), and secondly for synthesis of secretions or body building (*structural metabolism*). Most of the food we eat is for use as an energy source, only a small amount is required for structural purposes.

Energy for bodily processes is obtained by breaking down the *proteins, fat* and *carbohydrates* in the diet to simpler molecules with a lower energy content. The amount of energy released from each is called the *biological fuel value* (Table 6.7). Proteins and carbohydrates produce about 4 kcal/g (17 kJ/g) and fats about twice that at 9 kcal/g (37 kJ/g).

The amount of energy used by the body may be assessed by the heat output of the body, for all energy is eventually degraded to heat. The heat output of a man can be measured directly in a *calorimeter*, but is more often done by indirect means (because of the difficulty of making a whole body calorimeter). *Indirect calorimetry* is done by measuring oxygen consumption, for it is known that when the body is burning a normal mixture of carbohydrate, fat and protein one litre of oxygen is consumed for every 4·82 kcal (20 kJ) produced.

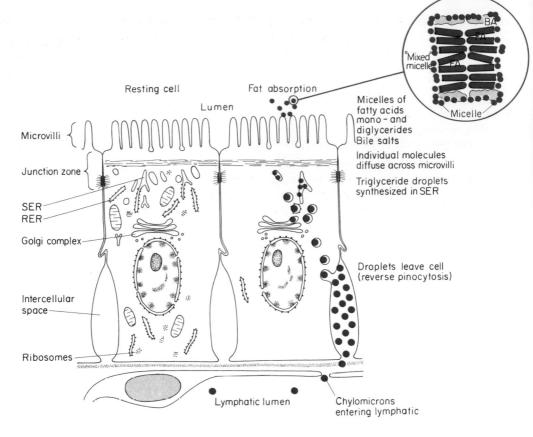

Fig. 6.19. Absorption of fat. The pancreatic lipases split the 1-ester linkage releasing 2-monoglycerides, glycerol and free fatty acids of differing sizes. Cholesterol, lysolethicin and the longer free fatty acids form micelles. All movement is by passive diffusion down concentration gradients.

Table 6.7. Biological fuel values of fats, proteins and carbohydrates (approximate); with those for alcohol and glucose for comparison.

The energy equivalent of 1 litre of Oxygen during the combustion of a mixed diet is around 20 kJ (5 kcal), but depends on the RQ (after Passmore & Robson).

	kJ/g	kcal/g	RQ
Protein	17	4	0·81
Carbohydrate	16	4	1·00
Fat	37	9	0·71
Glucose	16	4	1·00
Alcohol	29	7	0·66

To measure the heat production exactly we need to know the actual amounts of carbohydrate, fat and proteins being burnt at the time, for each uses a different amount of oxygen during degradation. This can be calculated as follows:

1 the *respiratory exchange ratio* (previously known as the respiratory quotient or RQ) is measured by the ratio of the volume of CO_2 produced to the volume of O_2 consumed when a substance is oxidised. For carbohydrate the reaction is:

$$C_6H_{12}O_6 + 6O_2 - 6H_2O + 6CO_2$$

so the RQ for carbohydrate is 1. The RQ for fat is about 0·7 and for protein approximately 0·82;

2 the amount of nitrogen excreted in the urine gives the average amount of protein used during the experiment.

By combining these two measurements an accurate assessment of the heat production, and hence energy usage, of the body can be made. Usually this is measured under resting conditions, when it is known as the *basal or resting metabolic rate*; in an adult male this is about 2000 kcal (8·4 MJ) per day. This is often expressed as per *surface area* of the body. Although the surface area is not easily measured, it can be estimated from the weight and height as follows:

$$72 \times weight^{0·4} \times height^{0·7}$$

The rule of 9 gives a rough idea of the distribution of this area: 9 per cent for each arm, 9 percent for the head and neck, 18 percent for each leg and 36 percent for the trunk.

The metabolic rate increases by about 6 percent after eating food (diet induced thermogenesis, previously the *specific dynamic action*). This is because energy is required to handle the conversion of food to fuel in the body. This effect is greatest after a protein meal, for protein requires more energy for this conversion than the others.

Proteins consist of amino acids and are mostly used in making and repairing the structures of the body (*structural metabolism*). Certain amino acids cannot be made in our bodies (*essential amino acids*) and so must be supplied in the diet. For these reasons protein cannot be replaced entirely by fats and carbohydrates. Proteins are found in meat, fish, milk, eggs and dairy products and there are variable amounts in all kinds of vegetables (Table 6.8). Amino acids are not stored in the body to any great extent, so a proper mixture must be eaten in each meal. This occurs naturally if animal products are eaten, but needs some planning if a purely vegetarian diet is eaten.

Fats consist mainly of a mixture of triglycerides, which are triesters of glycerine and fatty acids; they are mostly used to provide energy. Fatty acids may have all their bonds saturated or be unsaturated to a variable extent (i.e. have free bonds to which hydrogen atoms can be attached); vegetable fats are more unsaturated than animal ones. Certain unsaturated fatty acids cannot be synthesised and so must be eaten (*essential fatty acids*). Fats in food may be visible as layers in meat, oils etc., or concealed as small droplets. Modern farming methods encourage the production of food with concealed fats; this partly explains the high fat intake of Western man.

Table 6.8. Energy value and nutrient content of some common foods. The composition of meat and sausage is very variable, depending on the amount of concealed fat (from McCance & Widdowson).

	Energy		Water	Protein	Fat	Carbohydrate	Alcohol
	kJ	kcal	g	g	g	g	g
Whole wheat flour	1420	339	15	14	3	69	—
White bread	1020	243	38	8	1	53	—
Rice, raw	1510	361	12	6	1	87	—
Milk, fresh, whole	276	66	87	3	4	5	—
Butter	3320	793	14	1	85	tr	—
Cheese, Cheddar	1780	425	37	25	35	tr	—
Beef steak, fried	1140	273	57	20	20	0	—
Haddock, fried	733	175	65	20	8	4	—
Potatoes	293	70	80	3	tr	16	—
Peas, canned	360	86	73	6	tr	17	—
Cabbage, boiled	38	9	96	1	tr	1	—
Orange, with peel	113	27	65	1	tr	7	—
Apple	197	47	84	1	tr	12	—
White sugar	1650	394	tr	tr	0	105	—
Beer,* bitter	130	31	97	1	tr	2	3
Spirits* (gin, whisky 70% proof)	929	222	64	tr	0	tr	32

*Values per 100 ml; the others are per 100 g.

Carbohydrates consist of simple sugars (monosaccharides) and compounds of two or more simple sugars (di-, oligo- or polysaccharides). They are used primarily as a source of energy, but partly as building blocks for other compounds (Fig. 11.32). Most of our carbohydrate is from vegetables and is in the form of plant starch. Many vegetables contain large amounts of cellulose fibre, which is indigestible but may be of importance in gut function.

It is obviously useful for animals to have a means of *storing energy* for lean times. Fat gives the most energy per gram and so is the lightest way of carrying fuel; consequently it is the main means of storage. A small amount of carbohydrate is stored in the liver and muscle as glycogen (about 8 hours supply), but no protein is stored. In prolonged starvation structural protein is broken down; so, for example, muscles can then do less work (Fig. 6.20).

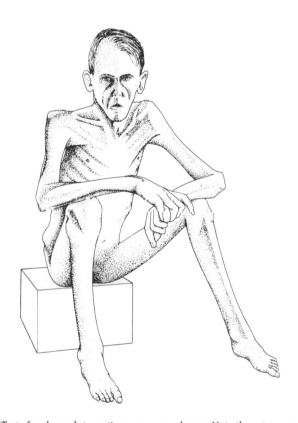

Fig. 6.20. Effect of prolonged starvation on a normal man. Note the extreme muscle wasting.

Vitamins are organic substances necessary in small quantities for the normal functioning of the body, which the body cannot make enough of for itself; their energy content is negligible. They are classified as fat-soluble or water-soluble (Table 6.9). The first ones discovered were named after letters of the alphabet, now most are given chemical names. The chemical structure of the vitamins is very diverse; their action is

Table 6.9. Vitamins involved in human nutrition. The water-soluble vitamins are easily absorbed; the fat soluble ones (A, D, E, K) require the presence of dietary fat and bile or pancreatic lipase (partly after Ganong).

Water Soluble				Fat Soluble			
Vitamin	Action	Deficiency Symptoms	Sources	Vitamin	Action	Deficiency Symptoms	Sources
B complex Thiamine (vitamin B_1)	Cofactor in de-carboxylations.	Beriberi, neutritis	Liver, unrefined cereal grains	$A(A_1, A_2)$	Visual pigments maintain epithelia.	Night blindness, dry skin	Yellow vegetables and fruit
Riboflavin (vitamin B_2)	Constituent of flavoproteins.	Glossitis, cheliosis	Liver, milk	D group	Increase intestinal absorption of calcium and phosphate (Chapter 21).	Rickets	Fish liver
Niacin	Constituent of NAD^+, $NADP^+$.	Pellagra	Yeast, lean meat, liver				
Pyridoxine (vitamin B_6)	Forms prosthetic group of certain decarboxylases and transaminases.	Convulsions, hyperirritability	Yeast, wheat, corn, liver	E group	Antioxidants; cofactors in electron transport in cytochrome chain?	Muscular dystrophy and fetal death in animals	Milk, eggs, meat, leafy vegetables
Pantothenic acid	Constituent of CoA.	Dermatitis, enteritis, alopecia, adrenal insufficiency	Eggs, liver, yeast	K group	Catalyse γ carboxylation of glutamic acid residues to complete synthesis of prothrombin. Also catalyse synthesis of several other clotting factors in the liver.	Hemorrhagic phenomena	Leafy green vegetables
Biotin	Catalyses CO_2 "fixation" (in fatty acid synthesis, etc)	Dermatitis, enteritis	Egg yolk, liver, tomatoes				
Folates (folic acid) and related compounds	Coenzymes for "one-carbon" transfer; involved in methylating reactions.	Sprue, anemia	Leafy green vegetables				
Cyanoco-balamin (vitamin B_{12})	Coenzyme in amino acid metabolism. Stimulates erythropoiesis.	Pernicious anaemia	Liver, meat, eggs, milk				
C	Necessary for hydroxylation of proline and lysine in collagen synthesis.	Scurvy	Citrus fruits, leafy green vegetables				

usually highly specific in cellular metabolism; they are often catalytic elements in enzyme systems.

Water makes up more than 50 percent of most foods (Table 6.8) and together with that drunk and that produced by metabolism constitutes the water intake (see Fig. 14.2).

The most important *salts* in the diet are the cations sodium, potassium, calcium and magnesium and the anions chloride and phosphate. A diet of 'natural' substances contains much potassium, but little sodium so 'salt' (NaCl) was a prized substance in primitive societies. Modern man is almost addicted to NaCl; a circumstance which has been implicated in high blood pressure.

The *trace elements* are present in minute quantities in the diet. They can be considered as those:
1   with known or suspected function, e.g. iron, fluorine, iodine, copper, manganese, molybdenum and zinc;
2   with known toxic effects, e.g. antimony, arsenic, lead, cadmium, mercury and thallium;
3   with no known function, e.g. aluminium, boron, silver and tellurium.

Various aromatic substances are added to food to affect its taste and smell. These *condiments* make food more palatable, and also affect the production of the gastrointestinal secretions.

Foods contain various *residues* remaining from their production and handling, such as drugs, metals, flavourings, pesticides etc.

Thirst and hunger

Thirst and hunger are examples of *general sensations*, i.e. those which arise within the body itself but which we cannot ascribe to any particular sense organ or part of the body. Tiredness, shortness of breath and sexual appetite are other examples. General sensations are inborn and so need not be learned, though they are modified by life's experiences; such modifications occur at various points in the process. They have obvious survival value.

These general sensations are produced by stimuli acting on various receptors (known or unknown) within the body. In addition to producing these sensations, these stimuli induce a drive to correct the deficiency. For example water deficiency leads both to the sensation of thirst and to the search for water to drink and so remove the sensation.

Thirst

There are about 40 litres of water in a 70 kg man, distributed between the various compartments (p. 10). In ideal conditions this fluctuates within a range of about ± 250 ml (a cupful), due to the interplay of water lost and gained during the day (p. 413). This variation is only about 0·5 percent of the body weight. We feel thirsty when our body water drops by about this amount, and have a diuresis when it rises by this amount.

Regulation of body water is achieved both by negative feedback mechanisms and by *feedforward* systems (see p. 13). *Feedback control* is exercised by a reflex involving detectors, various neural pathways in the central nervous system and control of water intake or loss. Thus water loss from the body leads to a higher osmolality of that left; whereas excess water leads to a lower tonicity. The signals for the reflexes correcting these changes depend on these changes. These signals are thought to arise from (Fig. 6.21):

1 The volume or osmolality of the cells. The osmoreceptors which detect changes in cellular osmolality are dotted throughout areas in the front part of the hypothalamus; the hypothalamus itself seems to process this information.

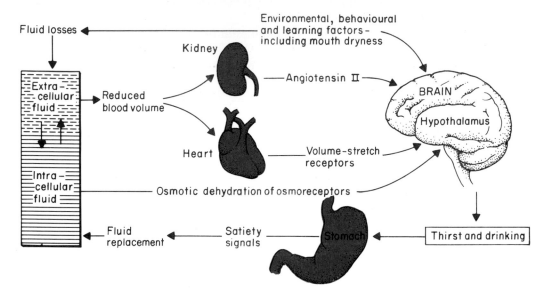

Fig. 6.21. Relationships between changes in body fluids, environmental, other influences and thirst. 'Satiety signals' is to indicate that drinking stops before the osmotic pressures return to normal (see text) (after *New Scientist*, 1983).

2 The volume or tonicity of the extracellular fluid. It is likely that the low pressure stretch receptors in the veins near the heart are involved in this regulation, but the information is scanty. The hormones renin and angiotensin II may also be involved. The hypothalamus again acts as a processing centre.

It was previously thought that the dryness of the mouth and throat which arises from a reduction of salivary secretion was also a signal for thirst, but it is now thought to be a symptom of thirst rather than a signal. The false thirst arising from a dry mouth after speaking etc. is satisfied by wetting the mouth, whereas real thirst due to dehydration is not.

The sensation of thirst does not adapt, i.e. a slowly rising dehydration leads to the same water intake as a rapidly produced dehydration of the same degree.

Feedforward regulation. Thirst is, of course, quenched by drinking water but there are a number of interesting features to this. There is a delay between drinking water and the correction of osmolality in the body fluids, for the water needs to be absorbed. It is always found that dehydrated animals stop drinking water when the correct amount has been drunk, but before the osmolality of the body fluids has had time to settle to the normal level. The body can therefore measure and calculate the expected effects of the water drunk with great accuracy. The way this is done is not clear; although it does depend on signals from the GI tract. The system is even more complex than this in that we, and other animals, tend to anticipate thirst and drink to prevent it; e.g. if eating a salty meal we drink more water. If this system was not present excess water would be absorbed, which would then need to be excreted, leading to a larger oscillation of body water than occurs.

Once thirst is satisfied there is a slow loss of body water until the next time water is drunk. After about 250 ml is lost we feel thirsty, i.e. a thirst threshold is passed, so we drink again. The body water therefore fluctuates between a maximum following drinking and a minimum just before drinking again. In man the fluctuations are greater than this, because of our social habits; this is called *secondary drinking* compared to the *primary drinking* following thirst.

Hunger
The body weight of man and animals is kept remarkably constant over very long periods of time, e.g. a 70 kg man eats about a ton of food each year and yet may stay the same weight for decades. It is probable that this is done primarily by controlling the food intake by hunger signals, rather than by controlling the efficiency of usage once the food is absorbed. The system which does this has both longer-term and shorter-term components, the latter copes with day to day changes in food intake and work loads. Hunger, like thirst, is regulated both by negative feedback and by feedforward mechanisms; in man of course there are also strong social factors involved. The central processing regions for hunger and satiety are in the hypothalamus, perhaps with separate areas for each.

Feedback regulation. Hunger is the generalised sensation which is related to the feedback control of food intake; we 'feel' it in the region of the stomach. Factors involved in hunger are:
1   the availability of glucose to the cells (*glucostatic hypothesis*) appears to be crucial; receptors for this are thought to be in the diencephalon, liver, stomach, and small intestine;
2   the amount of fat in the body appears to be monitored by liporeceptors (*lipostatic hypothesis*), this is probably most important in long-term regulation;
3   it is a common observation that we eat less in hot climates, and more in cold ones; it is thought that the general level of body temperature may be integrated and lead to hunger signals (*thermostatic hypothesis*).

It used to be thought that hunger arose primarily from the contractions of an empty stomach, but denervation of the stomach does not much affect it so stomach contractions probably only play a small part in it.

We stop eating when *satiated*. Part of this generalised sensation is related to the events of *post-absorption*. It is thought that the enterorereceptors discussed above in hunger regulation probably provide the signals for this. This part of the satiety mechanism prevents us from eating again immediately after a meal.

Feedforward regulation. Hunger is assuaged by eating, but we stop eating before most of the food reaches the body cells; this is the *pre-absorptive phase* of satiety. It is related to processes involved in food intake. The information for this is thought to arise from chemo- and mechanoreceptors in the nose, mouth, throat, oesophagus, stomach and upper small intestine. Circulating gut hormones may also be involved.

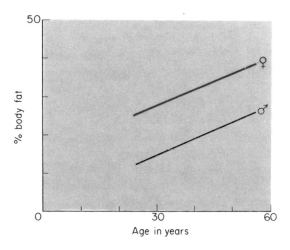

Fig. 6.22. Body fat at different ages in males and females. An index of fatness is given by

$$W/H^2$$

where W is the body weight in kg, and H is height in metres. The optimum body weight for maximum longevity, based on life-insurance statistics, for adult men is 20–25, and for adult women is 19–24. (Graph from data of Durnin & Womersley, equation from Garrow.)

We also eat when we are not hungry. This is partly for social reasons but partly in anticipation of becoming hungry before the next meal and the possible energy expenditure expected before then. This is another example of a predictive regulatory system.

Nutritional requirements

In a steady state the amount of food eaten needs to balance the energy expenditure; various factors such as work-load, sex, age etc., are involved in this equilibrium. Deficiencies arise when the intake is too small for the current amount of work done or is not properly balanced. In western societies many people eat excessive amounts of food, so that their body weight rises to a new steady state with large depositions of fat (Fig. 6.22). It is thought that this predisposes these people to various degenerative diseases.

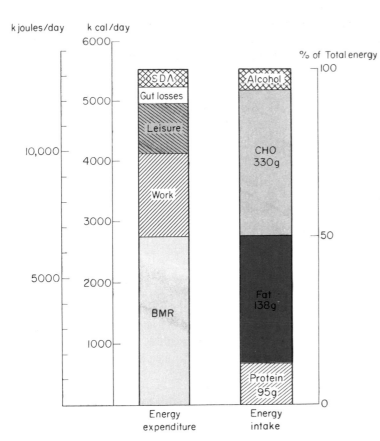

Fig. 6.23. The daily energy balance of a middle-aged man doing moderately hard work. The food intake shown is that eaten by an average male European, and is not that considered to be ideal. An ideal diet should contain more carbohydrate and less fat; the proportion due to fat should be less than 30 percent instead of the 40 percent actually eaten. Gut loss refers to incomplete absorption, SDA to specific dynamic action; both are around 6 percent.

A certain minimum amount of each of the foodstuffs must be eaten for replacement of bodily tissues. The minimum amount of protein is determined by this replacement rate: that of fat largely by the dissolved vitamins; that of carbohydrate to supply the brain with glucose. The ideal protein and fat intakes can be remembered as about a gram per kilogram body weight per day, with carbohydrate making up the rest (around 350 g/day). Fig. 6.23 shows the energy balance of a middle-aged man doing moderately hard work. Details of actual diets can be found in books on dietetics.

Chapter 7
Respiratory System

Respiration is the transfer of gases between body cells and the environment. It goes hand in hand with metabolism, the process by which food produces energy for the needs of the body. In metabolism oxygen is taken from the environment and carbon dioxide added to it; in our bodies these processes proceed at $37°C$ due to the presence of catalysts—the enzymes. Lavoisier thought that similar processes occurred in a fire and in animal metabolism as in both O_2 is absorbed, CO_2 given off and energy produced. We now know that, though the end results of these processes are similar, the detailed reactions are very different; in metabolism hydrogen is removed from substrates whereas in a fire oxygen is added to them. The reason for these differences is probably that animal metabolism evolved before the atmosphere contained oxygen to act as an electron acceptor; so metabolism predates combustion.

For a list of symbols used in respiratory physiology see Table 7.2 at the end of the chapter.

Diffusion of gases in respiration
Most living animals ultimately gain their O_2 and lose their CO_2 by diffusion. This is the process whereby a net movement of molecules occurs from an area of higher concentration to one of lower concentration. This was discussed fully on p. 14 and will only be summarised here. The rate at which substances diffuse increases with the driving force and the area available for diffusion and decreases with the distance. Diffusion in gases is much faster than in liquids because the molecules in a gas are less tightly packed. Such a transfer of substances is fast over short distances but very slow over large distances. As animals increase in size there are two reasons why the supply of gases by diffusion becomes inadequate. The first is that the distance for diffusion becomes too great and the second is that the surface area of the animals becomes relatively smaller than the mass of the metabolising tissue. With spherical animals this point is reached at sizes of about 1 mm. All animals with sizes greater than this require a specialised respiratory system together with an associated circulatory system (Fig. 7.1). In man this respiratory system consists of *lungs* in which gas movement occurs by a bellows action of the chest wall, which creates a negative pressure in the chest and so draws air in. As we require a large gaseous exchange with the environment the surface area of our lungs is very large due to their division into hundreds of millions of small air spaces, the alveoli (Fig. 7.3).

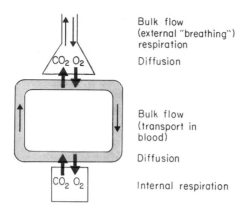

Bulk flow
(external "breathing")
respiration

Diffusion

Bulk flow
(transport in
blood)

Diffusion

Internal respiration

Fig. 7.1. Schematic diagram of lung function. Air is conveyed to and gas away from the gas reservoir through the conducting airways. The forces doing so are partly mechanical—by the bellows action of the chest—and partly by diffusion. Exchanges between the gas reservoir and the blood occur across two layers of cells, some $0.3\ \mu m$ thick, driven by diffusion only (after Bell, Davidson & Emslie-Smith).

Ventilation of the lungs

Distribution of gas and blood to the lungs. Air is filtered by hairs in the nose—which remove the larger dust particles—moistened and warmed by the mucosa of the nose and pharynx so that by the time it reaches the trachea it is saturated with water vapour and is at body temperature. The hairs in the nose only filter out gross particles from the inspired air, and particles smaller than $1\ \mu m$ stay in suspension and are breathed out again. Dust of intermediate size, that is from 10 down to $1\ \mu m$, is removed from the air during its passage along the respiratory tract, the larger particles in the nasopharynx and upper respiratory tract and the smaller in the bronchial tree itself. An important factor in this filtration is the presence of mucus on the surface of the respiratory tract epithelium. Turbulence occurs in the air flow as it changes direction, particles impinge on the walls and are trapped in the mucus. This particle-laden mucus must then be disposed of and this is the task of the ciliated epithelium. The epithelium of the respiratory tract is ciliated, that is it has minute hair-like projections from its surface on which the sheet of mucus is carried. These cilia move in a coordinated fashion with a fast stroke in one direction and a slow one in the other.* This has the effect of carrying the sheet of mucus, with its entrapped particles, in one direction, downwards in the nasal mucosa towards the pharynx and upwards in the bronchial tree to the pharynx. Once in the pharynx the mucus is unconsciously swallowed.

The ciliated epithelium ends at the terminal bronchioles, so what happens to particles which are deposited in the respiratory bronchioles and the alveoli? In the main they are mopped up by wandering alveolar macrophages which get to the mucociliary 'escalator' and so to the

---

* This movement is due to the presence of axial microfilaments containing actin and myosin.

pharynx, or are carried away in lymph vessels or are sequestered in lung tissue by a tissue reaction.

The trachea divides into right and left branches and then continues dividing into smaller and smaller bronchi, finally reaching the terminal bronchioles which lead to the alveoli. The incoming air thus eventually reaches the alveoli where the main exchange of CO_2 and O_2 occurs. These alveoli are some 100–300 μm in diameter and consist essentially of a bundle of capillary vessels suspended in air. Most of the alveolar surface is covered by these blood vessels, which are some 1500 miles in length spread out over the 40–60 m² of diffusing surface. At rest the capillary blood is renewed every second and during exercise every third of a second or less. To do so the total blood flow is about 5 l/min at rest and during exercise about 20–30 l/min. The air flow to the alveoli is about 4 l/min at rest and up to 100 l/min during exercise.

The details of the blood and air flow to the lungs will be considered in more detail, separately.

Pulmonary circulation

The pulmonary circulation, being in series with the systemic one, receives the total cardiac output (Fig. 7.2). The right ventricle pumps this blood—mixed venous blood—through the pulmonary artery to the lungs, and its branches accompany the bronchioles to the terminal bronchioles where they break up into the capillary bed (Fig. 7.3).

The main difference between the pulmonary circulation and the systemic one is that it is a low-pressure system with a systolic/diastolic pressure of 25/10 mmHg (3·3/1·3 kPa) and a mean pressure of 15 mmHg (2 kPa). The left atrial pressure is about 6 mmHg (0·8 kPa), so

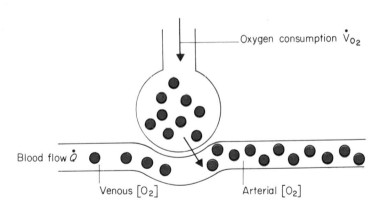

Fig. 7.2. Measuring cardiac output by the Fick principle. Blood leaving the lungs contains more oxygen than blood entering the lungs. This difference in amount is equal to the oxygen added by the lungs. So

$$O_2 \text{ removal from lungs} = \text{arterial–venous oxygen concentration} \times \text{blood flow}$$

or

$$\text{blood flow} = \frac{O_2 \text{ uptake}}{\text{arterial–venous oxygen concentration}}$$

all per unit of time.

If O_2 usage is 250 ml/min and arterial and venous contents are 200 and 150 ml/l the cardiac output = 5 l/min. This neglects the nutritional blood flow.

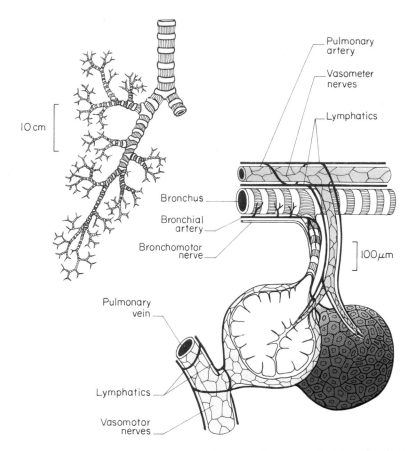

Fig. 7.3. Composite diagram of the lung structures. The diagram on the left is taken from a cast of the airways of a human lung, showing airways from the trachea to the right terminal bronchioles (the alveoli are removed). The diagram on the right continues the airways down to the alveoli. Note the difference in scales.

The alveolar sac on the right is shown covered by a network of capillaries. As the blood moves across this network (< 1s) it gains O_2 and loses CO_2 to the gas in the alveolus. During respiration, this whole alveolus expands and contracts as the size of the chest changes. Gas movement is by bulk flow to about the level of the respiratory bronchioles and by diffusion beyond them.

the driving force for blood flow through the pulmonary circulation is only some 10 mmHg (1·3 kPa) compared to about 80 mmHg (10·6 kPa) for the systemic one. As the same volume of blood flows through both circulations this means that the resistance to flow in the pulmonary circulation is much less than in the systemic one; this is because the walls of the pulmonary arterioles are much thinner and contain less smooth muscle than those elsewhere.

When cardiac output rises, as in exercise, the pulmonary artery pressure does not rise and this must mean that resistance to blood flow through the lungs decreases. There is both an increase in the diameter of pulmonary vessels and an opening up of some which have been closed under resting conditions; this results in a decreased resistance to flow, so enabling the pressure to remain low.

The pulmonary circulation is also different from the systemic one in other respects:

1 The capillary network is more dense in the lung than elsewhere (Fig. 7.3) in order to give a large interface for gas exchange.

2 The hydrostatic pressure in lung capillaries is much lower than the oncotic pressure of the plasma protein tending to reduce filtration but the interstitial pressure is also lower than elsewhere due to lung factors, so net filtration is quite high (Fig. 7.4).

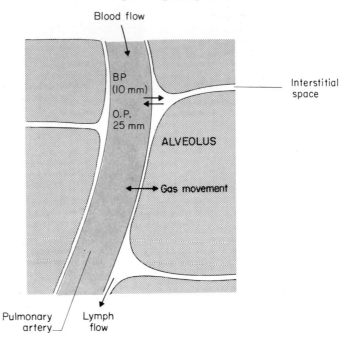

Fig. 7.4. Forces and fluid movement in lung capillaries. Bulk flow of plasma filtrate occurs into the interstitial space, as elsewhere; it is then either reabsorbed or forms lymph. Little fluid passes into the alveoli mainly because of the presence of tight junctions between the alveolar epithelial cells, the presence of surfactant and an active transport mechanism. Gas exchange occurs at the 'thin regions' shown.

3 The pulmonary arterioles constrict to a fall in Po_2 or a rise in Pco_2, the opposite of what happens in the systemic circulation. This has the 'useful' effect of diverting blood away from poorly ventilated segments where less exchange of gases could occur.

4 Because mean pressure is so low the difference due to gravity above and below heart level is significant, particularly in the upright posture. Blood is therefore normally distributed preferentially to dependent parts (see p. 192 for a consideration of the effects of this).

Mechanics of respiration

Air moves into and out of the lungs as a result of changes in pressure within the chest cavity, the lungs themselves expanding and shrinking passively. The energy for these changes comes from contraction of the diaphragm at the bottom of the chest cavity and the muscles of the chest wall. The lungs do, however, contribute an elasticity to the system by virtue of their tendency to collapse (Fig. 7.5). Normally they are prevented from doing so by adhering to the inside surface of the barrel-like thoracic cage.

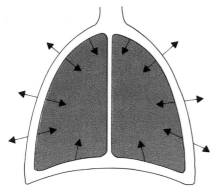

Fig. 7.5. Diagram showing the tendency of the lungs to collapse and the tendency of the chest wall to expand. At the end of a normal expiration these forces are balanced.

About 500 ml of air are taken into the chest during each breath as a result of descent of the diaphragm and the elevation of the rib cage. The diaphragm is dome-shaped and so it descends on contraction; the rib cage is altered mainly by the contraction of external intercostal muscles. During quiet respiration movements of the diaphragm account for 75 percent of the air movement. Because of the way the ribs are articulated, the breadth as well as the depth of the chest is increased during a breath.

At the end of a normal quiet inspiration the nervous activity driving the muscles of the diaphragm and chest wall falls off, and the chest size decreases due to the elastic recoil of the lungs. So at the end of a normal expiration the elasticity of the lungs tending to cause them to collapse is exactly balanced by the elastic tendency of the chest wall to expand. During vigorous respiration this passive collapse of the chest and diaphragm is too slow, and active expiration occurs. This involves various thoracic and abdominal muscles which pull the chest wall down and push the diaphragm up, thus increasing the rate of expiration.

Quantitative information on the mechanical factors modifying breathing has been obtained by measuring pressures in various parts of the system, together with the volume of gas breathed into and out of the chest and the rate of gas flow (Fig. 7.6). The important pressures are:

1 The intrapleural or intrathoracic pressure, which is measured in the 'space' between the pleural membrane lining the inside of the chest wall and that lining the outside of the lung. (These layers are normally only separated by a thin film of pleural fluid.)

2 The intrapulmonary pressure, which is the pressure inside an airway. The place where it is measured must be specified as there is a pressure gradient between the mouth and the alveoli, with the alveolar pressure most negative during inspiration. It is this pressure gradient which 'pulls' the gas along the airway.

3 Atmospheric pressure is the reference level to which the others are compared.

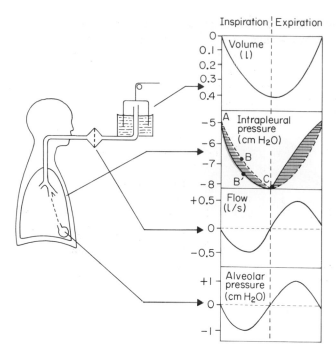

Fig. 7.6. Alveolar pressure, intrapleural pressure and volume of air breathed, plotting during one respiratory cycle. Numerical values shown are for a normal subject. If there was no airways resistance, alveolar pressure would remain at zero and intrapleural pressure would follow the broken line, ABC, which is determined by the elastic recoil of the lung. Airways (and tissue) resistance contributes the hatched portion of intrapleural pressure (see text) (after West).

The effect of the inspiratory effort is to reduce the intrathoracic pressure, which causes a reduction in the alveolar pressure and thus causes a gas flow to occur (Fig. 7.6).

The intrepleural pressure may be measured by inserting a hollow needle, connected to a manometer, through the chest wall so that the needle tip lies between the two pleural layers. Normally it is negative (or subatmospheric) throughout the cycle with a resting value of about $-5\,cmH_2O\,(-0.5\,kPa)$, at the end of a deep inspiration it may become $-20\,cmH_2O\,(-2\cdot0\,kPa)$, and during forced expiration against a resistance, for example, blowing bagpipes, it may reach $+70\,cmH_2O$ ($+6\cdot9\,kPa$) or so. The relation of the change in intrapleural pressure to the volume of air inhaled is about $0\cdot2\,l/cmH_2O$ (i.e. if the pressure drops by $1\,cmH_2O$, $0\cdot2$ litre of air are inhaled).

Air flows through the airways because there is a pressure difference between the alveoli and the atmosphere. This circumstance may be used to calculate a resistance to flow, using Ohm's law. In a normal subject this is about $2\,cmH_2O\,l^{-1}\,s^{-1}$, i.e. if air is breathed in at a rate of $1\,l/s$, there is a $2\,cmH_2O$ drop of pressure between the alveoli and the atmosphere.

During one respiratory cycle the events are quite complex. Fig. 7.6 shows these various pressures, volume changes and rate of gas flow during one breath. Air flow into and out of the lungs depends on the pressure difference between atmospheric air and that in the alveoli. The alveolar pressure depends on the algebraic sum of the pressure created by the lung elasticity and the superimposed pressure due to respiratory muscles. Pressure swings are much greater during heavy breathing leading to more rapid flows of air. To create a pressure difference requires work and larger pressure differences require more work than small ones. During vigorous breathing the work of breathing is therefore greater than at rest.

The elastic behaviour of lung is made up of two different components; that due to the elasticity of the lung tissue and that due to the surface tension of the thin layer of fluid lining the alveoli. These may be shown by inflating an isolated lung with air—which measures both components—and then with saline, which removes the surface tension effect and so measures that due to the tissue elasticity. The elastic property of lung tissue is fairly constant with distension (Fig. 7.7), whereas that due to lung and surface tension is complex and differs depending on whether the lung is being inflated or deflated, i.e. shows hysteresis. This complexity is due to the effects of the surface tension of the fluid lining the alveoli and small bronchioles.

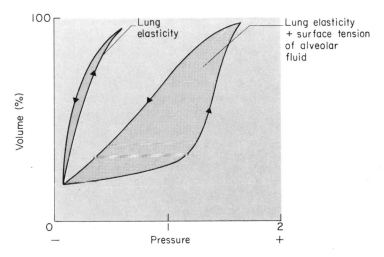

Fig. 7.7. The behaviour of isolated cat lungs to changes in pressure. The lung elasticity was obtained by inflation with saline whereas the lung + surface tension was obtained by air inflation.

These curves show the stretchability or compliance of the tissues. This is measured as the ratio of volume change/pressure change. Lungs in the body are never deflated as far as shown (after West).

We all know that it is more difficult to start inflating a balloon than to continue inflating it once it is big. This is partly because the wall becomes thinner as the size increases, but largely because of the physical properties of spheres. This can be studied most easily in gas bubbles which have a constant wall thickness; small bubbles require a greater force to stop them collapsing than larger bubbles. This can be shown by connecting a large and small bubble together, the small one collapses and empties into the larger one. It would clearly be inconvenient if small alveoli in the lungs collapsed and emptied into larger ones. This effect is counteracted in normal lungs by the presence of surfactant, which not only diminishes the surface tension but also diminishes it more in the smaller alveoli than in the larger ones. Surfactant therefore adjusts the surface tension of the wall of an alveolus to correspond with its size. Figure 7.8 suggests how this happens. Reducing surface tension also decreases the force causing the exudation of fluid from the capillaries (see Fig. 7.4).

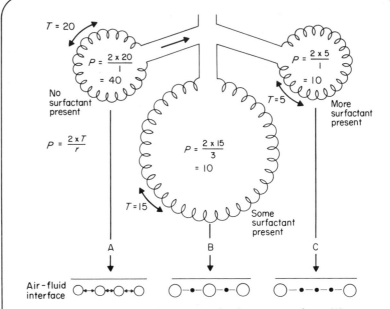

Fig. 7.8. Surface tension is due to intermolecular attraction forces (A). Surfactant (the black circles) coats the water molecules (open circles) with a layer of chemical groups which have a lesser attractive force between them, so reduces the surface tension (B); the smaller the surface area, the less is surface tension (C). The higher pressure in the alveolus (A) would cause air to pass into the other two, with collapse of A.

The alveoli can be considered as spheres of various sizes. The relationship between the internal pressure (P) applied to them, the wall tension (T) and radius (r) is described by Laplace's law (p. 105 also):

$$P = \frac{2T}{r}$$

The work of breathing is of two kinds: non-elastic required to move air through the airways; and elastic required to overcome the elasticity of the lungs and chest wall and to stretch the air:water boundary in the alveoli. The non-elastic work depends on the rate of breathing, for as the rate increases the velocity of air flow in the airways increases and so there is an increase in frictional work, whereas the elastic work depends on the depth of breathing for as the depth increases the lungs need to be stretched more. The amount of air breathed depends both on rate and depth, so that any particular ventilation can be obtained by various combinations of these. If the total work of breathing is measured over a range of different combinations of rate and depth at constant ventilation then a minimum value is found (Fig. 7.9). In natural breathing, rate and depth are set to this minimum value. It is an interesting but so far unresolved question as to how the body works this out; perhaps it is related to volume information via the vagus and length–tension information from respiratory muscles. It is likely that the total work of breathing rarely exceeds 3 percent of the total energy expenditure of the body.

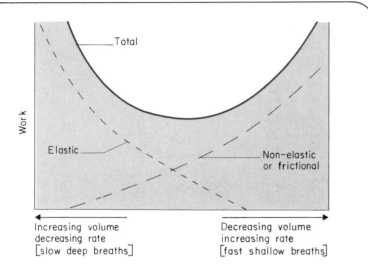

Fig. 7.9. The work of breathing at constant total ventilation but differing combinations of rate and volume.

The same total ventilation can be obtained by slow deep breaths or fast shallow breaths; with slow deep breaths the main work is against the elasticity of the lungs (static), with many shallow breaths the work is mainly against the turbulence of the air flow (dynamic). At some combination of rate and depth the work done is at a minimum; this is the state during normal breathing at rest.

The non-elastic frictional work is small during quiet respiration, for the frictional resistance to air moving through the airways is then not very great. The friction is enough, however, to mean that the air flow, and so the change in volume, lags slightly on the pressure so that if a

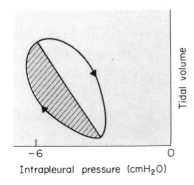

Fig. 7.10. Pressure–flow relationship in an intact chest during quiet respiration. If air flow followed pressure change instantly then this diagram would be reduced to the centre sloping line; but air flow into and out of the lungs lags behind the pressure change producing it so giving the curves shown.

pressure–volume curve is plotted it has the form shown in Fig. 7.10. The shaded area is the frictional or non-elastic work required to do this. This increases as the air flow increases

Lung volumes

The total lung capacity of a 30-year-old 70-kg man is about 6 litres. This is divided into the various components shown in Fig. 7.11.

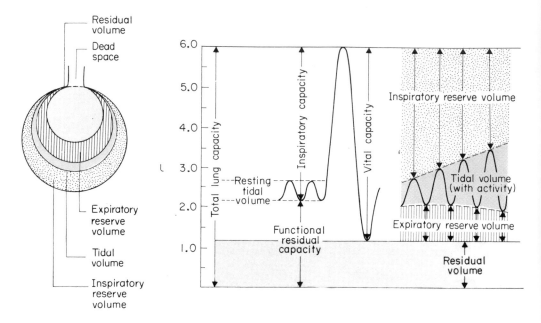

Fig. 7.11. The lung volumes and capacities. Left: illustrates the four primary lung volumes and their magnitude; right: typical record obtained by spirometry.

Gas mixtures and pressures

Certain physical properties of gases must be considered in relation to the movement of gases in the body. These physical properties determine the molecular concentration of oxygen and carbon dioxide and so control their diffusion between various parts of the body and their reaction with chemicals in the blood.

The concentration of a gas in a mixture of gases may be indicated in a number of different ways, for example moles per litre or volumes percent. When a gas has access to both gas and liquid phases it is necessary to use the concepts gas tension and partial pressure. The relationship between pressure and volume of gases is given by Boyle's law and between volume and temperature by Charles's law. Thus the pressure of a gas at a given temperature defines its concentration.

$$PV = nRT$$

where n = number of moles of gas, V = volume, T = absolute temperature and R = gas constant

$$\text{Concentration } (n/V) = \frac{P}{RT}$$

With mixtures of gases the *partial pressure* of an individual component is defined as the contribution of a gas to the total pressure. The composition of dry air is

	percent
O_2	20·93
N_2	79·04
CO_2	0·03

The partial pressures of the gases, indicated by the symbols Po_2, PN_2 and Pco_2 are derived from these proportions. Thus the partial pressure for oxygen will be given by

$$\frac{20·93}{100} \times 760 = 159 \, \text{mmHg} \, (21·2 \, \text{kPa})$$

Similarly, the partial pressure for nitrogen will be 600·7 mmHg (80·09 kPa) and Pco_2, 0·23 mmHg (0·03 kPa). When the air is moist then water vapour contributes its partial pressure to the whole. At 37°C the partial pressure of water vapour in saturated air is 47 mmHg (6·27 kPa). So air entering the alveoli of the lung has a lower partial pressure of O_2 than dry air. For example,

$$\frac{20·93}{100} \times (760 - 47) = 149 \, \text{mmHg} \, (19·9 \, \text{kPa})$$

However, the actual partial pressure of O_2 in alveolar air is less than inspired air, even making allowances for water vapour pressure, as O_2 is removed by the blood and a steady state is reached with a partial pressure of around 100 mmHg (13·3 kPa).

It will be noticed that alveolar gas differs from inspired air in having less O_2 and more CO_2 (Fig. 7.12); this is due to O_2 diffusing from the alveoli into the blood and CO_2 diffusing from the blood into the alveoli. The normal average values of Po_2 of 100 mm (13·3 kPa) and Pco_2 of 40 mm (5·3 kPa) are remarkably constant in resting man; and the respiratory control system seems to be 'set' to a Pco_2 value near to 40 mm (5·3 kPa).

When gases are dissolved in a liquid the molar or molecular concentration depends on the partial pressure of the gas in the liquid (which at equilibrium is the same as that of a gas in contact with it) and the solubility of the gas in the liquid. Commonly we use the partial pressure of the gas (tension) as a measure of the molar concentration; this is convenient and correct for a single gas but some care is necessary when comparing different gases. CO_2 is 23 times more soluble than O_2 in water so that equal molecular concentrations occur with a partial pressure of CO_2 23 times less than that of O_2.

Oxygen lack at high altitudes can be explained by these considerations, for with air breathing at the reduced atmospheric pressure the alveolar Po_2 falls to levels too low to adequately oxygenate the haemoglobin in blood.

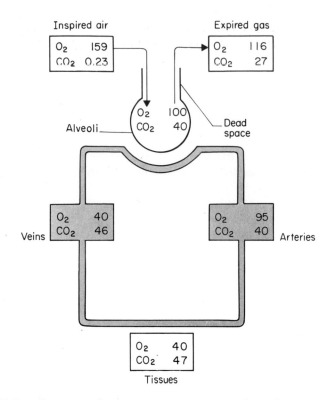

Fig. 7.12. Partial pressures of gases in mmHg in respiration. The total pressure in the tissues and veins is subatmospheric due to less CO_2 being produced than O_2 used. The tissue O_2 is often less, and the CO_2 greater, than shown here but is not easy to measure. The normal percentages of O_2, CO_2 and N_2 in air are 20·93, 0·03 and 79·04 respectively (after Ganong).

Respiratory dead space and alveolar ventilation

Because the alveoli are situated deeply within the body at the end of the conducting airways (Fig. 7.3) not all the air breathed in reaches them. These conducting passages constitute the 'anatomical dead space' of the respiratory system and, by their presence, reduce the efficiency of respiration. The anatomical dead space may be measured in a cadaver by filling the air passages with water, and in life by continuously monitoring the N_2 concentration of expired air after a single breath of pure O_2 and measuring the volume exhaled before the N_2 concentration rises due to alveolar gas. Normally, the anatomical dead space is about 150 ml and so represents about one-third of the volume of an average breath at rest.

The total, or physiological, dead space includes both gas in the anatomical dead space and that ventilating alveoli with relatively poor or no blood flow, so it is larger than the anatomical dead space and more relevant to a consideration of the efficiency of breathing. If an alveolus receives no blood supply then the gas in it will be unchanged and so, when exhaled, will be like atmospheric air. If an alveolus receives an adequate blood supply the gas within it will equilibrate with the blood and so have a similar gas composition to blood. The composition of expired air could thus vary between that of the inspired air (= no blood supply to the lung) and that of the arterial blood leaving the lungs (= all the alveoli receive an adequate blood supply and there is no dead space). Bohr's formula for calculating the physiological dead space is based on this idea:

$$\frac{\text{physiological dead space volume}}{\text{tidal volume}} = \frac{\text{arterial } P_{CO_2} - \text{expired gas } P_{CO_2}}{\text{arterial } P_{CO_2}}$$

The result is expressed as a ratio with a normal value of 0·3.

The volume of air which enters the alveoli per minute and so causes effective ventilation is therefore the product of the breathing rate × (the tidal volume − physiological dead space). This is the alveolar ventilation, V_A, and varies from some 4 l/min at rest up to 100 l/min during severe exercise.

Gas exchange between the alveoli and the blood occurs continuously, whereas that between the alveoli and the air is intermittent. Therefore the partial pressures of gases in the alveoli vary around a mean value during respiration. This variation is kept to a value of ±4 per cent of the mean because the total alveolar volume is much larger than each single breath. The overall effect of the respiratory control systems discussed later is to minimise changes in this mean level during varying metabolic activity of the body.

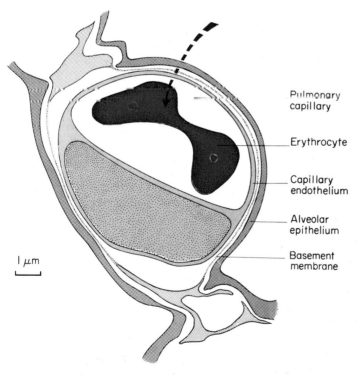

Pulmonary capillary

Erythrocyte

Capillary endothelium

Alveolar epithelium

Basement membrane

1 μm

Fig. 7.13. Pulmonary capillary in the wall of an alveolus. The arrow indicates the diffusion path from alveolar gas to the interior of the erythrocyte which includes the layer of surfactant (not shown in the preparation), alveolar epithelium, interstitium, capillary endothelium and plasma. Note the extremely thin blood–gas barrier of less than 0·5 μm.

The transfer of gases between alveoli and blood

Although it was at one time thought that O_2 was 'secreted' by the lungs, it is now clear that the movements of O_2 and CO_2 between the alveoli and blood are determined only by the pressure differences between them and the nature of the intervening membranes. The diffusion path between alveolar air and blood is short (Fig. 7.13), and consists of two layers of cells with very little interstitial fluid in the thinnest parts. The net transfer of a gas from the alveolus to capillary blood continues so long as there is a pressure gradient. For O_2 under normal resting conditions this only occurs in the first third of the capillary network which it reaches; thereafter no net transfer occurs as there is no pressure gradient (Fig. 7.14). An individual erythrocyte (and the plasma around it) spends

Capillary

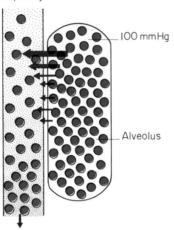

100 mmHg

Alveolus

Fig. 7.14. Net O_2 transfer between an idealised alveolus and lung capillary under resting conditions. Incomplete oxygenation may occur if the O_2 tension in the alveolus is reduced (for example, high altitude) or the membrane is thickened (by disease), particularly if the time available for equilibration is decreased (in exercise). The horizontal arrows indicate the relative size and direction of O_2 transfer.

about 1 second in an alveolar capillary under resting conditions and perhaps a third of this time during vigorous exercise. At normal oxygen tensions, even this shortened time is sufficient for full oxygenation to occur, but at reduced oxygen tensions or with thickening of the membrane (for example, at altitude or in disease) this may not be sufficient and so less than equilibrium oxygenation occurs.

Gravity effects on blood and gas flow to the lungs

For most efficient transfer of gases between alveoli and blood, all alveoli should receive an adequate air and blood supply and these should be matched properly. In fact both the air and blood supply to alveoli varies greatly from one part of the lung to another. The main reason for this variation is the effect of gravity, which, in man standing upright, causes a greater air and blood flow to the base than to the top of the lungs (Fig. 7.15).

The partial pressure of oxygen and carbon dioxide in the alveoli of any zone of the lung and of the blood leaving that zone is determined by the ratio of ventilation to perfusion. Theoretically, if the ratio decreased to zero one would get the situation seen in Fig. 7.16 where the alveolar gas partial pressures of oxygen and carbon dioxide are the same as that

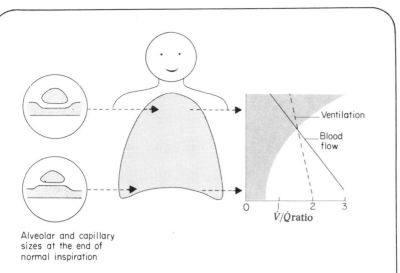

Fig. 7.15. The ventilation ($\dot{V}$), blood flow ($\dot{Q}$), and the ventilation/perfusion ratio ($\dot{V}/\dot{Q}$ in red) in the lung of erect man. The inserts show the relative sizes of the alveoli and capillaries at apex and base.

For reasons related to gravity the ventilation and blood flow are both greater in the lower than the upper parts of the lung but the ventilation difference is less than the blood flow difference.

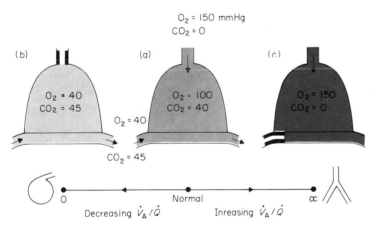

Fig. 7.16. Effect of alterations of the ventilation/perfusion ratio on the Po_2 and Pco_2 in a lung unit. Common causes of a changed $\dot{V}/\dot{Q}$ ratio are airway and blood flow restrictions (shown in black) (after West).

of mixed venous blood. With an increasing $\dot{V}/\dot{Q}$ ratio, the extreme situation would be of an alveolus with no blood supply; the partial pressures of oxygen and carbon dioxide in the lung unit would be the same as atmospheric air. Obviously, mismatching of ventilation and perfusion could result in any ratio between these two extremes.

How do variations in the ratio affect blood gas levels? Blood leaving an area with a low $\dot{V}/\dot{Q}$ ratio will have a lower Po_2 than normal and as more blood comes from the bases in the upright lung this will tend to decrease the arterial blood Po_2. This is not compensated for by the areas which have a high $\dot{V}/\dot{Q}$ ratio. This makes little difference in normal circumstances, but if exaggerated it results in a lowering of arterial saturation (Fig. 7.17).

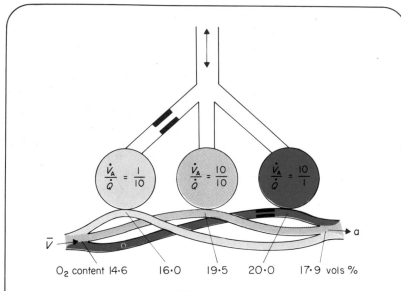

Fig. 7.17. The effect of various $\dot{V}/\dot{Q}$ ratios on the saturation of arterial blood. The $\dot{V}/\dot{Q}$ ratios shown are exaggerated, and would only be seen in diseased conditions. The normal $\dot{V}/\dot{Q}$ ratio at the base of the lungs of around 0·7 gives a saturation of over 19 ml percent. These results can be explained by the shape of the oxygen dissociation curve. Common causes of a changed $\dot{V}/\dot{Q}$ ratio are airway and blood flow restrictions (shown in black) (after West).

As the CO_2 dissociation curve is more linear over its physiological range, the high $\dot{V}/\dot{Q}$ areas do tend to compensate for the low $\dot{V}/\dot{Q}$ areas in CO_2 carriage. Further, any remaining increase in arterial Pco will stimulate the overall ventilation and return the Pco_2 to normal levels, without adding much oxygen to blood. The final effect will be to lower the blood oxygen level slightly in normal lungs, though $\dot{V}/\dot{Q}$ ratios can be grossly disturbed in lung disease. The difference between top and bottom is much greater for blood than for air so that the ratio of ventilation to perfusion changes from 3 at the apex to less than 0·5 at the base with a mean overall value of 0·8 for the whole lung.

Carriage of O_2 and CO_2 by the blood

In man at rest, about 250 ml of O_2 and 200 ml of CO_2 is carried per minute between the lungs and the tissues. The main substance responsible for O_2 carriage is haemoglobin, which, for osmotic reasons, is carried within the erythrocytes. Haemoglobin combines reversibly with oxygen in the reaction

$$4Hb + 4O_2 \rightleftharpoons (HbO_2)_4$$

to carry over 98 percent of the oxygen present in the blood.

A smaller proportion (25 percent) of the CO_2 is carried in direct combination with haemoglobin as a carbamino compound, but haemoglobin is also important in buffering CO_2 carriage as bicarbonate.

Oxygen dissociation curve (see also p. 68)

One gram of haemoglobin can combine with 1·34 ml of oxygen. So 100 ml of blood, which normally contains 15 g of haemoglobin, can carry about 20 ml of oxygen when saturated. This degree of saturation occurs at the oxygen tension of 100 mmHg normally found in the alveoli.

The relationship between the partial pressure of oxygen and the quantity of it combined with Hb is sigmoid shaped (Fig.7.18) because of the interaction between the four oxygen binding sites on the haemoglobin molecule. Once a single molecule of O_2 has been bound, the affinity of the other three sites is increased and so on (p. 69). This sigmoid shape is of clear physiological 'usefulness' in that (1) the amount of O_2 carried is not very sensitive to changes in Po_2 in the lungs over the range of 100 down to 60 mmHg; (2) at tissue Po_2 levels the curve is very steep so that small changes in Po_2 below 60 mmHg lead to a large release of oxygen to tissues. In myoglobin—a molecule found in muscle containing a single haem and a polypeptide chain—the relationship is hyperbolic, as expected when no molecular interactions can occur (Fig. 7.18).

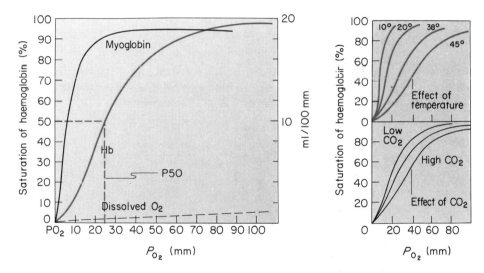

Fig. 7.18. Relationship between partial pressure of oxygen and the oxygen carried by haemoglobin, myoglobin and a simple solution in the plasma. The inset shows the effect of CO_2 and the temperature variations on the steep part of the Hb curve. Myoglobin is found in muscle; it has a simple hyperbolic type of uptake because it only binds one O_2 molecule per molecule. The $P50$ (shown as red dotted line) is the Po_2 at which 50 percent of the Hb is saturated.

The carriage of oxygen is related to that of CO_2 in that with increasing concentrations of CO_2 more O_2 is released at a given tissue Po_2 level (the Bohr effect). Increases in temperature have a similar effect. Both effects are clearly 'useful' adaptations in that in active tissues producing more

CO$_2$ and having a higher temperature more oxygen will be given off. In active muscle this effect may be very large.

The carriage of CO$_2$ and the CO$_2$ dissociation curve

Considerably more CO$_2$ is held in blood than the small amounts which are exchanged between the tissues and the lungs on each circulation (Fig. 7.19 and Table 7.1). Our primary concern here is only to consider the carriage of this 'extra' CO$_2$ from tissues to lung consequent on tissue respiration. Although most of this CO$_2$ is carried as bicarbonate, with a lesser amount combined with protein, we shall see that this carriage is greatly influenced by the presence of haemoglobin.

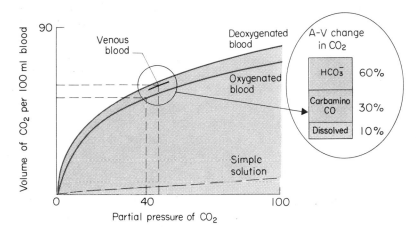

Fig. 7.19. CO$_2$ dissociation curve for whole blood. The arterial and venous points indicate the total CO$_2$ content of normal blood. The inset shows the fate of the extra CO$_2$ which enters venous blood.

Table 7.1. The amounts of CO$_2$ carried in 100 ml of venous and arterial blood.

	Arterial blood	Mixed venous blood	A-V difference	
			Units	Percentage
Pco$_2$ (mmHg)	41	47·5	6·5	
Dissolved (ml/100 ml)	2·7	3·1	0·4	8
As H$_2$CO$_3$ (ml)	43·9	47·0	3·1	62
As carbamino compound (ml)	2·4	3·9	1·5	30

Carbon dioxide diffuses from the tissue cells into the blood and thence mainly into the erythrocytes, driven by its concentration gradient. The most important reactions which occur are that CO$_2$ in the erythrocytes forms bicarbonate and also combines with the local proteins (mainly haemoglobin) as a carbamino compound (Fig. 7.20). The result of these reactions is that much CO$_2$ can be taken up by the erythrocyte without a large rise in the CO$_2$ tension. Lesser reactions also occur.

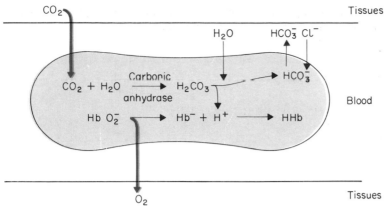

Fig. 7.20. Formation of bicarbonate from CO_2 in erythrocytes. For each 1 mmol/l of O_2 released 0·7 mmol/l of CO_2 and hence H^+—can be taken up by the Hb. A small amount of bicarbonate is also produced in the plasma.

The formation of bicarbonate in the erythrocyte is more complex than might appear at first sight and will now be considered. CO_2 forms bicarbonate as follows:

$$CO_2 + H_2O \rightleftharpoons H_2CO_3 \rightleftharpoons H^+ + HCO_3^-$$

The equilibrium point for this reaction is such that little bicarbonate can be formed unless the reaction products—H^+ and HCO_3^- are removed. In the erythrocyte most of the H^+ is removed by combination with ionised haemoglobin to give unionised haemoglobin, i.e.

$$H^+ + Hb^- \rightarrow HHb$$

This reaction can proceed because reduced haemoglobin is a weaker acid than oxygenated haemoglobin (due to a molecular rearrangement consequent on O_2 removal, see p. 69) and so is less ionised. It is clearly very 'useful' that the change in haemoglobin which occurs when O_2 is given up 'mops up' the H^+ produced by an uptake of CO_2.

The bicarbonate produced as a result of CO_2 entering the erythrocyte diffuses into the plasma in exchange for Cl^- (the chloride shift—see below), so removing this reaction product from the cell. The whole reaction by which CO_2 is taken up by the erythrocyte is speeded up because of: (a) the enzyme carbonic anhydrase, present in erythrocytes which speed up CO_2 hydration; and (b) the presence of a chloride/bicarbonate carrier system in the erythrocyte membrane which allows the chloride shift to occur rapidly. These factors ensure that the reaction goes to completion in the short time the erythrocyte spends in the tissue and lung capillary.

Carbon dioxide reacts with the terminal amine groups of haemoglobin and other proteins to produce a carbamino compound. Reduced haemoglobin again binds more CO_2 than oxyhaemoglobin. The total amount of CO_2 bound as carbamino is not large, but 30 percent of the carriage of CO_2 from the tissues to the lungs is by this method (see Fig. 7.19).

If the tissue respiration increases then there is a greater concentration difference between tissues and blood and so more CO_2 diffuses into, and

O_2 out of, the blood. This automatic adjustment of the rates at which substances are transferred is an important consequence of the laws of diffusion. It continues to operate so long as the capillary blood flow is sufficiently large to keep the blood CO_2 tension lower than the tissue CO_2 tension.

REGULATION OF RESPIRATION

Control system

Respiration occurs due to rhythmic discharge of the motor neurones in the cervical and thoracic spinal cord which supply the respiratory muscles. This rhythm of discharge is dependent on impulses which come down from higher centres; neither the spinal cord, nor the respiratory muscles in the chest, abdominal walls or in the diaphragm have any inherent rhythmic activity.

Respiration is controlled by two separate but interacting neural mechanisms: a voluntary system, located in the cerebral cortex, and an automatic system, located in the pons and medulla. The voluntary system is used in behavioural activities such as speaking, whereas the automatic or involuntary system matches respiration to the metabolic needs of the body. The voluntary system sends nerve impulses to the spinal motor neurones via the corticospinal tracts, whereas the automatic system sends impulses to the same neurones via the reticulo-spinal tracts.

As with other activities which depend on opposing sets of muscles, there is a reciprocal innervation which ensures that the inspiratory muscles are inhibited when the expiratory ones are working, and vice versa. These probably work via collaterals from excitatory pathways which synapse on inhibitory interneurones, and then send impulses to motor neurones supplying antagonistic muscles.

Brainstem centres. At one time it was thought that a discrete respiratory 'centre' was located in the brainstem with separate inspiratory and

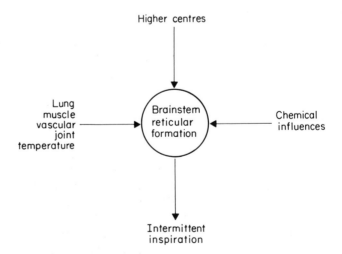

Fig. 7.21. Factors altering the rhythmic activity of the brainstem respiratory centre.

expiratory components. The current view, based on further experiments, is that neurones with activity related to the respiratory cycle are more diffusely located throughout the reticular formation. We think it convenient to retain the term respiratory centre, but with the new meaning of a diffuse region in the brainstem with a respiratory rhythm. This rhythm, producing inspiratory and expiratory impulses, is usually modified by various influences from the rest of the body. The respiratory control system is thus rather like the pacemaker region of the heart, it has an inherent rhythm which is normally modified by external influences (Fig. 7.21).

Regulation of the respiratory centre—introductory outline. The activity of the respiratory centre is altered by two kinds of stimuli, chemical and non-chemical. The chemicals with this effect are the respiratory gases, O_2 and CO_2, and H^+ ion. A rise in its P_{CO_2} or H^+ concentration or a drop in its P_{O_2} increases respiratory centre activity, while changes in the opposite direction cause a decrease in activity. These changes are mediated by two sets of chemoreceptors; situated either in the medulla or peripherally. These chemical substances act as mediators between the metabolic activity of the body and the respiratory activity necessary to maintain the body fluids in a constant state. The non-chemical influences are mainly afferent nerve inputs from various other systems. These two kinds of stimuli will now be discussed in more detail.

Non-chemical influences on respiration

Higher centres. The normal automatic pattern of breathing can be interrupted by pathways from higher centres. This voluntary control of respiration occurs, for example during speech and playing certain musical instruments; in both of which lungs are used as air reservoirs to be emptied at appropriate rates. In these activities there is a surprising amount of modulation of the respiratory muscles during stressing of words and musical notes. This kind of control is integrated with the normal automatic control to keep alveolar P_{CO_2} normal. Pain and emotional events also affect respiration.

Proprioceptors. The main proprioceptor input is from the lungs. In many animals a maintained inflation of the lungs delays the appearance of the next inspiration, whereas a maintained deflation increases the rate and depth of inspiratory efforts (the inflation and deflation reflexes respectively). These two reflexes—the Hering–Breuer reflexes—are mediated by vagal afferent fibres which run from stretch receptors in the smooth muscle of the conducting airways to the brainstem. In adult man there is some doubt as to their normal role in ventilation, though it is known that afferent impulse patterns change continually during the breathing cycle. In certain animals, in young babies and in diseases which increase the stiffness of the lungs the Hering–Breuer reflexes are probably important.

Joint movements affect respiration. This occurs via the proprioceptor input from joints and tendons and helps to control ventilation during exercise.

Irritation of air passages. Irritation of these passages leads to coughs and sneezes. Coughs are due to a large build up of air pressure in the chest against a closed glottis, followed by sudden opening of the glottis. The resultant blast of air helps to dislodge the irritant.

Gut reflexes. During swallowing and vomiting material passing along the oesophagus is prevented from entering the lungs by closure and elevation of the glottis.

Spinal reflexes. Respiratory muscles, like other striated muscle, contain muscle spindles which monitor the length of the muscle (see p. 259). If an increased load, i.e. an increased resistance to air flow, is imposed on respiratory muscles, the decreased shortening, i.e. smaller tidal volume for the same effort, which occurs will be sensed by receptors on the spindles, resulting in an increased excitation of the alpha motor neurones and increased muscle contraction. Some control of ventilation can thus occur at spinal level.

Chemical control of respiration

The chemical regulatory mechanisms adjust breathing so that the P_{CO_2} of the arterial blood is kept constant, the effects of excess H^+ ions in the circulation are overcome, and the arterial P_{O_2} is increased if it should fall markedly. The total ventilation is proportional to the metabolic rate, but the link is CO_2, not O_2. A decrease in P_{O_2} causes respiratory stimulation via the peripheral receptors, but acts centrally to depress the respiratory centres, as it does on most neurones. An increase in P_{CO_2} stimulates respiration both via the peripheral chemoreceptors and centrally via the medullary receptors.

Chemoreceptor areas. The peripheral chemoreceptors are located in the aortic and carotid bodies (Fig. 7.22). They consist of glomus cells, arranged in islands, surrounded by sinusoidal vessels. The blood supply per unit of tissue is very large (2000 ml/100 g) compared to all other tissues. This means that the oxygen needs of the cells can be met largely by dissolved oxygen alone, without needing to draw on the oxygen stored in the haemoglobin. It used to be thought that these receptors were not much affected by decreases in 'chemically combined oxygen', such as occurs in anaemia and carbon monoxide poisoning, and only responded to dissolved oxygen, which is proportional to the partial pressure of oxygen in blood. However, recent experiments have cast doubt on this idea.

The central chemoreceptors, situated on the surface of the medulla, respond to changes in the $[H^+]$ in their immediate environment—the extracellular fluid. This is readily influenced by changes in blood P_{CO_2}, for CO_2 equilibrates rapidly with cerebral extracellular space and cerebrospinal fluid. These fluids are less well buffered than blood, so changes in P_{CO_2} produce larger changes in $[H^+]$ in brain fluids than in blood. Changes in P_{CO_2} produce only a small direct effect on these receptors in the absence of $[H^+]$ change.

The initial response to a rise in P_{CO_2} may be through the peripheral

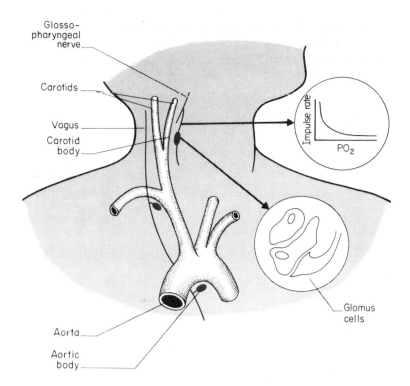

Fig. 7.22. The chemoreceptor areas in the right side of the human neck with insets showing the structure of the carotid body and their response to anoxia. In addition to the sensory nerves these chemoreceptor cells have motor nerve fibres running to them; at present the function of these fibres is unclear.

chemoreceptors but the major response is via the hydrogen ion receptors of the medulla—both seem to depend on CO_2 producing changes in the extracellular hydrogen ion concentration at the receptors.

Ventilatory response to pure changes in CO_2, O_2 or $[H^+]$

If the P_{CO_2} of venous blood is raised (for example, by transfusing blood containing a raised concentration of CO_2), the alveolar P_{CO_2} rises and the respiratory 'centre' is stimulated, causing an increase in ventilation which eliminates more CO_2 (Fig. 7.23). This brings the alveolar and arterial blood P_{CO_2} back towards normal. It does not quite return to the normal value but reaches a new equilibrium level near it. In this way, a rise in venous P_{CO_2} is reflected in only small changes in alveolar P_{CO_2}. A decrease in P_{CO_2} works the other way in that it decreases or inhibits respiration.

When the alveolar P_{O_2} falls there is stimulation of ventilation but this effect is very small until levels of about 60 mmHg are reached, when it becomes marked. This non-linearity of response to changes in P_{O_2} is in marked contrast to the linear response to changes in P_{CO_2} (Fig. 7.23). The reasons for the small response to changes in P_{O_2} down to 60 mmHg are not clear; probably the chemoreceptors show a similar non-linear response. An increase in P_{O_2} has little effect on respiration.

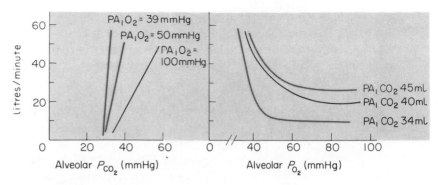

Fig. 7.23. Ventilatory responses to O_2 and CO_2 changes from normal. The black lines in each figure are the responses to changes in one gas at the normal alveolar concentration of the other. The red lines are the responses to changes in one gas at the altered concentrations of the other. Increased ventilation occurs due to increases in both rate and depth of ventilation.

If the blood pH falls for metabolic reasons then ventilation increases. This is mediated by the peripheral receptors only, because hydrogen ions do not enter the CSF readily. This mechanism is effective around the normal body pH of 7·4.

Ventilatory response to combined O_2, CO_2 and [H⁺] changes

If the alveolar concentrations of both O_2 and CO_2 change, the ventilatory changes which result are complex. The CO_2 response curve becomes more steep as Po_2 decreases and minor degrees of hypoxia now stimulate respiration in the presence of increased Pco_2 (Fig. 7.23). Therefore, when both gases change the resultant change in ventilation is greater than the sum of each separate change. This complex response presumably matches ventilation to metabolism more precisely during periods of very vigorous metabolism, for example exercise, than would otherwise be the case.

Changes in the H⁺ ion concentration of blood due to metabolic causes, on the other hand, produce effects which are simply additive with Pco_2 changes, i.e. the slope of the Pco_2 curve does not alter.

Non-respiratory features of pulmonary circulation

Apart from its main function of gas exchange, the pulmonary circulatory system serves several other important functions within the cardiovascular system.

1 At any instant, the pulmonary vessels contain about 1 litre of blood, of which only 100 ml are within the capillaries. This volume of blood within the vessels can vary under different conditions, increasing, for example, in changing from the upright to the horizontal position. It is thought to be one of the important reservoirs for blood in the circulatory system, from which blood is immediately available to the left ventricle for increasing the cardiac output.

2 The pulmonary circulation acts as a filter for small thrombi, preventing them from reaching the systemic circulation and so obstructing the blood supply to other vital tissues.

Various constituents of blood are modified in their passage through the lung. As the total cardiac output passes through the capillary bed every minute, and as the blood volume is approximately equal to the cardiac output at rest, all of the blood is exposed to this large endothelial surface in this time. Vaso-active substances such as bradykinin, 5-hydroxytryptamine, prostaglandins, histamine and noradrenaline may be taken up or released from the lung. One example of this type of activity is the conversion of the polypeptide angiotensin I to angiotensin II by a converting enzyme from lung epithelium (Fig. 11.28).

4 The large number of mast cells in lung containing heparin are now thought to play an important part in the clotting mechanisms of blood.

Nutritional blood supply to the lungs

The lungs, like any other organ, require a nutritional blood supply. This is carried by the bronchial artery which arises from the aorta, supplies the needs of the tissue in the bronchial tree, and then largely drains back into the left atrium via pulmonary veins. This flow is about 1 percent of the total blood flow to the lungs; when this venous blood is mixed with the oxygenated blood from the alveolar capillaries it causes a small drop in the Po_2 of the arterial blood. These blood vessels are systemic ones and so respond to hormones and circulating metabolites like systemic vessels, rather than like pulmonary vessels.

Table 7.2. Symbols used in respiratory physiology (for reference mainly).

Primary symbols		*Secondary symbols*	
P	pressure or partial pressure	1	Gas phase
V	volume of gas	I	inspired air
$\dot{V}$	volume of gas per unit time	E	expired air
F	fractional concentration in dry gas	A	alveolar air
Q	volume of blood	T	tidal air
$\dot{Q}$	volume of blood per unit time	D	dead space air
C	concentration of gas in blood	B	barometric
S	percentage saturation of haemoglobin with O_2	2	Blood phase
		a	arterial blood
		v	venous blood
		c	capillary blood
		$\bar{v}$	mixed venous blood
		c′	end capillary blood

Examples

Partial pressure of CO_2 in mixed venous blood Pv, CO_2

Fractional concentration of N_2 in inspired gas $Pı, N_2$

O_2 concentration in arterial blood Ca, O_2

Chapter 8
The Kidney and Micturition

The main job of the kidneys is the differential excretion of various substances so that the chemical composition of the blood plasma and hence the extracellular space is kept constant. What we eat has much less influence on the composition of our bodies than what we excrete. Substances excreted in the urine are in two different categories:

1 *Waste or end products of metabolism:* (a) the nitrogenous end products: urea, uric acid, and creatinine, which are of no further use and are produced in too large amounts to be excreted in any other way are excreted in the urine; if allowed to accumulate (e.g. as in kidney failure) they gradually poison the body; (b) many foreign substances, including *drugs*, are excreted in the urine.

2 *Physiologically essential substances* such as sodium ions, water etc., the kidneys regulate the water and electrolyte balance (p. 413) and the acid-base status of the body (p. 411) and so maintain the ionic composition, osmotic concentration and pH of the body fluids; the rates of excretion of these substances are regulated by various hormones.

The kidneys have an *endocrine function*, in that they release various hormones and alter others. The main hormone systems involved are the *renin-angiotensin system, Vitamin D metabolism, and the production of erythropoietin.* In addition large amounts of *prostaglandins* are released.

Structure
The nephron consists of a filter, the *glomerulus*, leading to a long duct, the *tubule*. The detailed structure of the nephron is complex and needs to be considered along with its blood supply (Fig. 8.1).

Features of renal function
The 'primary urine' is produced as a filtrate of plasma in the glomerulus and is then modified by adding or removing substances as it passes down the tubule; only a few percent of it remains to enter the bladder as urine. You can consider the nephron as a much modified capillary, whose arterial end becomes the glomerulus and venous end the tubule (see Fig. 1.11).

The tubules *reabsorb* most of the filtrate into the blood of the peritubular capillaries, but also *secrete* some substances into the filtrate. The substances in the urine therefore consist of those filtered plus those secreted minus those reabsorbed (Fig. 8.2).

The *energy* for producing the glomerular filtrate comes from the capillary pressure produced by the heart. As it is a purely mechanical process it is not substance-specific, so that any foreign substance (if it is small enough) will be filtered and so may be excreted. The transport of substances across the walls of the tubules is roughly like that across other

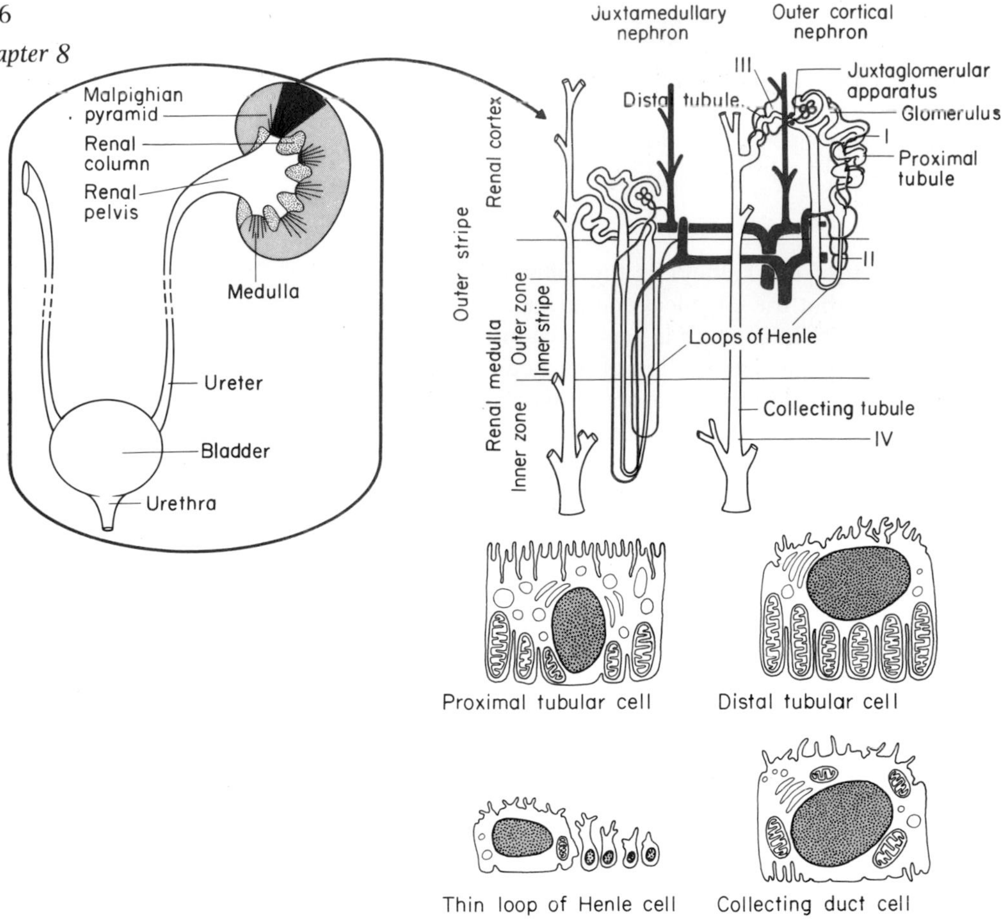

Fig. 8.1 The inset shows the gross structure of the kidney and urinary system. The main figure shows the cortical and juxtamedullary nephrons with vascular supply. The lower insets show the tubular cells in 4 regions.

membranes (p. 20). Thus *passive transport* occurs down concentration gradients, the rate depending on the *lipid solubility, molecular weight and charge; active transport* occurs into or out of the tubule and depends on the presence of specific carriers for each substance.

The concept of clearance

The kidneys remove various substances from the blood plasma and concentrate them in the urine. It is obvious that the amount of any substance removed from the plasma will depend on the way the tubule handles it; thus the rate of excretion of a substance which is both filtered and secreted will be greater than that of one filtered and reabsorbed. A convenient way of describing the differing rates of excretion is to calculate the volume of plasma which is completely *cleared* of the substance. The method of doing so is based on *Fick's principle* (p. 14).

The amount of a substance removed from the plasma during its passage through the kidney must equal that appearing in the urine, i.e.

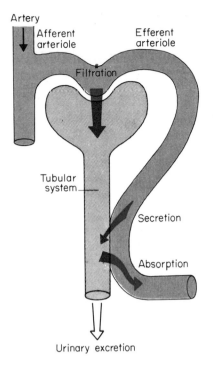

Artery

Afferent arteriole

Efferent arteriole

Filtration

Tubular system

Secretion

Absorption

Urinary excretion

Fig. 8.2. A diagram illustrating the factors which control the formation of the urine (adapted from Vander *et al.*).

plasma concentration × volume of plasma cleared = urine concentration × urine volume

or $P_s \times C_s = U_s \times V$

where C_s is the clearance; P_s and C_s are the concentrations of s in plasma and urine; V is the urine volume per minute.

so the clearance $C_s = \dfrac{U_s \times V}{P_s}$ with units of ml/min.

Fig. 8.3 shows the various common clearances.

Renal circulation

The kidneys receive around 20 per cent of the cardiac output, which gives an average *specific perfusion rate* of about 4 ml/g of kidney per minute; an amount much larger than that of other big organs (e.g. heart, liver, brain) in the body. This is mainly to produce a high *glomerular filtration rate* rather than to provide O_2 for the metabolism of the kidney cells; one consequence is that the O_2 extracted is small, i.e. the arterio-venous O_2 difference is only 1–2 percent.

The blood supply to all parts of the kidney is not the same however, for it decreases from the outside inwards: the flows are in the proportion of $100:25:6$ for the cortex : outer medulla : inner medulla (i.e. $4:1:0.2 \text{ ml g}^{-1} \text{ min}^{-1}$). The reason for the reduced flow in the medulla is the presence of the second set of capillary loops (the vasa recta see Fig. 8.1) which have a high resistance. The medullary flow is least when a concentrated urine is being produced, and rises during diuresis.

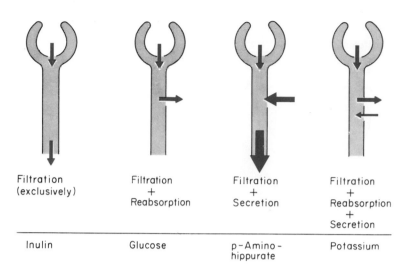

Filtration (exclusively)	Filtration + Reabsorption	Filtration + Secretion	Filtration + Reabsorption + Secretion
Inulin	Glucose	p–Amino–hippurate	Potassium

Fig. 8.3. The four possible modes of handling excretion: filtration alone (example: inulin), filtration with reabsorption (example of complete reabsorption of the amount filtered: glucose), filtration with secretion (example of massive secretion: p-aminohippuric acid) and all three together (example: potassium ions)

Clearance is expressed as millilitres of plasma cleared per minute. U and V refer to urinary concentrations and volume of the substance respectively and P is the plasma concentration.

The glomerular filtration rate (GFR). Inulin is filtered by the glomeruli but neither secreted nor absorbed by the tubules, therefore

$$\text{GFR} = \text{inulin clearance} = \frac{U_I V}{P_I}$$

Renal plasma flow. P-aminohippuric acid (PAH) is both filtered by the glomerulus and excreted by the tubules and is completely extracted by a single passage through the kidneys. Therefore

the effective renal plasma flow (RPF) = PAH clearance = $\dfrac{U_{PAH} V}{P_{PAH}}$

(figure after Schmidt & Thews).

Autoregulation of renal blood flow. The cortical blood flow is hardly affected by the general processes of circulation control, and remains remarkably constant over the blood pressure range of 80–180 mmHg (Fig. 8.4). This constancy of flow remains after denervation, in isolated perfused kidneys and transplanted kidneys so that it is presumed to be an intrinsic renal mechanism. It probably arises from a direct response of the smooth musculature in the renal resistance vessels to changes in pressure (mainly the afferent arterioles). Autoregulation of renal cortical flow is necessary for the autoregulation of the glomerular filtration rate.

Autoregulation in the cortex does not occur under all circumstances, however, for both neural and hormonal influences do alter blood flow, e.g. it is decreased in exercise and increased in some fevers. After haemorrhage renal blood flow drops, and physicians then monitor urine production carefully in order to assess the patient's general condition. There are also *circadian effects* on blood flow as on other kidney parameters.

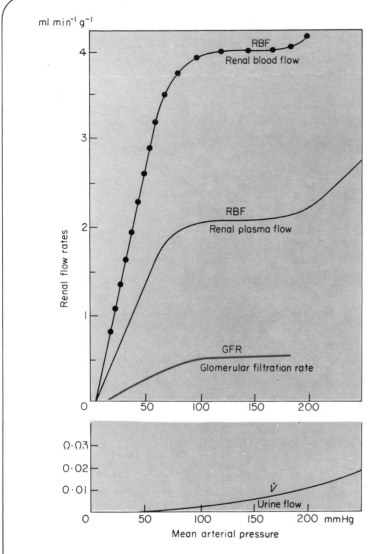

Fig. 8.4. The autoregulation of renal blood flow and glomerular filtration rate (after Shipley & Study).

The rate of oxygen consumption of the kidneys is quite high compared to other organs ($55\,\mu l\,g^{-1}\,min^{-1}$). This is mainly to provide energy for the sodium pumps, necessary for reabsorbing the large amount of sodium filtered: it is therefore found that an increased reabsorption causes increased oxygen consumption.

GLOMERULAR FILTRATION

The *glomerular filter* consists of about 30 capillary loops enclosed by the inner layer of Bowman's capsule (Fig. 8.5). The *glomerular membrane* is

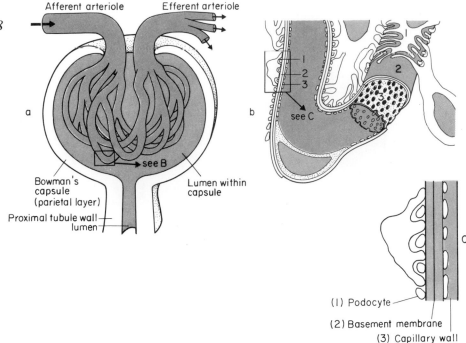

Fig. 8.5. The glomerulus and the glomerular filter membrane. (a) The 20 to 40 capillaries arise from the afferent arteriole, and rejoin to form the efferent arteriole; note that this then breaks up again to form a second capillary network. (b) Each capillary is covered by the inner layer of Bowman's capsule, and (c) together with the podocytes forms the filter (after Schmidt & Thews).

made up of three layers: the capillary wall, the basement membrane and the inner layer of Bowman's capsule. Of these only the basement membrane has a structure capable of acting as a filter, for the capillary wall is full of holes (i.e. fenestrated) and the capsule has large interdigitating processes. The structure of the basement membrane is probably made up of a network of collagen fibres.

Permeability of the glomerular membrane. The net effect of this arrangement is that a much larger fraction of water and small molecules leave the plasma in the glomerular capillaries than leaves normal capillaries. Whether a molecule is filtered or not depends largely on its size but partly on its shape and charge. Thus molecules below 7000 daltons pass easily, larger ones are progressively retarded until those over 70,000 daltons are excluded (Fig. 8.6). Negatively charged molecules are filtered less than neutral ones, possibly due to negative charges in the basement membrane. From this it will be seen that the glomerular filtrate is almost protein free and so has a negligible oncotic pressure.

The composition of the filtrate is determined by the size of the holes in the glomerular membrane, whereas the rate of filtration depends on the *effective filtration pressure* (i.e. the net driving force) and on the area available for filtration (the *filtration coefficient*). The much larger filtration which occurs across these capillaries compared to 'normal' capillaries

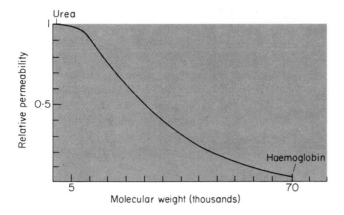

Fig. 8.6. The permeability of the glomerular filter to molecules of different sizes.

elsewhere is due to both of these factors being larger: firstly the effective filtration pressure is around 20 mmHg (Fig. 8.7) compared to the 7 mmHg (32–25) in normal capillaries, and second the capillary permeability is some × 100 that of normal capillaries.

The current view is that the capillary pressure P_{cap} remains high and constant in the glomerulus due to the resistance provided by the efferent arteriole, so that the whole capillary behaves like the arteriolar end of a

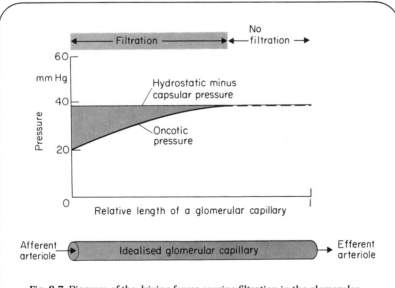

Fig. 8.7. Diagram of the driving forces causing filtration in the glomerular capillaries. Net filtration proceeds until the capillary hydrostatic pressure—the capsular pressure = the oncotic pressure. Beyond this point exchanges occur but there is no net filtration. Changes in the rate of blood flow affects the equilibrium point but equilibrium is probably always reached; this feature can be regarded as providing a high 'safety factor' for GFR. As 20 percent of the blood is usually filtered the blood flow determines the GFR. (after Schmidt & Thews).

normal capillary. This pressure is opposed by the pressure in Bowman's capsule (P_{Bow}; 15 mmHg) and by the oncotic pressure of the plasma proteins (P_{co}), the effective filtration presure (FP_{eff}) is then:

$$FP_{eff} = P_{cap} - P_{Bow} - P_{co}$$

As the blood travels along the glomerular capillary and filtration proceeds, the protein left in the plasma becomes more concentrated so the oncotic pressure rises; at the point where it equals the net hydrostatic pressure, filtration stops (Fig. 8.7). This model of the glomerulus depends on the glomerular membrane having a very high permeability: other models also exist.

The glomular filtration rate (GFR). The renal blood flow in a 70 kg man is about 1100 ml/min, so the renal plasma flow (for a haematocrit of 55 percent) is 600 ml/min. About 20 percent of this plasma is filtered at the glomerulus, giving a filtrate of about 120 ml/min (180 l/day); the remainder continues on to the efferent arterioles. Since the total plasma volume is around three litres, this means that the total blood volume of a 70 kg man is filtered about 60 times a day or once every half hour. This allows a very precise control of the internal environment of the body.

The *composition of the filtrate* depends on the permeability properties of the glomerular membrane, so that it contains all the small molecules at around the same concentration as in plasma, but only traces of plasma proteins. You need to remember that if small molecules are bound to plasma proteins then they will not be filtered; thus various hormones (e.g. T3) and ions (e.g. half the Ca^{2+}) are not much filtered. Binding can be considered as a mechanism for 'saving' substances from excretion.

The tubular transport processes

An approximate description of the nephron is to state that the tubular cells of the proximal tubule 'process' and reabsorb over 80 percent of the glomcrular filtrate, and that the subsequent parts of the nephron are concerned with the 'fine control' of the excretion of electrolytes, water, and hydrogen ions in order to ensure homeostasis.

As well as being classified into those with long and short loops of Henle, nephrons probably differ in other characteristics. The properties of a tubule also vary along its length; the tight junctions (p. 35) near the glomerulus are fairly 'loose' and so have a high permeability to water, they become progressively 'tighter' towards the collecting duct and so the permeability decreases. The high permeability in the proximal tubule means that the filtrate is always in osmotic equilibrium with the plasma; the low permeability further down the nephron means that large osmotic gradients can, and are, set up. The ability of the tubular cells to pump Na^+ and glucose also decreases along it's length, whereas the secretion of organic acids (e.g. PAH) increases.

Tubular reabsorption and secretion of organic substances. The tubular transport processes of secretion and reabsorption can be active or passive. The following processes are *active*.

1   Normally the urine contains no glucose because it is completely reabsorbed in the first part of the proximal tubule. Raising the plasma, and so the filtrate concentration, above about 10 mmol/l (180 mg percent) causes glucose to appear in the urine (*glycosuria*), this is because there is a *threshold* above which reabsorption stops (Fig. 8.8). Glucose leaves the tubule on a carrier with sodium, the energy coming from the sodium gradient (Fig. 8.10).

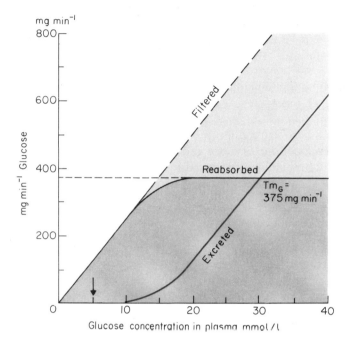

Fig. 8.8. The reabsorption and excretion of glucose at varying plasma glucose concentrations. At normal concentrations (arrow) all filtered glucose is reabsorbed; at higher concentrations the threshold for reabsorption is exceeded and glucose appears in the urine. Note that the threshold (Tm_G) is an amount presented to the tubules not a concentration.

2   *Amino acids* are reabsorbed in the same kind of way as glucose but on different carriers. The threshold for reabsorption is so high that it is rarely exceeded.

3   The small amounts of protein normally filtered are reabsorbed by *pinocytosis* in the proximal tubule.

4   There are three separate systems in the proximal tubule which *secrete* substances into the tubule. The substances secreted by these systems are *organic acids* (e.g. PAH, penicillin etc); *strong organic bases* (e.g. tetra ethylammonium-TEA); and *EDTA*. These show saturation at high concentrations.

The following processes are *passive*:

1   The urea concentration in the filtrate is the same as in plasma. As sodium and water are reabsorbed, urea becomes concentrated and so diffuses passively back into the plasma. The urea permeability is not as high as that of water (p. 218), so urea reabsorption cannot 'keep up'

with water reabsorption and therefore the tubular concentration remains higher than that of the plasma. The faster the urine flow rate the greater the urea excretion, i.e. the *urea clearance depends on the rate of urine flow*. Normally about half the filtered urea is excreted.

We shall see later that urea is most concentrated in the fluid in the collecting tubule. In this region urea diffuses into the interstitial space and then back into the urine in the loop, so it creates a *urea cycle*. You can think of this as a way of retaining urea in the system in order to increase the osmotic pressure.

2 A number of weak acids and bases are reabsorbed by a process of *non-ionic diffusion*. These substances are reabsorbed in their undissociated form through the lipids of the cell membranes, and excreted when ionised. So weak acids are reabsorbed in acid urine, and excreted in alkaline urine. Weak bases are the opposite. Drug poisoning can sometimes be treated by adjustment of the urinary pH.

Tubular reabsorption of water and electrolytes.

The final urine volume produced by the kidneys ranges from around $\frac{1}{2}$ percent of that filtered during periods of water conservation (anti-

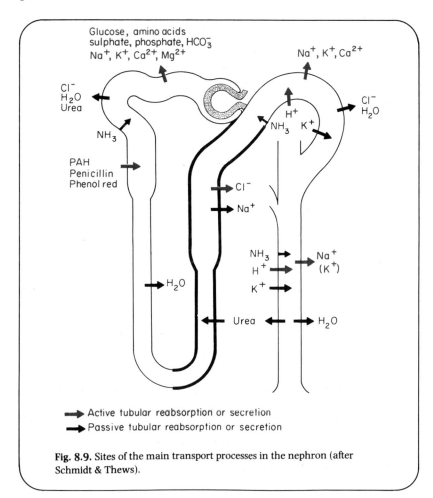

Fig. 8.9. Sites of the main transport processes in the nephron (after Schmidt & Thews).

diuresis), to 15 percent of that filtered following large water intakes (*diuresis*). So tubular reabsorption of water ranges from over 99 percent to 85 percent of the GFR. The main *site of water reabsorption* is the proximal tubule (80 percent in man, Fig. 8.9); this does not vary much during various states of diuresis. The fine control over the rest occurs in the distal tubule; so the homeostatic control is exercised there.

The *reabsorption of sodium* in the proximal tubule is based on *active transport of sodium* back into the plasma with water following passively (Fig. 8.9); this keeps the osmotic pressure of the tubular fluid the same as plasma. Fig. 8.10 shows a more detailed account of this which suggests that some sodium also moves passively. This can be thought of as a 'clever' way of saving energy, for each free sodium ion in the body is reabsorbed in the kidney about 60 times per day.

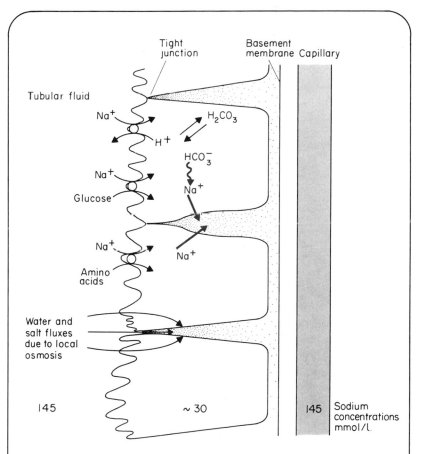

Fig. 8.10. Model of the isotonic reabsorption of fluid in the proximal tubule. Upper part: sodium enters passively through the luminal membrane in exchange for H+, with glucose or with amino acids; it is then actively transported into the lateral spaces with bicarbonate. Lower part: local osmosis brings water and dissolved solutes across the tight junctions and the cells into the lateral space compartment. (These processes occur in all the cells of course). Note Na+ concentrations.

About 15 percent of the *loops of Henle* dip into the medulla; the further inwards they go the greater the osmotic pressure. This rise in osmolality is ultimately due to an active outward transport of NaCl in the ascending limb together with a low water permeability.

The distal convolution starts with properties like that of the ascending limb of the loop of Henle and ends with those of the collecting tubule. The 'passive' permeability of the distal tubule is much lower than that of the proximal tubule (i.e. has tighter junctions), so that the active sodium transport which occurs there can create a high sodium gradient between the tubular urine and the plasma. This in turn creates a potential between the lumen and the plasma (lumen negative), which can be used for the secretion of cations (e.g. potassium). The distal convolution is the site of action of *antidiuretic hormone* which regulates water excretion, and of the *mineralocorticoids* which conserves sodium.

Water handling in the distal nephron, unlike that in the proximal tubule, is largely independent of ion transport; it depends on the water permeabiltiy (controlled by ADH) and on the osmotic gradient set up by the counter-current mechanism.

Hormonal influences on water and ion transport in the nephron

Aldosterone increases sodium reabsorption by the pumps of the distal nephron, and increases potassium and hydrogen ion loss due to the larger potential gradient. Potassium handling is, however, rather complex. The action of aldosterone is probably to increase the sodium leak into the tubule cells, so that more is presented to the pumps along the lateral space (see pp. 354–356; Fig. 8.10); hence more is reabsorbed.

Antidiuretic hormone (ADH or vasopressin) saves water by reducing the urine volume and increasing the concentration of its contained solutes. When water is in short supply the kidneys can, under the influence of ADH, excrete all the necessary substances in a minimum volume of water of $\frac{1}{2}$ to 1 litre per day. The main effect of ADH is to increase the fraction of water reabsorbed in the distal tubule and collecting ducts; an effect brought about by an increase in the water permeability of the apical part of the cells, due to an action of cyclic AMP inserting more 'pores' into the membrane. Another effect of ADH is to make the fluid leaving the loop of Henle isotonic with plasma rather than hypotonic.

When ADH is absent the distal nephron becomes impermeable to water and so the urine becomes hypotonic. The maximum excretion of water is then around 15 percent of the GFR, about 25 litres per day. Thus 15 percent of the filtered water is under hormonal regulation (*facultative reabsorption*), whereas the other 85 percent is always reabsorbed (*obligatory reabsorption*).

An abnormal type of diuresis can be produced if the filtrate contains a large amount of a substance difficult to reabsorb. An example of this is the diuresis produced by infusions of mannitol into the blood and in diabetes mellitus. The mannitol remains in the tubular urine and overwhelms the loop of Henle upsetting the normal concentrating system.

Fig. 8.11. Sequence of events in establishing a concentration gradient in the kidney. We assume (1) that active NaCl transport only occurs in the ascending limb of the loop of Henle and is capable of producing an osmotic difference of 200 mosmol/kg (2) The ascending limb is impermeable to water; the descending limb is highly permeable to water; the transition occurring at the tip of the loop. (3) The descending limb is sparingly permeable to NaCl.

The ascending limb secretes NaCl so that the fluid left in the ascending limb becomes hypotonic, and that in the interstitium hypertonic. As the walls of the descending limb are highly permeable to H_2O its contents come into equilibrium with the interstitium mainly by the exit of water. Some solute entry may occur. For simplicity we can represent the system as a transfer of NaCl from the ascending to the descending limbs of the loop of Henle as shown.

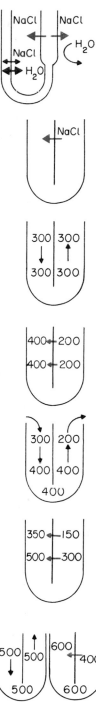

Now stop the NaCl pump, and allow more isotonic fluid to enter the lumen.

Now stop the flow and restart the NaCl pump. At equilibrium a concentration difference of 200 mosmol/kg is again produced.

Now allow more isotonic fluid to enter the top of the descending limb, so that hypotonic fluid is ejected from the top of the ascending limb (again stop the pumps).

Now allow the NaCl pumps to operate on the stationary fluid, producing the concentrations shown.

Repeating this sequence increases the concentration gradients to those shown (the upper figures are omitted in this figure). (After Lote).

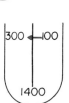

Parathormone increases the excretion rate of phosphate, and decreases that of calcium. *Calcitonin* increases the excretion of phosphate and that of calcium; it also increases the excretion of NaCl.

Producing a concentrated urine by the counter current system of the renal medulla.

The concentrating mechanism in the kidney depends on establishing a large osmotic gradient between the cortex and medulla. This gradient is set up because the long loops of Henle of the juxtamedullary nephrons and the collecting ducts act with their associated blood vessels (the vasa recta) as a *counter current system*. The osmotic gradient in the kidney is established in the extracellular space, in the loop of Henle, in the collecting duct and in the vasa recta by a system which involves the recirculation of both sodium and urea round these structures (sometimes called the *sodium and urea cycles*). The process is aided by the anatomy of the region, for each segment of the kidney is wedge-shaped so that the volume involved is quite small; the circulation in the inner medulla is also rather sluggish so that the osmotic gradient is not 'washed away'.

The system is complex and although its general working is understood, the details are not yet clear. Fig. 8.11 shows a simplified version of it.

ADDITIONAL POINTS

1   The current view is that in the loop of Henle the active NaCl excretion only occurs in the thick segment of the ascending limb. Table 8.1 shows current values for the various permeabilities and Fig. 8.12 an overview.

2   *Urea excretion.* The isosmotic reabsorption of sodium and water in the proximal tubule raises the concentration of the other solutes, including urea, so that concentration gradients are established favouring passive reabsorption. By the end of the proximal tubule, about 50 per cent of the urea filtered has been reabsorbed. By the time the filtrate reaches the collecting ducts, the urea concentration is high; some there-

Table 8.1.   Transport and permeability properties of parts of the nephron.

Segment	Active salt transport	Permeability		
		H_2O	NaCl	Urea
Loop of Henle				
Descending	nil	high	low	very low
Ascending (thin)	nil	low	high	very low
Ascending (thick)	high	low	low	low
Collecting ducts				
Cortex/outer medullary	very low	high (if + ADH)	low	low
Inner medullary	very low	high (if + ADH)	low	high

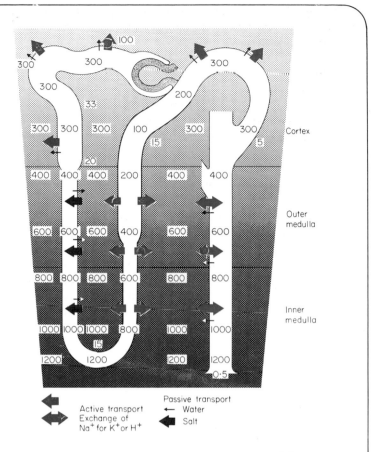

Fig. 8.12. An alternative view of the properties of the juxtamedullary nephron during the production of a concentrated urine. The spaces between the tubules contain the vasa recta; this is considered (together with the interstitium) to be a single compartment. The *black* numbers indicate the osmotic concentration of the tubular or peritubular fluid; the *red* numbers show the percentage of the glomerular filtrate still present at successive points in the nephron. Compare the descending limb with Table 8.1 (after Kokko & Rector).

fore diffuses out (especially in the presence of ADH) and raises the interstitial concentration so contributing to the osmotic gradient. The urea then diffuses back into the loop of Henle forming the 'urea cycle'. The final excretion of urea by the kidney is usually only about 50 percent of that filtered.

3   The *vasa recta* are capillaries which have a 'hairpin' arrangement and run beside the loops of Henle (see Fig. 8.1); they are derived from the efferent arterioles. This anatomic arrangement means that as a RBC moves along the capillary loops it is subjected to a progressive rise in osmotic pressure (and shrinks) until it reaches the tip of the loop and then it swells again as it returns from the loop. As the capillary loop ends in interstitial fluid of the same osmolarity as it started, blood flow

does not dissipate the osmotic gradient. The blood flow of course does supply nutrients and removes metabolites in the usual way, and allows some lateral stirring of the compartment.

4   The block of kidney tissue comprising the loops of Henle, vasa recta, and collecting ducts can be considered as a closed compartment, so that the input and outputs must balance. If the output of the collecting ducts to the urine is hypotonic, then the vasa recta removes hypertonic fluid.

5   *Long and short loops of Henle.* In man, only about 15 percent of the nephrons (the *juxtamedullary nephrons*) have loops of Henle which are long enough to pass deeply into the medulla. The other 85 percent of the nephrons (the *cortical nephrons*) have short loops which only reach the edge of the medulla. Only the long loops produce the medullary hypertonicity, but of course, the collecting ducts from all the nephrons are affected by the gradient as they pass through it.

6   *Actions of antidiuretic hormone* (ADH). ADH reduces the volume of urine flow by causing the reabsorption of water in the distal parts of the nephron, so that a highly concentrated urine is produced. The effect is mediated by cyclic AMP produced in the cells of the distal convoluted tubule and collecting duct by the action of ADH on adenylcyclase receptors. The cyclic AMP also escapes into the urine. Fig. 8.13 and Table 8.2 show the difference caused by ADH on the volume of fluid in various parts of the nephron.

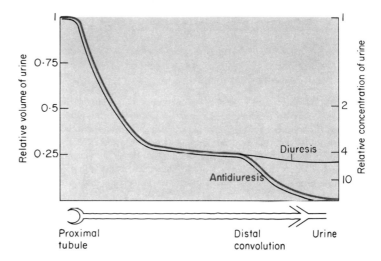

Fig. 8.13. Relative volume flow and concentration in various parts of the nephron in the presence and absence of ADH. With ADH absent, the distal nephron is virtually impermeable to H_2O, so that a hypotonic urine is produced; with ADH present the water permeability increases and water reabsorption occurs in all distal parts. About 15 percent of the glomerular filtrate is altered by ADH. (Based on Dufjen).

Regulatory functions of the kidneys

The kidneys are the effector organs for the regulatory mechanisms which keep the ionic composition and volume of the extracellular fluids constant. Their role in maintaining the osmolarity and volume of the extracellular fluids is discussed on pp. 413 and 416. Here we are concerned with acid-base balance.

Table 8.2 Changes in water metabolism produced by ADH at a constant osmotic load of 700 mosmol/day. Specific gravity (SG) and osmolality are not always related to each other because SG depends on the nature as well as the number of molecules present (after Ganong).

	GFR (ml/min)	Percentage of filtered water reabsorbed	24-hour urine vol. (litre)	Urine concentration (mosmol/l)	Specific gravity
Urine isotonic to plasma	125	98·7	2·4	290	1·010
ADH present	125	99·7	0·5	1400	1·035
No ADH (diabetes insipidus)	125	88	23·3	30	1·002

Role of the kidney in the regulation of acid-base balance.

Adults on a *normal diet* produce a surplus of about 60 mmoles of acid per day, derived from the breakdown of protein etc. When this, or any other acid (e.g. HA) appears in any of the tissue spaces the homeostatic response occurs in three stages (Fig. 8.14):

1   an almost instantaneous compensation by the various intra- and extracellular buffering systems;

2   a rapid compensation by respiratory excretion of CO_2 which lowers the HCO_3^- by replacing it with the non-bicarbonate base (this used to be called fixed base);

3   a slower 'cleaning up' operation by the kidneys. This involves a compensatory excretion of the non-bicarbonate bases and return of HCO_3^- to the blood.

The presence of excess alkaline materials leads to opposite effects, i.e. an increase in the plasma bicarbonate and excretion of fixed acids. This

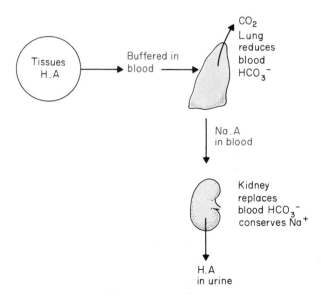

Fig. 8.14. Regulatory response of the body to addition of an acid (HA).

general system is discussed more fully in Chapter 13 (p. 407); here we outline the way in which the kidneys excrete the fixed (i.e. non–volatile) acids and bases from the body.

Disturbances of our acid-base status actually occurs most frequently towards the acid side rather than the alkaline side, for we normally produce a surplus of acid from our food (p. 221); this means of course that an acid urine must be excreted to balance this. Those living on a largely vegetarian diet produce a surplus of alkali (fixed cations), excrete these with bicarbonate and consequently have an alkaline urine. Physiologically we are more able to cope with an excess of H^+ (i.e. fixed base) than an excess of alkali (i.e. fixed acids), because the equilibrium point of our blood buffers (CO_2-bicarbonate system) is at a pH of 6·1 compared to the blood pH of 7·4.

The kidney excretes the fixed base (A^-) in a most economic way for it exchanges it with HCO_3^- and saves the associated Na^+ by excreting H^+ instead. So an acid (e.g. HA) from the tissues is excreted as HA, although in between times it takes other forms.

Renal excretion of hydrogen ions. Over 99·9 percent of the 60 mmol of H^+ which the kidney has to excrete per day occurs by binding the H^+ to an 'acceptor' in the urine; only the tiny 0·1 percent remaining is excreted as free H^+. (The reason for this is that the pH of the urine can only fall to 4·5, sufficient to 'carry' only 60 μmol, i.e. 0·1 percent, of H^+ per day). These acceptors are the buffer anions in plasma and ammonia (NH_3) produced for the purpose by the kidney itself.

Hydrogen ions are *secreted* in both the proximal and distal parts of the nephron.

1 Hydrogen ions enter the proximal tubule in exchange for sodium, combine with filtered HCO_3^-, form CO_2 which then diffuses back into the plasma (Fig. 8.15). This represents over 90 percent of the H^+ excreted

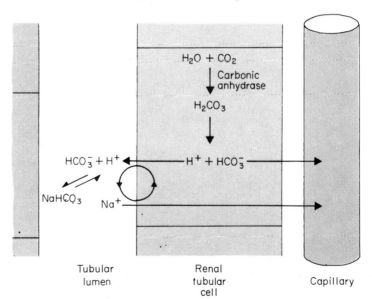

Tubular
lumen

Renal
tubular
cell

Capillary

Fig. 8.15. Reactions involved in H^+ secretion and bicarbonate reabsorption by proximal tubular cells.

by tubule cells, some 4500 mmol/day. It is, however, simply a recovery system for HCO_3^- not a way of excreting H^+. The kidneys need this kind of system for the recovery of HCO_3^- because the permeability of cell membranes for ions is too low to reabsorb bicarbonate by simple diffusion; the high permeability for CO_2 allows fast rates of reabsorption. Bicarbonate behaves in this system as a threshold substance (Fig. 8.8), in that it only appears in the final urine if its plasma concentration exceeds about 28 mmol/l. The threshold concentration is thus about the same as that of the normal blood concentration.

2 K^+ competes with H^+ in the Na^+ exchange process in the distal nephron, i.e. a Na^+–K^+ exchange is added here (Fig. 8.16). The relative rates of each process varies according to the state of the cell, which in turn reflects the needs of the body.

3 Ammonia is produced from glutamine in all the tubule cells, and passes easily into the lumen of the tubule through the lipids of the membranes. It combines with H^+ by the reaction (Fig. 8.16):

$$NH_3 + H^+ = NH_4^+$$

For the reasons already discussed the ionised NH_4^+ does not diffuse back into the tubular cells, and so is excreted.

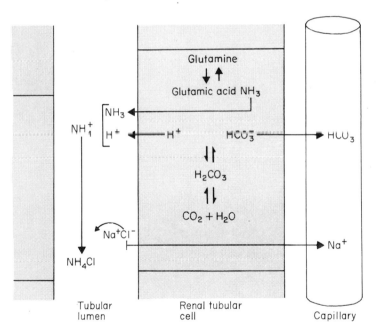

Fig. 8.16. Mechanism of ammonia excretion in tubular cells. The ammonia diffuses in the non-ionic form (as NH_3) because the membrane has a low permeability to the ionic form (NH_4^+).

Mechanism of micturition

Urination is a local spinal reflex which is influenced by higher centres in the brain. Urine collects in the bladder by peristaltic contractions of smooth muscle in the ureteral walls. The bladder walls are made up of thick layers of smooth muscle supplied with stretch receptors and a rich parasympathetic nerve supply. The afferent neurons of the stretch receptors in the bladder wall synapse in the spinal cord with these parasympathetic nerves. There is also a circular layer of voluntary skeletal muscle around the base of the bladder called the external urethral sphincter. This sphincter is able to hold the urethra closed even against strong bladder contractions. Stimulation of the bladder stretch receptors activates the parasympathetic nerves causing bladder contractions and simultaneously inhibits the somatic motor nerves innervating the external urethral sphincter resulting in urination (Fig. 8.17). In adults de-

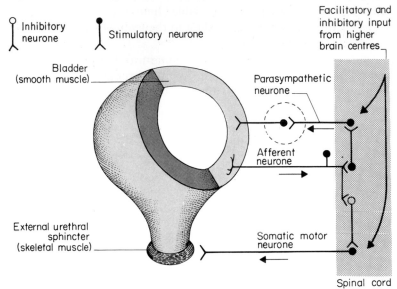

Fig. 8.17. Structure and nerve supply to bladder (after Vander *et al.*).

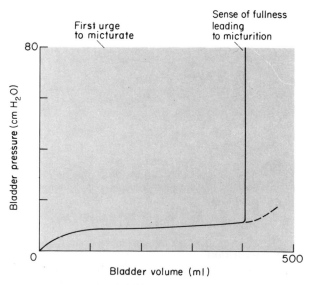

Fig. 8.18. The volume–pressure relationship (cystometrogram) of the bladder of a normal person. The dotted line indicates the further increase in pressure if micturition had not occurred (after Ganong).

scending pathways from the cerebral cortex inhibit the bladder parasympathetics, resulting in stimulation of the motor nerves to the external sphincter. This opposes the synaptic input from the bladder stretch receptors and allows urination to be delayed at will. Control of the local spinal reflex for urination is a learned process and children must master the control of these pathways during 'toilet training'.

When urine enters an empty bladder there is a small rise in pressure (Fig. 8.18); as the bladder fills up there is little further rise in pressure until the micturation reflex is initiated. So there is not much rise in pressure until the bladder is well stretched. This is partly due to the normal 'plastic' behaviour of the smooth muscle of the bladder wall, i.e. when it is stretched, the tension produced initially is not maintained; but partly due to the Law of Laplace ($P = 2T/R$, p. 105) for as the bladder expands, wall tension has less effect on the internal pressure in the organ.

Part 3
Communications and Central Control

Chapter 9
The Nervous System

Nerve cells

Nerve cells or neurones are specialised for receiving and transmitting electrical signals, often working over long distances. Classically the

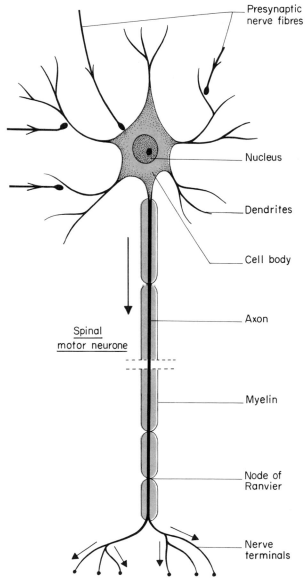

Fig. 9.1. Spinal motor neurone. The terminals of presynaptic neurones form synapses upon the cell body and dendrites of the motor neurone. Impulses are conducted along the axon, which is surrounded with myelin formed by Schwann cells. The motor neurone nerve terminals form synapses upon skeletal-muscle fibres.

neurone can be considered to consist of a receiving zone, a conducting zone and an output zone. The dendrites of neurones act as receiving zones and are contacted by other nerve cells at junctions known as synapses. Signals which are sufficiently strong to excite the neurone pass down the conducting portion of the cell, the axon, and finally reach the output zone of the cell, the nerve terminals, which make synaptic contact with other cells. Each neurone has an enlarged portion, the cell body, in which the nucleus is located. Figure 9.1 shows the structure of a spinal motor neurone. This type of neurone is directly responsible for the activation of muscles and so the synapses formed by motor neurone terminals are upon muscle fibres. Recently, however, the division of neurones into receiving, conducting and output zones has become blurred, since it has become apparent that the dendrites of a number of

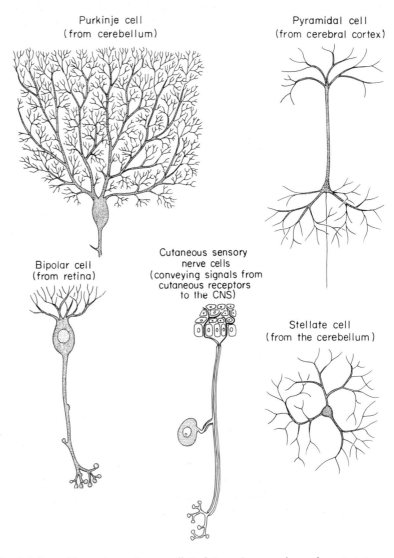

Purkinje cell
(from cerebellum)

Pyramidal cell
(from cerebral cortex)

Bipolar cell
(from retina)

Cutaneous sensory
nerve cells
(conveying signals from
cutaneous receptors
to the CNS)

Stellate cell
(from the cerebellum)

Fig. 9.2. Some different types of nerve cell. Each type of neurone has a characteristic appearance which distinguishes it from that of other types and reflects its function.

different types of neurone are capable not only of receiving signals from other neurones, but can also transmit signals themselves to other cells. Therefore, some dendrites can be involved with transmitting as well as receiving information.

The structure of different types of neurone can vary considerably, so that the size of the dendritic tree relative to the length of the axon and the number of terminals can be large or small. Depending upon the type of neurone, the axon can be a number of metres long or as short as a fraction of a millimetre. The position of the cell body of a neurone can also vary. The structures of several different types of neurone are shown in Fig. 9.2.

Satellite cells

Although in the majority of descriptions of the nervous system most attention is paid to neurones, they are, in fact, outnumbered approximately 10:1 by satellite cells; these, although smaller than neurones, occupy about half the volume of the central nervous system (CNS). Satellite cells within the CNS are termed neurologia or simply glia and, like neurones, are embryologically derived from neural ectoderm. Glia can be classified into the following types: oligodendrocytes, astrocytes and ependyma. Oligodendrocytes and astrocytes are found both in the grey and white matter of the CNS, although oligodendrocytes are found predominantly in the white matter. Ependyma are found on the inner surfaces of the CNS, lining the cerebral ventricles (see Fig. 9.25).

A major function of at least some oligodendrocytes is to form myelin around the larger nerve fibres of the CNS. This role is similar to that of Schwann cells in the peripheral nervous system, although the embryological origins of Schwann cells and glia are different. Other functions of glia are not so clearly defined; it has been suggested that they may supply nutrients to neurones, provide them with mechanical support, and may also be involved in the control of the fluid-bathing neurones. In addition to true glia the CNS contains 'microglia'. These are not true glia since they invade the CNS during fetal life and are of mesodermal origin. These cells may be important following damage to the CNS, since they become phagocytic and may remove tissue debris. The structure of the different types of satellite cell is shown in Fig. 9.3.

Synapses

Synapses, the junctions between neurones, can be divided into two basic types, electrical and chemical, depending upon whether the effects of activity in one neurone (the presynaptic neurone) are passed on to the postsynaptic neurone directly by the flow of current or indirectly by the release from the presynaptic neurone of a chemical known as a transmitter substance which effects the activity of the postsynaptic neurone. Transmission from nerve to muscle has already been discussed in Chapter 2.

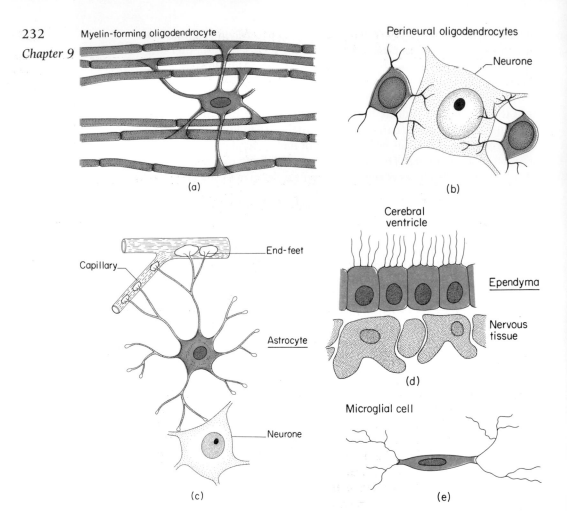

Fig. 9.3. Satellite cells. (a) Myelin-forming oligodendrocytes are found mainly in white matter of the CNS where they wrap around and ensheath axons; (b) perineuronal oligodendrocytes are closely associated with the cell bodies of neurones; (c) astrocytes have many branches, some of which may terminate on neurone cell bodies and others which may contact blood capillaries; (d) ependyma line the ventricular system of the brain; (e) microglia appear to act as scavengers, especially after damage to the CNS.

Electrical synaptic transmission

The nerve impulse travels along an axon by means of local currents, which depolarise the region of the axon in front of the impulse (Fig. 9.4a). At the terminals of such an axon, the local currents may flow into and depolarise the postsynaptic neurone if two conditions are met. First, there must be a low resistance to current flow between the pre- and postsynaptic neurones. Second, there must be only a relatively small loss of current through the external solution in the gap between the pre- and postsynaptic membranes (Fig. 9.4b). If these conditions are met transmission may be electrical, if not (Fig. 9.4c), potential changes in the presynaptic neurone can only be passed on to the postsynaptic cell indirectly through the action of a synaptic transmitter substance.

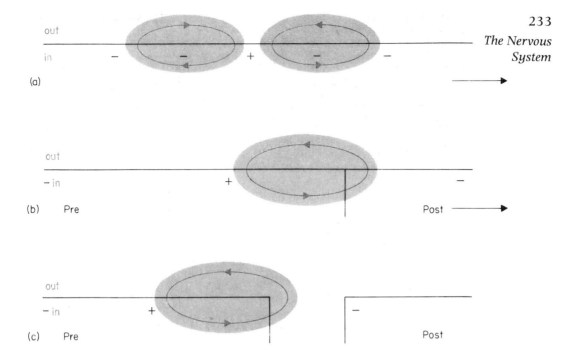

Fig. 9.4. Flow of electrical signals in nervous tissue. (a) The nerve impulse is transmitted along an axon by flow of current away from the active zone; this depolarises and excites the adjacent regions of the axon. (b) If the electrical resistance between two cells is low and their membranes are close together, current can flow from one cell to another and produce excitation. (c) If the gap between two cells is large, current will flow into the intercellular space and there can be little or no direct flow of current from presynaptic to the postsynaptic cell; under these circumstances excitation must be mediated through a chemical transmitter substance

It is now known that transmission across a number of synapses is mediated electrically. Electrical transmission is correlated with the presence of 'gap' junctions between neurones (p. 36). At these junctions, the membranes of the electrically coupled cells come very close together and are separated by a gap of only 2 nm (instead of the 20 nm gap seen at chemically mediated synapses).

Until recently electrical transmission was considered relatively rare and unimportant in the vertebrate cns. However, careful observation has led to an increase in the number of electrical synapses which have been demonstrated in vertebrates. Electrical synapses have no synaptic delay and are resistant to most drugs and therefore can easily escape detection. Chemical synapses in a given cns pathway have often been detected either by observing a delay in conduction (transmission across chemical synapses is associated with a delay of about 0·5 ms) or by modifying synaptic transmission with drugs.

Chemical synaptic transmission
In this type of transmission between nerve cells a substance, termed a chemical transmitter substance, is released by a nerve impulse from the

terminals of a presynaptic neurone. The transmitter substance diffuses across the synaptic gap and, by interacting with specific molecules (receptors) in the postsynaptic membrane, brings about a change in its permeability to certain ions. Depending upon the ions whose permeability is increased, the postsynaptic membrane potential can be either clamped, depolarised (made less negative) or hyperpolarised (made more negative). The direction of the shift in membrane potential will determine whether the synapse has an excitatory or inhibitory action on the excitability of the postsynaptic neurone. So far, because of technical difficulties, only a few relatively low molecular weight trans-

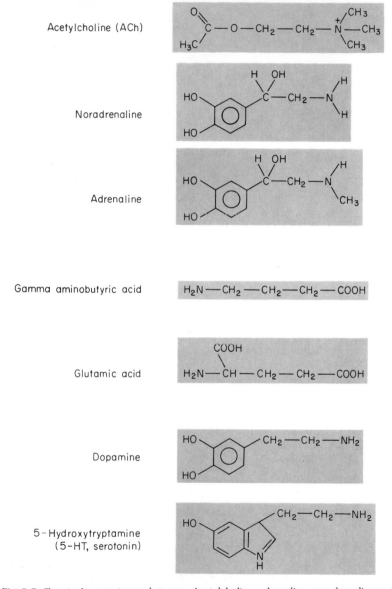

Fig. 9.5. Chemical transmitter substances. Acetylcholine, adrenaline, noradrenaline and gamma aminobutyric acid are known to be transmitter substances; glutamic acid, dopamine and 5-hydroxytryptamine are likely candidates.

mitter substances have been identified with certainty. These include acetylcholine (ACh), adrenaline (epinephrine), noradrenaline (norepinephrine) and gamma aminobutyric acid (GABA), although there are other likely candidates, such as dopamine, 5-hydroxytryptamine (5-HT, serotonin) and glutamic acid (Fig. 9.5).

Figure 9.6b shows the structure of the chemical synapse as it appears under the electron microscope. The features of different chemical synapses are relatively constant. The presynaptic terminals contain mito-

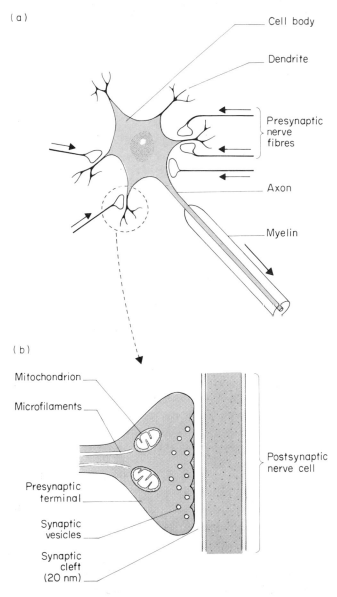

Fig. 9.6. Chemical synapse. (a) Presynaptic nerve fibres form terminal buttons on the cell body and dendrites of a neurone. (b) Enlarged inset. The presynaptic and postsynaptic membranes are separated by a 20-nm gap. Terminal buttons normally contain mitochondria, microfilaments and synaptic vesicles. The synaptic membranes are frequently thickened and modified.

chondria, microfilaments, smooth endoplasmic reticulum and aggregates of vesicles close to the synaptic cleft. The vesicles vary between about 50 and 500 nm in diameter at different junctions and may have dense cores. (Dense-cored vesicles are found in terminals known to release catecholamines such as noradrenaline.) The pre- and postsynaptic cells are separated by a gap or synaptic cleft of approximately 20 nm.

The transmitter substance is stored in the presynaptic terminal within the synaptic vesicles. When a nerve impulse arrives at the presynaptic terminal it causes an influx of calcium ions, which triggers the release of transmitter substance into the synaptic cleft. For this to occur some of the vesicles in the terminal fuse with the surface membrane of the terminal and release their contents by a process known as exocytosis (Fig. 9.7). Having released their contents, the vesicles are then thought

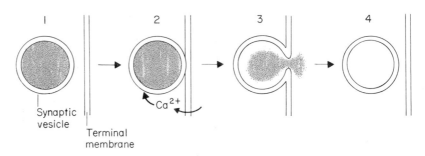

Fig. 9.7. Release of chemical transmitter substance. (1) Some synaptic vesicles come to lie close to the synaptic membrane. (2) Some calcium ions enter the presynaptic nerve terminal during an impulse; this triggers fusion of vesicles with the synaptic membrane. (3) The vesicle opens to the exterior and releases transmitter substance into the synaptic cleft. (4) The vesicle then separates from the synaptic membrane and is ready to be refilled with transmitter substance.

to be pinched off from the surface membrane for refilling with transmitter substance in a recycling process. Although nerve terminals are capable of synthesising some materials, such as the transmitter substance ACh, they are ultimately dependent on the transport of materials, such as enzymes from the cell body of the neurone, since nerve terminals themselves contain no ribosomes and so are incapable of protein synthesis.

Recently, it has become clear that, in addition to the above low molecular weight substances, peptides may also serve as transmitter substances in the central nervous system. Among these peptides are many which also act as hormones of the hypothalamic–pituitary axis (e.g. adrenocorticotrophic hormone, thyrotropin-releasing hormone, somatostatin, luteinising hormone-releasing hormone, and vasopressin) or the gut (e.g. insulin, glucagon, gastrin/cholecystokinin, secretin and vasoactive intestinal polypetide). Perhaps the most celebrated of the neurally active peptides are the encephalins, which are involved in modulating the perception of pain.

On the outer surface of the postsynaptic membrane are situated receptor molecules, which can interact specifically with a given transmitter substance. The interaction of a transmitter substance with its receptors may be pictured as operating gates on channels which span the cell membrane and are permeable to particular types of ion (p. 39). At excitatory synapses the transmitter substance causes a simultaneous increase in permeability of the postsynaptic membrane to sodium and potassium ions. The equilibrium potential for sodium is about $+50$ mV, while that for potassium it is about -100 mV; the increase in the permeability of the membrane to sodium and potassium induced by the transmitter substance drives the membrane potential towards a level intermediate between the equilibrium potentials for each ion (approximately -10 mV). The neuronal membrane has, however, a relatively

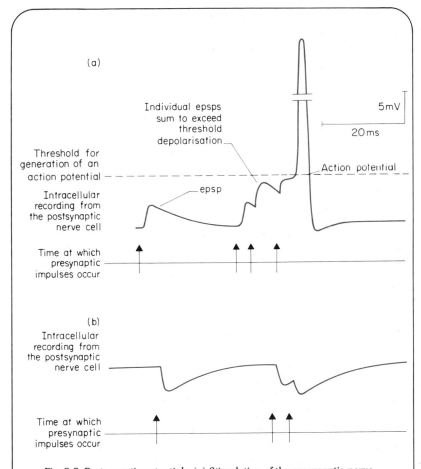

Fig. 9.8. Postsynaptic potentials. (a) Stimulation of the presynaptic nerve fibre at an excitatory synapse results in a depolarisation of the postsynaptic membrane; this is known as an excitatory postsynaptic potential (e.p.s.p.). A single e.p.s.p. normally produces insufficient depolarisation to reach threshold for generation of an action potential; to produce a postsynaptic impulse, several e.p.s.p.s must therefore add together. The arrows in the lower trace in (a) show the time at which presynaptic impulses occur. (b) Stimulation of the presynaptic nerve fibre at an inhibitory synapse results in a hyperpolarisation of the postsynaptic membrane; this is known as an inhibitory postsynaptic potential (i.p.s.p.). The effect of i.p.s.p.s is to move the membrane potential away from the threshold for excitation, thus opposing the effect of e.p.s.p.s.

high resting potassium permeability (see p. 27) which partially 'short-circuits' the effects of the transmitter substance. This tends to 'clamp' the membrane potential and to limit the amplitude of the response to only a few millivolts depolarisation. This depolarising response to an excitatory transmitter substance is known as an excitatory postsynaptic potential (e.p.s.p.) (Fig. 9.8a). The amplitude of e.p.s.p.s produced by successive presynaptic action potentials may fluctuate. Such amplitude variations are not continuously graded, but vary in steps. For this reason the amplitude of e.p.s.p.s is said to alter in 'quantal' fashion. This occurs because the transmitter substance is released in 'packets' rather than as individual molecules. Each 'packet' of transmitter substance is thought to correspond to the release of the contents of a single synaptic vesicle. At synapses in the mammalian CNS there is an interval of about 0·5 ms between the arrival of the impulse at the presynaptic terminal and the onset of the postsynaptic response. This is known as the synaptic delay and results mainly from the time taken for the presynaptic impulse to trigger release of the transmitter substance.

From the above description, it may be seen that the e.p.s.p. differs from the nerve impulse first in that the increase in membrane permeability to sodium and potassium is simultaneous and not sequential. In addition, the e.p.s.p. is local and not regenerative (see p. 30) and is not followed by a refractory period. Thus, postsynaptic potentials themselves are not propagated along the axon of neurone. Normally a single e.p.s.p. produces insufficient depolarisation to exceed the threshold for action potential generation. For this reason it is necessary for several e.p.s.p.s to add together (summate) in order to evoke an action potential (see p. 239 and Fig. 9.8a) If the e.p.s.p. showed a refractory period summation of successive e.p.s.p.s would be impaired.

THE CENTRAL NERVOUS SYSTEM

Inhibition

The CNS contains many synapses which, when activated, cause inhibition of the postsynaptic neurone. At these synapses the receptor molecules of the postsynaptic membrane which interact with the inhibitory transmitter substance cause an increase in the permeability of the postsynaptic membrane either to chloride ions, potassium ions or to both. Since the equilibrium potentials for both chloride and potassium ions in neurones are usually more negative than the resting potential, such a permeability increase causes membrane potential hyperpolarisation. Such a hyperpolarising response is known as an inhibitory postsynaptic potential (i.p.s.p.). (Fig. 9.8b.) The i.p.s.p. moves the membrane potential away from the threshold for excitation and so has an inhibitory action. Like e.p.s.p.s, i.p.s.p.s can summate.

Whether a given transmitter substance produces an e.p.s.p. or an i.p.s.p. in a given postsynaptic neurone is dependent upon the ionic channels opened by the postsynaptic receptors for that transmitter substance. Therefore, although a single presynaptic neurone contains only one transmitter substance, it can produce either e.p.s.p.s or i.p.s.p.s in different postsynaptic cells, depending on the type of ionic channel opened by the receptor molecules on each of these neurones.

Presynaptic inhibition

Postsynaptic inhibition is produced by the membrane hyperpolarisation. The result of this is to reduce the effectiveness of all excitatory synaptic input to that neurone. In presynaptic inhibition terminals of the inhibitory neurone form synaptic connections on to presynaptic terminals of other neurones (Fig. 9.9a). The effect of inhibition on to these terminals is to reduce the number of synaptic vesicles released by each impulse from the inhibited terminals. In this way it is possible to achieve more selective inhibition. That is to say, the effectiveness of a given presynaptic neurone can be reduced while leaving other inputs unaffected (Fig. 9.9a).

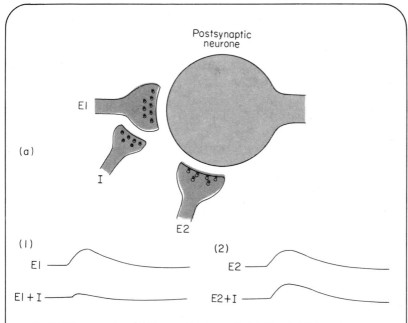

Fig. 9.9. Presynaptic inhibition. In this form of inhibition, inhibitory synapses are formed upon the terminal buttons of some presynaptic neurones; thus nerve fibre I forms a synapse upon the excitatory nerve terminal E1. The effect of stimulating the nerve fibre I is to reduce the amount of transmitter substance released from the terminal E1, thus reducing the amplitude of the e.p.s.p. produced in the postsynaptic neurone by E1 (trace 1). The inhibitory nerve fibre has no effect on the nerve terminal E2 (trace 2).

Functions of synapses

One of the roles of synapses is to perform the integrative functions of the CNS. Since in most instances a single e.p.s.p. is insufficient to initiate an action potential, it is necessary for a number of e.p.s.p.s to summate to produce excitation. This may result either from the summation of many successive e.p.s.p.s produced by activity in a single presynaptic neurone (temporal summation), or from the summation of e.p.s.p. produced by activation of different presynaptic neurones (spatial summation) (Fig. 9.10). Thus a single neurone may be excited either by a high level of activity in relatively few of its presynaptic neurones or by a lower level of activity in a larger number of its presynaptic neurones. The effectiveness of e.p.s.p.s, however, will be reduced by i.p.s.p.s which may sum-

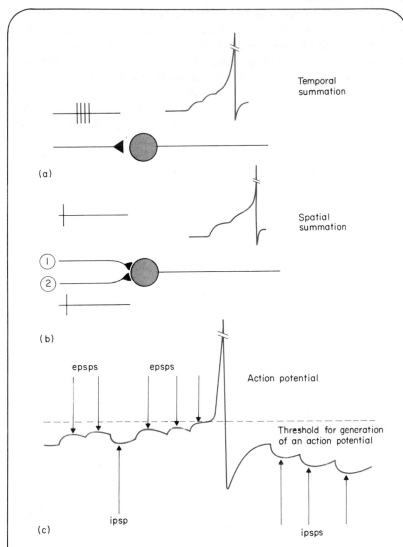

Fig. 9.10. Summation. In order to produce excitation of a neurone, e.p.s.p.s must add together (summate) to reach the threshold for generation of an action potential. (a) A burst of impulses in a single presynaptic nerve fibre can produce a series of e.p.s.p.s capable of summation (= temporal summation). (b) Impulses in several presynaptic nerve fibres can also produce summation of e.p.s.p.s (= spatial summation). (c) Normally a nerve cell may receive a constant barrage of e.p.s.p.s and i.p.s.p.s; the level of excitation or inhibition of the postsynaptic neurone will depend on the overall balance between excitatory and inhibitory influences at any one time.

mate as do e.p.s.p.s. Therefore, the number of impulses produced by a given neurone will depend not only upon the level of activity of its excitatory inputs, but also upon the balance between this and the level of inhibition (Fig. 9.10c).

Most chemicals and some electrical synapses allow signals to pass from the presynaptic nerve terminal to the postsynaptic cell but not in the reverse direction. Without this property of synapses it would be difficult to channel signals through the relevant nervous pathways.

Chemical synapses are able to act as amplifiers. An example of this can be seen at the nerve–muscle junction, which is a specialised type of excitatory synapse at which no integration occurs. Instead, each nerve impulse generates an action potential in the muscle fibres it innervates. The nerve terminals of motor neurones are very small relative to muscle fibres. For this reason, even if the nerve terminal and the muscle fibre formed an electrical synapse, the local currents produced by impulses in such small terminals would be insufficient to cause excitation of the muscle fibre. The presynaptic impulse, however, releases sufficient ACh to excite the postsynaptic muscle fibre with a considerable safety margin. Thus, the presynaptic impulse is amplified by the release of a transmitter substance.

One major advantage of electrical over chemical synapses is that they have negligible synaptic delay. For this reason electrical synapses are frequently found in neurone circuits producing escape responses, where great speed may mean the difference between life and death.

Another role often played by electrical synapses is to synchronise the activity of groups of neurones; if all the members of the group of neurones are interconnected by electrical synapses, then they will all tend to become active simultaneously.

Sensory receptors

Sensory receptors are the specialised structures which supply us with information about both the outside world and the state of the body. There are different types of sensory receptor, each of which is specialised to respond to a particular type of stimulus. Sensory receptors may be classified as follows:

1 Mechanoreceptors, which respond to mechanical deformation of tissues (for example muscle spindles, Golgi tendon organs, labyrinthine hair cells, cochlear hair cells, Pacinian corpuscles).

2 Chemoreceptors, which respond to certain chemicals with which they come in contact (for example, receptors mediating smell and taste and those which detect chemicals such as carbon dioxide in the blood).

3 Thermal receptors, which respond to temperature changes. These can mainly be divided into cold receptors which respond to a fall in temperature and warm receptors which respond to temperature rises.

4 Photoreceptors, which respond to light falling upon the retina. These are known as rods and cones.

5 Pain receptors, which, for a number of technical reasons, are hard to study. For this reason the identity of pain receptors is not known with certainty.

Figure 9.11 shows some of the more common types of sensory receptor

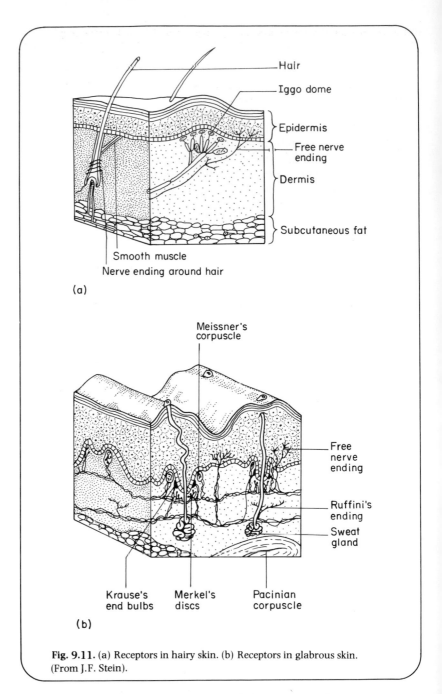

Fig. 9.11. (a) Receptors in hairy skin. (b) Receptors in glabrous skin. (From J.F. Stein).

found in the body. The functions of the different types of sensory receptor found throughout the body are listed in Table 9.1.

Signals from sensory receptors are conveyed to the CNS via sensory nerves. Each sensory-nerve fibre normally innervates a number of sensory receptors which are distributed within a relatively small area of the body. The area innervated by a sensory nerve and within which stimulation of the appropriate receptors causes it to be excited is known as the receptive field of that sensory-nerve fibre (Fig. 9.12). Receptive fields of

Table 9.1. Function of some types of sensory receptor found throughout the body.

Mechanoreceptors	Thermal receptors	Pain receptors

Widely distributed
  Free nerve endings
Glabrous skin
  Meissner's corpuscles
  Merkel's discs
  Krause's end-bulbs
Hairy skin
  Hair receptors
  Ruffini end-organs
  Touch domes
Deep structures
  Pacinian corpuscles
  Muscle spindles
  Golgi tendon organs
	Probably free nerve
	  endings
		Probably free nerve
		  endings

neighbouring sensory-nerve fibres normally overlap, so that stimulation of these zones of overlap will excite more than one sensory-nerve fibre (Fig. 9.12). The precision with which a stimulus may be localised will be dependent both upon the size of the receptive fields of individual sensory-nerve fibres and upon the extent to which these receptive fields overlap.

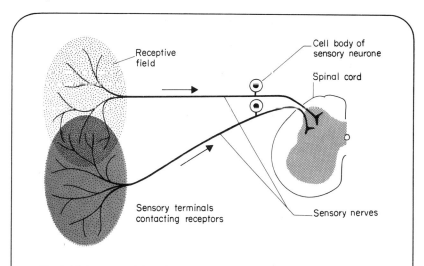

Fig. 9.12. Receptive fields of sensory nerve fibres innervating the skin. The receptive field of a neurone is the areas of skin within which stimulation can influence the activity of the neurone. Receptive fields of adjacent neurones normally overlap; stimulation of such overlap zones may excite several sensory neurones (after Vander *et al.*).

The sensation produced by excitation of a particular sensory receptor depends upon the nerve fibre through which this signal reaches the CNS and not upon the pattern of impulses. In this way a blow in the eye makes one 'see stars', since it stimulates fibres of the visual pathway.

Similarly, a person who has had a limb amputated may suffer from sensations which feel as if they arise in that limb. Such sensations are known as 'phantom limb' phenomena. They result from impulses in damaged sensory-nerve fibres within the limb stump. These nerve fibres originally innervated the amputated limb and so, when active, give rise to a particular sensation apparently originating in that limb. Thus, each of the nerve fibres which enters the CNS is labelled both in terms of the sensation it conveys and the area of the body it represents.

The CNS must be capable not only of recognising which sensory receptor has been activated, but also of detecting the intensity of the particular stimulus. Since sensory receptors can only communicate with the CNS by means of action potentials of fixed amplitude, stimulus intensity can only be coded in a single nerve fibre by varying the impulse frequency.

Stimuli generate impulses in sensory-nerve fibres by increasing the permeability of the receptor-cell membrane. This causes a depolarisation of the sensory-nerve terminals known as a generator potential (Fig. 9.13). The generator potential is graded according to the strength of the stimulus and is not propagated along the sensory-nerve fibre. If it is sufficiently large it will initiate one or more action potentials in the sensory nerve. The action-potential frequency in a given sensory-nerve fibre is proportional to the amplitude of the generator potential (Fig. 9.14). If a prolonged constant stimulus is applied to a sensory receptor,

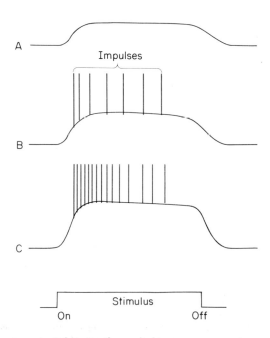

Fig. 9.13. Generator potential. A stimulus applied to a sensory receptor produces a depolarisation of the receptor cell membrane—the generator potential. The generator potential is graded; as the stimulus strength is increased (*A–C*), the amplitude of the generator potential increases. If the generator potential is sufficiently large, it evokes impulses in the sensory nerve (*B* and *C*).

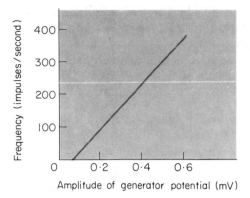

Fig. 9.14. Relationship between the amplitude of the generator potential and the frequency of impulses in the sensory nerve supplying a muscle spindle (after Katz).

the action-potential frequency declines. This process is known as adaptation. Sensory receptors in which there is only a small fall in the frequency with which impulses are generated in the sensory-nerve fibre during a prolonged stimulus are called 'slowly adapting' receptors, while those which produce only a short burst of action potentials during prolonged stimulation are called 'rapidly adapting' receptors and are

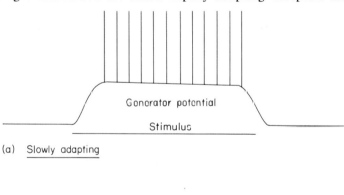

(a)   Slowly adapting

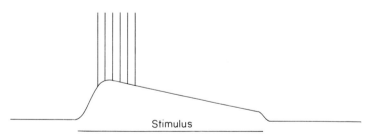

(b)   Rapidly adapting

Fig. 9.15. Adaptation of sensory receptors. (a) A constant stimulus applied to a slowly adapting sensory receptor produces a generator potential which shows only a small decline in amplitude with time. (b) A similar stimulus applied to a rapidly adapting sensory receptor produces a generator potential which shows a relatively rapid decline in amplitude; because of this fast fall in amplitude, a rapidly adapting stretch receptor will only evoke a short burst of impulses in its sensory nerve (after Eyzaguirre).

best suited to changing rather than constant conditions in the environment. The reason for the decline in the impulse frequency in the sensory-nerve fibre is that, after reaching a peak, the amplitude of the generator potential falls during a prolonged stimulus. The rate of decline determines the rate of adaptation (Fig. 9.15).

Neurone circuits

There is a great deal of information available about the connections made between different parts of the CNS by large fibre tracts, but because of the incredible complexity of the human brain (containing between 10^{10} and 10^{12} nerve cells), our knowledge of circuits at the cellular level is still relatively limited. For this reason we do not know the neuronal basis of even simple forms of human behaviour. Certain types of circuits, however, seem to be fairly common in the CNS and are described below.

Convergence. Most neurones in the CNS receive signals from a large number of presynaptic nerve cells (Fig. 9.16a). In this way a single postsynaptic

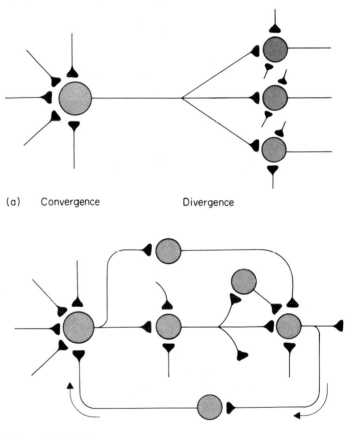

(a)   Convergence   Divergence

(b)   Feedback loop

Fig. 9.16. Neurone circuits. (a) *Convergence.* Many presynaptic neurones form synaptic connections with a single postsynaptic cell. *Divergence.* A single neurone normally forms synaptic connections with a large number of postsynaptic cells. (b) *Feedback loop.* The output of electrical activity from a chain or group of neurones may be fed back so that it can modify the subsequent activity of that group of neurones; thus, at any moment, the activity in the neurones of the group is influenced by their previous activity.

neurone samples the activity of a large number of presynaptic cells. Some presynaptic neurones may be excitatory while others are inhibitory.

Divergence. Each neurone not only receives information from many nerve cells, but is in turn itself presynaptic to a large number of other neurones. Thus a single neurone can influence the activity of many other cells. This is known as divergence (Fig. 9.16a). It must be remembered, however, that as information is passed from one neurone to another it is modified rather than relayed unchanged. A good example of the way in which signals are modified as they pass from one neurone to another is given in Chapter 10, page 303, in which the processing of visual information is described.

Feedback. Information flowing through a chain of neurones may be fed back to preceding neurones in that chain (Fig. 9.16b). Thus, the output from a group of nerve cells can be used to regulate their subsequent activity. If the feedback loop inhibits the activity of the group of neurones, it could have the effect of limiting the length of a burst of activity. If, on the other hand, the loop had an excitatory effect, then this would constitute a positive feedback loop and could cause the neurones in the chain to continue to be active long after the original input to the group of neurones had stopped.

Lateral inhibition. This is a form of contrast enhancement commonly found in sensory systems and within the central nervous system which may be illustrated by reference to mechanical stimulation of the skin (Fig. 9.17). If a blunt object is pressed against the skin, sensory receptors both under and around the object will be stimulated by tissue deformation. Lateral inhibition results from mutual inhibition between neurones with neighbouring receptive fields and enhances perception of the stimulus edge. Neurones with receptive fields at the centre of the stimulus will be strongly excited, but will also receive very powerful inhibition from their equally strongly excited neighbours; neurones with receptive fields at the edge of the stimulus, although also highly excited by the stimulus, will only receive strong inhibition from a proportion of their neighbours (those with receptive fields under the stimulus) and so will produce many action potentials (see right hand side of Fig. 9.17). Neurones with receptive fields around the stimulating object will produce few or no action potentials because they receive strong inhibition from neurones under the edge of the stimulus. The overall effect of lateral inhibition is that neurones with receptive fields just inside the edge of the stimulus produce most impulses so enhancing detection of the stimulus boundary.

General plan of the central nervous system
Generally, the terms used here to define the body axes are those of comparative anatomy. Thus, the terms 'dorsal' and 'ventral' will be used, rather than 'posterior' and 'anterior'. The most common terms used to describe orientation in the body are given in Fig. 9.18.

The CNS consists of the brain and spinal cord (Fig. 9.19). During the early

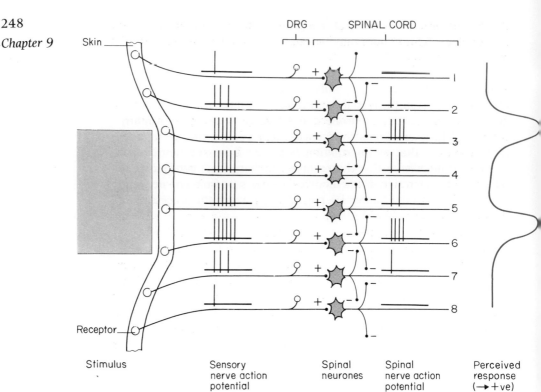

Fig. 9.17. Lateral inhibition. this is a mechanism for enhancing contrast and is widespread in the nervous system. An object pressed against the skin will cause tissue movement that stimulates sensory receptors both directly under and just around the object. Mutual inhibition between spinal neurones with adjacent receptive fields suppresses activity in neurones with receptive fields at the centre of the stimulus (numbered 4 and 5) as well as those surrounding it (numbered 1, 2, 7 and 8). Therefore spinal neurones with receptive fields at the edge of the stimulus produce most action potentials, thereby enhancing detection of the stimulus boundary. Plus and minus signs indicate respectively excitatory and inhibitory synapses.

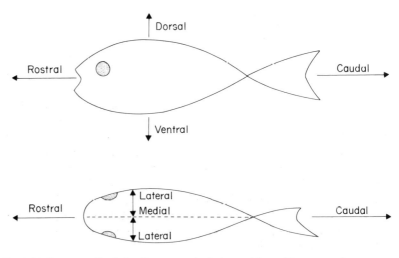

Fig. 9.18. Terms used to define the axes and relative positions of features in the nervous system.

stages of its formation the CNS is essentially a tubular structure, although the tubular origins of the brain become somewhat obscured during the course of development.

Spinal cord

In the adult human the spinal cord consists of a dorsoventrally flattened cylinder about the diameter of one's little finger running in the spinal canal of the vertebral column from the base of the skull to the upper lumbar region. The segmental organisation of the spinal cord can be seen by observing the 31 pairs of spinal nerves which travel from the spinal cord to various parts of the body (Fig. 9.19a). Because, during growth, the vertebral column continues to elongate at a greater rate than the spinal cord, the lumbar and sacral spinal nerves travel caudally in the spinal canal, to emerge at the appropriate level of the vertebral column (Fig. 9.19b). Close to the spinal cord, spinal nerves divide into two branches, known as the dorsal and ventral roots (Fig. 9.20). Sensory or afferent (incoming) nerve fibres enter the spinal cord via the dorsal roots, while efferent (outgoing) nerve fibres leave the spinal cord in the ventral roots. On each of the dorsal roots there is an enlargement, known as the dorsal-root ganglion. This ganglion contains the cell bodies of the sensory-nerve fibres in that segment. Cross-sections of the spinal cord show that it is divided into areas of grey and white matter. The white matter occupies the periphery of the spinal cord and consists of nerve fibres which carry signals to and from different parts of the CNS, including the brain. The grey matter contains the cell bodies of neurones and their synapses and occupies the central areas of the spinal cord. The overall outline of the spinal grey matter roughly forms an 'H', the upper branches of which are termed the dorsal horns, while the lower branches are known as the ventral horns.

Sensory-nerve fibres entering the spinal cord through the dorsal roots have many terminals in the dorsal horns of the grey matter, where they form synaptic connections with interneurones (nerve cells restricted to the CNS without branches in peripheral nerves—also known as inter-nuncial cells). The spinal cord contains many times more interneurones than motor neurones, reflecting the complex information processing the spinal cord can perform. The cell bodies of motor neurones supplying skeletal muscles are found in the ventral horns. Some sensory nerve fibres have terminals in the ventral horn where they make direct synaptic contact with motor neurones. In the regions of the spinal cord supplying fibres to the thoracic, upper lumbar and mid-sacral spinal nerves, the central grey matter has a lateral horn between the dorsal and ventral horns. This area contains the cell bodies of preganglionic autonomic neurones (see Autonomic Nervous System, p. 283). Since sensory-nerve fibres form synaptic contacts in the dorsal horn of the spinal grey matter, while the cell bodies of autonomic preganglionic fibres are in the lateral horn and those of motor neurones to skeletal muscle are located in the ventral horn, the spinal cord may be functionally divided as shown in Fig. 9.21.

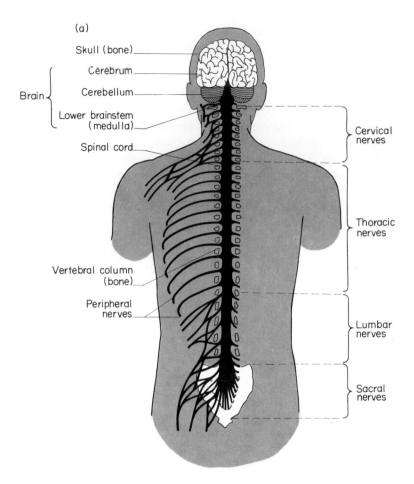

(a)

Skull (bone)

Cerebrum

Cerebellum

Brain

Lower brainstem (medulla)

Spinal cord

Vertebral column (bone)

Peripheral nerves

Cervical nerves

Thoracic nerves

Lumbar nerves

Sacral nerves

Fig. 9.19. The central nervous system. (a) The central nervous system consists of the brain and spinal cord. The spinal cord runs in the vertebral canal from the base of the skull to the upper lumbar region.

The brain

The brain can be divided into three major regions. These are the brainstem (consisting of the medulla, pons and midbrain), the cerebellum, and the cerebrum (Fig. 9.22a). The general pattern of organisation seen in the spinal cord (Fig. 9.21) is also present in the brainstem. In the brainstem, however, the arrangement is modified by specialised functions of particular areas. In this way, in the medulla the general spinal plan is modified by an enlargement of the visceral sensory and motor areas. These areas are enlarged as a result of the role played by the medulla in the control of basic systems such as the cardiovascular system, respiratory system and the gut.

The brain, like the spinal cord, has a number of peripheral nerves; these are known as the cranial nerves. They convey information to the brain from the special senses (smell, sight, hearing, balance and taste) and from general senses. They also control the voluntary muscles of the eyes, mouth, face, pharynx and larynx as well as many visceral structures (via the cranial outflow of the parasympathetic nervous system). The cranial nerves are shown in Fig. 9.22b. Within the brainstem the

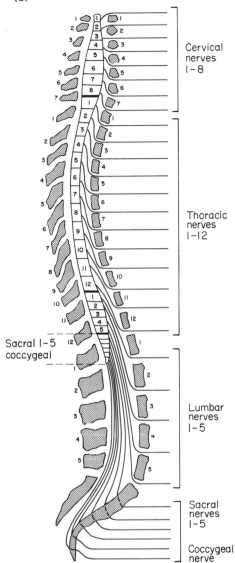

Cervical
nerves
1–8

Thoracic
nerves
1–12

Sacral 1–5
coccygeal

Lumbar
nerves
1–5

Sacral
nerves
1–5

Coccygeal
nerve

(b) Spinal nerves can be seen leaving the spinal cord along its length (facing page). (After Passmore & Robson.)

cell bodies of neurones associated with the cranial nerves are concentrated together into clusters known as nuclei. In this context a nucleus is an aggregate of neurone cell bodies.

The core of the brainstem contains a network of small nerve cells known as the reticular formation. In the caudal direction this system is continuous with interneurones of the spinal grey matter and rostrally extends into the cerebrum. This system is important in determining the level of wakefulness and alertness.

The cerebrum (cerebral hemispheres) consists of the cerebral cortex and a number of underlying nuclei (Fig. 9.23), including the basal

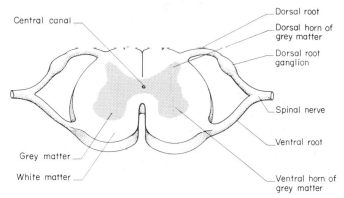

Central canal

Dorsal root

Dorsal horn of
grey matter

Dorsal root
ganglion

Spinal nerve

Ventral root

Grey matter

White matter

Ventral horn of
grey matter

Fig. 9.20. Transverse section of the spinal cord. The central grey-matter zone contains the cell bodies of neurones and their synapses; the surrounding white matter mainly consists of bundles of nerve fibres. Sensory nerve fibres enter the spinal cord in the dorsal roots of the spinal nerves, while motor nerve fibres leave in the ventral roots.

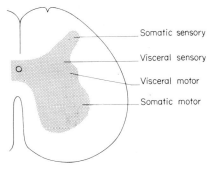

Somatic sensory

Visceral sensory

Visceral motor

Somatic motor

Fig. 9.21. Functional divisions of the spinal cord.

ganglia, thalamus and hypothalamus. The basal ganglia are important in the production of coordinated body movements, the thalamus relays and processes sensory information of all types except that associated with the sense of smell, and the hypothalamus is involved in the control of the body's internal environment and of several basic forms of behaviour. The cerebral cortex itself is a thin layer of grey matter containing an enormous number of nerve cells. The cerebral cortex is connected to other regions of the CNS by nerve fibres which form the white matter below the cortical surface. The cortex itself is thrown into many ridges (gyri) and grooves (sulci) and can be divided into several lobes (Fig. 9.24). The functions of the different areas of the central cortex will be described later.

The cerebellum is situated dorsal to the brainstem. The name means 'little brain' since anatomically it has similarities with the cerebrum. The cerebellum is not, however, involved in conscious sensation nor the initiation of behaviour. Instead it receives information primarily con-

Fig. 9.22. The brain. (a) There are three major brain regions: the brainstem (consisting of the medulla, pons and midbrain), the cerebellum and the cerebrum (consisting of the cerebral cortex and underlying nerve nuclei). (b) The cranial nerves are the peripheral nerves of the brain; they supply sensory and motor innervation to the head, the neck and some visceral structures (after Vander *et al.*).

(a)

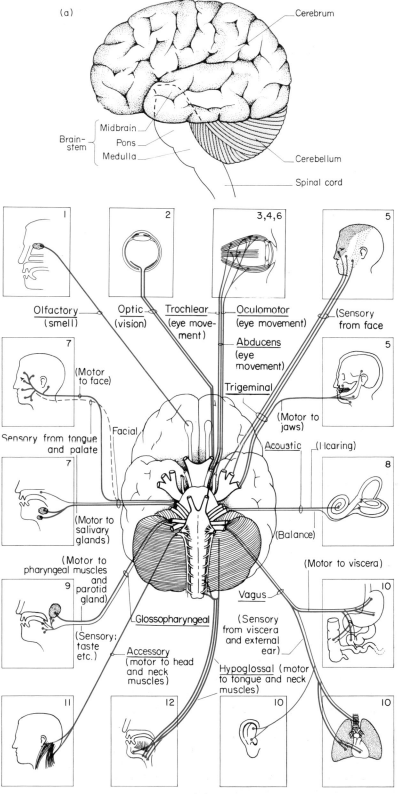

(b)

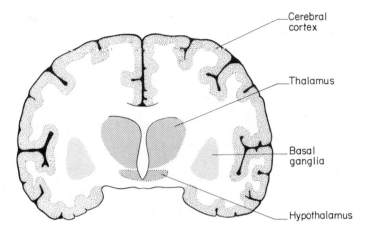

Fig. 9.23. Section of the brain showing the major features of the cerebrum.

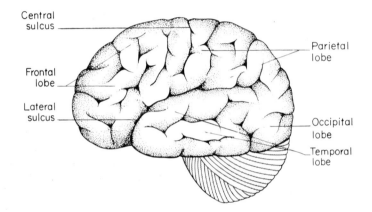

Fig. 9.24. The cerebral cortex. This superficial layer of the cerebrum is thrown into many folds; grooves (sulci) divide the cortex into four lobes.

cerning movement from a number of different sources. It uses this information to modify and coordinate body movements initiated by other brain regions.

Fluid circulation through the brain. The brain is perfused with two fluids: blood and cerebrospinal fluid (CSF).

The brain has an extremely rich blood supply which is essential to support its high level of metabolic activity. It is easily damaged by interruption of its blood supply; this is the cause of the symptoms which may occur after a stroke. The blood circulation through the brain and spinal cord is unusual since their cells do not become stained by dyes injected into the blood. This inaccessibility of the CNS to dyes and many other materials has given rise to the term 'blood–brain barrier'. The exact nature of this 'barrier' is unknown; it appears that it consists of both actual barriers to the penetration of some materials, and active transport systems which regulate the concentration of different substances in the brain. Nerve cells are extremely sensitive to a number of substances in the blood circulation and the 'blood–brain barrier' is essential for the maintenance of the constant environment necessary for the normal working of the nervous system.

Within the brain there are four interconnected fluid-filled spaces, known as the cerebral ventricles (Fig. 9.25). These ventricles are continuous with the central canal of the spinal cord. Within each ventricle there is a highly vascular structure known as a choroid plexus. The choroid plexuses are responsible for the production of cerebrospinal fluid, which fills the ventricles. Cerebrospinal fluid is isotonic with blood plasma, but is not merely produced by a process of filtration since it differs from plasma in several ways. It contains only a small amount of protein and less potassium and calcium than plasma. The concentration of chloride, on the other hand, is greater than that in plasma (Table 9.2). The choroid plexuses form approximately 300 ml of cerebrospinal

Table 9.2. Composition of cerebrospinal fluid.

Constituent	Plasma	CSF
Na^+	150 mM	147 mM
K^+	4·6 mM	2·9 mM
Cl^-	99 mM	113 mM
Ca^{2+}	2·3 mM	1·1 mM
Protein	6000 mg/100 ml	20 mg/100 ml

fluid per day. Once formed the cerebrospinal fluid flows through the ventricles and finally passes out of the external surface of the CNS via one medial and two lateral apertures (Fig. 9.25). The cerebrospinal fluid then circulates over and bathes the brain and spinal cord. Cerebrospinal fluid is finally returned to the blood circulation via 'valves' within arachnoid granulations on the upper outer surface of the brain (Fig. 9.25d). Because cerebrospinal fluid is being formed continuously obstruction of its circulation causes a rise in hydrostatic pressure within the ventricular system. This leads to the condition known as hydrocephalus which can produce severe brain damage. Cerebrospinal fluid apparently provides the CNS with mechanical support, acts as a buffer against mechanical shock, controls the chemical environment of the brain (together with the blood circulation), and exchanges nutrients and waste materials with the brain tissue.

Physiological organisation of spinal motor neurones
Spinal motor neurones are arranged in such a way that their position in the ventral horn of the spinal cord is related to the muscles they innervate. Thus motor neurones that supply the muscles of the arms (these neurones are located in the cervical segments of the spinal cord) are organised so that motor neurones supplying the upper arm are near the midline, while those supplying the arm extremities are located at the lateral edge of the ventral horn (Fig. 9.26). In addition, motor neurones are segregated so that flexors lie dorsal to extensors. This type of organisation in which there is a spatial separation of cells supplying different parts of the body is quite common throughout the CNS, (*cf.* Somatic sensory and motor areas of the cerebral cortex.)

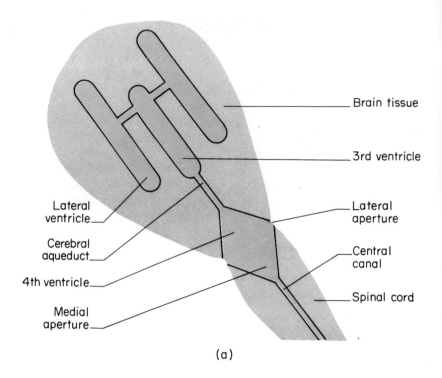

Brain tissue

3rd ventricle

Lateral
aperture

Central
canal

Spinal cord

Lateral
ventricle

Cerebral
aqueduct

4th ventricle

Medial
aperture

(a)

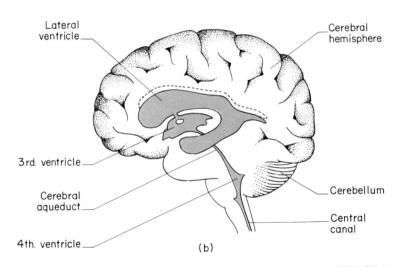

Lateral
ventricle

Cerebral
hemisphere

3rd. ventricle

Cerebral
aqueduct

Cerebellum

Central
canal

4th. ventricle

(b)

Fig. 9.25. The ventricular system of the brain. Within the brain is a series of fluid-filled spaces—the cerebral ventricles. (a) Diagrammatic representation of the ventricular system; these spaces are connected to the external surface of the brain and spinal cord by the medial and lateral apertures (after Bowsher). (b) Lateral view of the cerebral ventricles (after Netter). (c) Frontal view of the cerebral ventricles (facing page) (after Noback and Demarest). (d) Circulation of cerebrospinal fluid. The cerebral ventricles are filled with a liquid—cerebrospinal fluid—which is formed within the ventricles by the choroid plexes. Cerebrospinal fluid circulates through these spaces and then passes to the external surface of the CNS via the medial and lateral apertures; it is finally taken back into the blood through 'valves' in the arachnoid granulations on the upper surface of the brain (facing page) (after Netter).

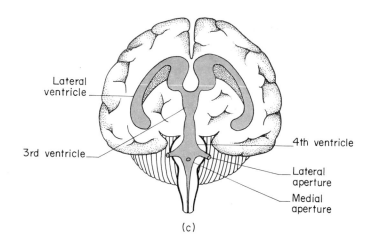

Lateral ventricle

3rd ventricle

4th ventricle

Lateral aperture

Medial aperture

(c)

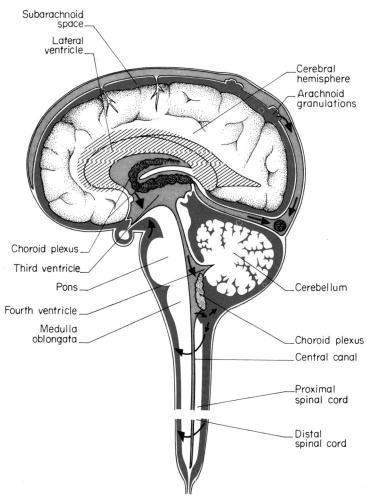

Subarachnoid space

Lateral ventricle

Cerebral hemisphere

Arachnoid granulations

Choroid plexus

Third ventricle

Pons

Fourth ventricle

Medulla oblongata

Cerebellum

Choroid plexus

Central canal

Proximal spinal cord

Distal spinal cord

(d)

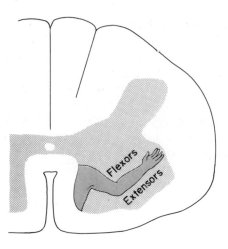

Fig. 9.26. Cross-section of a cervical segment of the spinal cord showing functional arrangement of spinal motor neurones. (The arms are innervated by motor neurones located in the cervical segments of the spinal cord.) Motor neurones supplying the arm extremities are located at the lateral margin of the ventral horn; motor neurones innervating the upper arm are at the medial margin of the ventral horn. Extensor motor neurones are found ventral to flexor motor neurones (after Netter).

Concepts of the physiological organisation of motor-neurone pools have changed in recent years. It is well known that the force of contraction in a given muscle can be graded both by increasing the impulse frequency in individual motor neurones and by recruiting successively more motor neurones with increasing demand. At one time it was thought that the order of recruitment of motor neurones supplying a given muscle was dependent upon the source of the stimulus. It has become apparent, however, that, with increasing demand the order in which motor neurones to a given muscle are recruited is always the same no matter what the driving stimulus. The smallest motor neurones to the muscle are always activated by the weakest stimuli. With increase in stimulus strength motor neurones of progressively increasing size become active. This system is efficient, since the smallest motor neurones and the muscle fibres they innervate are well adapted to sustained activity (pp. 55–57). Small motor neurones also innervate fewer muscle fibres than do large motor neurones. Thus, when only a low level of force output is required from the muscle small motor neurones are first brought into play. These activate only a relatively small number of muscle fibres and so generate relatively low muscle forces, making it possible to peform delicate manipulations. Because motor neurones are organised in this way a number of older concepts (such as occlusion, fractionation and subliminal fringes) can no longer be applied to groups of motor neurones supplying particular muscles.

Reflex activity of the spinal cord
Responses to many stimuli can be made without the need for conscious control of movement. Thus, if one's hand touches a hot object the hand is removed before we have had time to appraise the situation; only after the hand has been removed do the conscious actions take place. The

initial response is quite similar from one individual to another, while the responses which follow may vary a great deal! The initial, stereotyped response is an example of a reflex action. There is a wide range of reflexes of varying complexity. The simplest is the monosynaptic reflex (Fig. 9.27a). In this type of reflex stimulation a sensory receptor it produces a burst of impulses in a sensory nerve which synapses directly upon and activates motor neurones in the spinal cord. The motor neurones, in turn, cause muscular contraction and movement. This type of reflex is termed monosynaptic since it involves only one synapse within the CNS. In actual fact two synapses are involved, as motor neurones activate

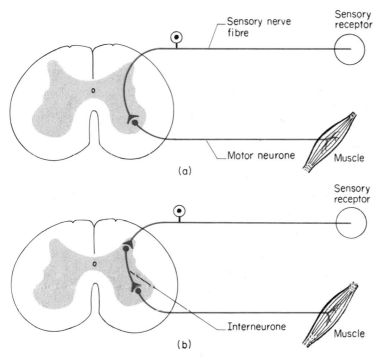

Fig. 9.27. Spinal reflexes. (a) Monosynaptic reflex. Sensory nerve fibres make direct connections with spinal motor neurones. (b) Polysynaptic reflex. At least one interneurone is interposed between the sensory nerve terminals and the motor neurones.

muscle fibres via nerve–muscle synapses. Reflexes involving more than one central synapse are termed polysynaptic reflexes (Fig. 9.27b). Other more complex reflexes may involve more than one spinal segment and are known as intersegmental reflexes.

In any reflex there is a delay (the reaction time) between the stimulus and the muscular response. Contributing to this delay is the time taken for conduction along nerve fibres to and from the CNS, transmission across the central and nerve–muscle synapses and activation of the contractile apparatus of the muscle fibres. This delay is approximately 20–24 milliseconds for the monosynaptic knee-jerk reflex.

Muscle spindles
If a normally innervated muscle is stretched it generates tension to resist the applied movement; this is known as a stretch reflex and is mono-

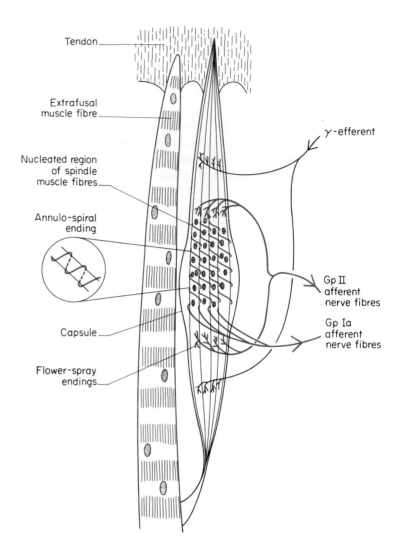

Fig. 9.28. Muscle spindle. Muscle spindles lie parallel with 'normal' (extrafusal) muscle fibres within the body of the muscle. Each spindle consists of a number of specialised (intrafusal) muscle fibres enclosed within a capsule.

The nuclei of the spindle muscle fibres are concentrated mid-way along the fibres; this region, unlike the two ends of the fibres, is unable to contract in response to stimulation. Annulospiral endings of Group Ia afferent nerve fibres wrap around the nucleated portion of the spindle muscle fibres (inset); flower-spray endings of Group II afferent nerve fibres are located just next to the nucleated region.

Spindle muscle fibres receive their motor supply from small γ-motor neurones whose terminals are situated on the contractile ends of the spindle muscle fibres.

synaptic. The sensory receptors responsible for this reflex are muscle spindles, which consist of a number of specialised muscle fibres enclosed in a capsule. The spindle muscle fibres (known as intrafusal muscle fibres) differ from 'normal' muscle fibres (or extrafusal fibres) in several ways: (1) they are smaller in diameter and their nuclei are aggregated about half way along their length, (2) their motor supply is provided by

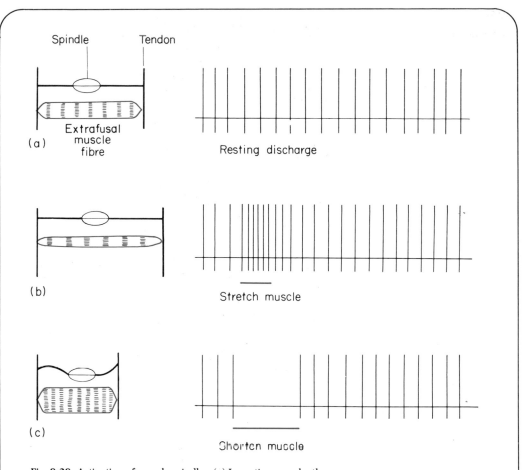

Fig. 9.29. Activation of muscle spindles. (a) In resting muscle, the sensory-nerve fibres of the muscle spindle produce a resting discharge of impulses. (b) If the muscle spindle is stretched, the central portions of the spindle muscle fibres are put under tension and their sensory-nerve terminals stimulated. (c) If the muscle shortens, the tension on the central regions of the spindle muscle fibres is reduced and the impulse frequency in their sensory-nerve fibres falls. Black bars under sensory nerve discharge pattern (b & c) indicate the duration of the muscle length change. (After Ganong.)

relatively small γ-motor neurones (extrafusal muscle fibres are innervated by α-motor neurones. (See p. 33 for classification of nerve fibres), and (3) intrafusal muscle fibres have a sensory nerve supply.

Spindles are provided with two types of sensory nerve fibre. Large Group Ia axons form primary (or annulo-spiral) endings, which wrap around the central nucleated portion of the spindle muscle fibres, whereas smaller Group II sensory nerve fibres form secondary (or flower-spray) endings. In the following description, the function of only Group Ia sensory fibres will be discussed, while the role of Group II fibres will be ignored.

Each skeletal muscle contains a number of muscle spindles which lie among and parallel to extrafusal muscle fibres, so that if the whole muscle is stretched, its muscle spindles will also be stretched; this depolarises the spindle sensory nerve endings and so increases the impulse frequency in sensory nerve fibres (Fig. 9.29b). If the muscle shortens (either passively or by contraction of extrafusal muscle fibres), the tension in the spindles is reduced, the sensory nerve endings are no longer excited and so the impulse frequency in their nerve fibres falls (Fig. 9.29c). Spindle sensory nerve fibres may produce impulses when their muscle is at normal resting length (Fig. 9.29a), therefore, spindle sensory nerve impulse frequency may be increased by stretching and decreased by shortening the muscle.

In the spinal cord, the sensory nerve from a muscle spindle form excitatory synapses on α-motor neurones of the same segment; these motor neurones in turn innervate extrafusal muscle fibres near to the original muscle spindle (Fig. 9.30). This circuit forms the basis of the monosynaptic stretch reflex. When a muscle is stretched a burst of

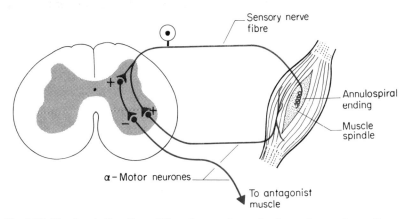

Fig. 9.30. Muscle spindle reflexes. When the annulospiral endings of a muscle spindle are stimulated a burst of nerve impulses travels to the spinal cord; the sensory-nerve fibres form excitatory synapses upon motor neurones supplying the muscle containing the activated spindle. Thus, impulses in the sensory-nerve fibres of a muscle spindle produce muscular contraction via a monosynaptic reflex—the stretch reflex. In addition to monosynaptic activation of motor neurones to their own muscle, the spindle sensory-nerve fibres also produce inhibition of motor neurones to antagonist muscles.

impulses in its spindle sensory nerve fibres causes excitation of α-motor neurones supplying the muscle which has been stretched. This muscle then contracts, relieves the tension on the spindle muscle fibres and allows the impulse frequency in the muscle-spindle sensory nerves to fall back to a lower level. The α-motor neurones are now no longer excited and contraction stops.

Muscle spindles not only produce monosynaptic excitation of α-motor neurones to their own muscle, but simultaneously cause reflex inhibition of α-motor neurones to antagonist muscles. This is produced by the activation of interneurones which inhibit the antagonist motor neurones (Fig. 9.30). This inhibition facilitates shortening of synergistic* muscles during the stretch reflex. The arrangement whereby opposing

*Synergistic muscles (or motor neurones) work together to produce a movement around a particular joint; antagonistic muscles (or motor neurones) have opposing actions.

effects are produced in neurones with antagonistic functions is known as reciprocal innervation and is found commonly in reflex and other actions of the CNS.

An example of the stretch reflex is the knee-jerk reflex, in which the patellar tendon is tapped with a hammer. This rapidly stretches the thigh muscle (and the spindles within it) and produces a knee jerk by stimulating contraction of the thigh muscles via their α-motor neurones.

Role of γ-motor neorones. Because muscle spindles are so extremely sensitive to changes in length they would be only capable of working effectively over a very narrow range of muscle lengths if they merely responded passively. γ-motor neurones, however, enable muscle spindles to work over the whole physiological range of muscle lengths without loss in sensitivity. The ends of spindle muscle fibres are contractile, while the middle, nucleated portion of these fibres (bearing annulospiral sensory endings) is not. Therefore, the effect of activity in γ-motor neurones

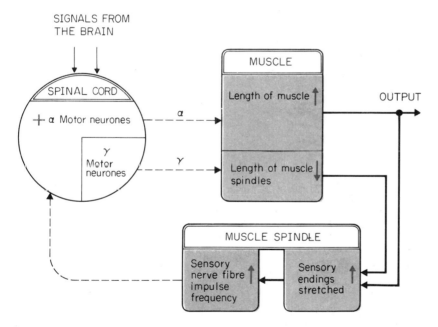

Fig. 9.31. Diagram showing the action of muscle spindles in controlling length of a skeletal muscle. γ-Motor neurones cause contraction of spindle muscle fibres and allow them to operate effectively when their muscle is being used at relatively short lengths (e.g. when the arm is bent rather than straight in the case of the biceps). In this way modulation of γ-motor neurone activity allows the muscle spindle sensory nerve fibres to maintain their sensitivity over a wide range of muscle lengths.

Normally commands from the brain produce simultaneous activation of the both α- and γ-motor neurones. Thus, as the whole muscle shortens, the spindle muscle fibres also shorten by the appropriate amount to enable them to work efficiently over the new range of muscle lengths. If, however, following the command from the brain, muscle shortening is prevented, spindle sensory nerve terminals will become over-extended and produce a burst of nerve impulses which excites α-motor neurones supplying the muscle; this will reinforce the initial command and cause more powerful muscular contraction. Acting in this way, muscle spindles can operate as load compensators.

is to shorten the ends of spindle muscle fibres and hence to put the middle portion under tension. Thus, if a muscle shortens, γ-motor neurone activity will 'take up the slack' in spindle muscle fibres and maintain the sensitivity of their sensory endings. If, on the other hand, a muscle is elongated, reduction in γ-motor neurone activity will prevent these sensory endings from being over-extended. In this way, modulation in the level of γ-motor neurone activity can be used to ensure that the sensitivity of muscle spindles is optimised for the range of lengths over which a muscle is operating at any given time. This is, in fact, just one example showing how the CNS can modulate the sensitivity of a peripheral sense organ.

Commands from the brain calling for limb movement usually produce simultaneous activation of α- and γ-motor neurones to the appropriate muscles, so that spindle muscle fibres shorten along with the extrafusal muscle fibres, keeping the spindles at their optimal sensitivity. This arrangement enables spindles to act as load compensators, e.g., if a heavy weight is to be lifted, α-motor neurone activation initially may be insufficient to cause muscle shortening (i.e. to lift the weight). Because the spindle muscle fibres are simultaneously activated via γ-motor neurones, under these circumstances their sensory endings are put under tension and they produce a burst of impulses in their sensory nerves. This further excites α-motor neurones to the muscle (via the monosynaptic stretch reflex) and so increases force generation (Fig. 9.31). It should be remembered, however, that even in this role, spindles are detecting muscle length rather than tension.

Golgi tendon organs

Information about muscle *tension* is provided by another type of sensory receptor, the Golgi tendon organ. The Golgi tendon organ consists of branched nerve endings embedded in the tendon of a muscle (Fig. 9.32). Golgi tendon organs are supplied with Group Ib sensory nerve fibres. When the muscle tendon is put under strain the nerve endings of its

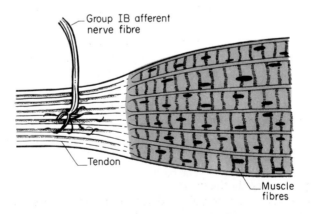

Fig. 9.32. Golgi tendon organ. Golgi tendon organs are located within the tendons of muscles and consist of many-branched nerve terminals; they respond to tension in the muscle. Different Golgi tendon organs have different thresholds and so work over different ranges of muscle tension.

Golgi tendon organs are stimulated. In this way, the CNS receives information about the force generated in a given muscle. If this force becomes so great that it may cause tissue damage, the Golgi tendon organs can mediate a protective reflex which inhibits the activity of motor neurones to the muscle with which they are in series; this results in a sudden loss of muscular tension. Thus, if an excessively heavy weight is applied to a muscle, the stretch reflex mediated by muscle spindles is overridden; the muscle no longer opposes extension and damage is prevented. Sensory-nerve fibres from Golgi tendon organs, on entering the spinal cord, form synaptic connections with inteneurones which inhibit motor neurones supplying their muscle (Fig. 9.33).

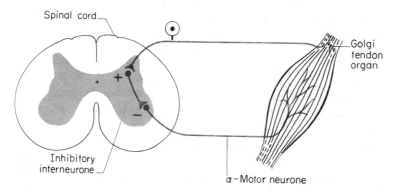

Fig. 9.33. Golgi tendon organ reflexes. Sensory-nerve fibres from Golgi tendon organs form synaptic contacts with interneurones in the spinal cord. These interneurones inhibit motor neurones supplying the muscle to which that Golgi tendon organ is attached.

Other reflexes

In the above descriptions we have seen that reflexes can mediate rapid responses which would be delayed if it was necessary consciously to decide what action to take. This allows us to devote our conscious thoughts to fewer matters while routine movements are delegated to automatic mechanisms.

The reflexes mediated by muscle spindles and Golgi tendon organs serve to illustrate some of the basic features of reflexes. However, there is a wide variety of reflexes of greatly varying complexity. Thus, if we move our bodies we have to make a large number of postural adjustments in order to maintain balance; these adjustments are largely produced by coordinated reflexes mediated by different types of receptor including the eyes and the vestibular apparatus of the inner ear as well as by muscle spindles and Golgi tendon organs.

Sensory pathways to the brain

Signals entering the spinal cord through nerve fibres from sensory receptors not only produce spinal reflexes but must also travel to the brain, where, among other things, they can produce conscious appreciation of the various stimuli which the body receives. The route by which sensory information travels from the spinal cord to the brain depends upon the degree of discrimination and the urgency with which information must reach the brain. Thus, information about joint position, for example, is transmitted rapidly to the brain by the dorsal column system (Fig. 9.34a). On the other hand, temperature sensation, which is less urgent and not precisely localised, is conveyed to the brain via the spinothalamic system (Fig. 9.34b).

Sensory-nerve fibres forming the dorsal column system enter the

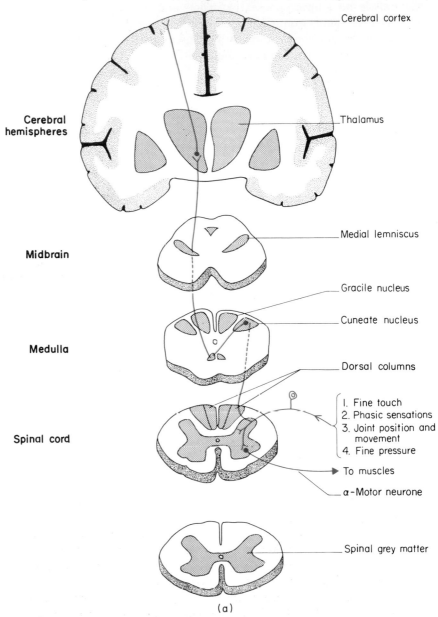

Fig. 9.34. Sensory pathways to the brain. (a) Dorsal column system. The nerve fibres of this system rapidly and precisely transmit sensory signals to the brain. (b) Spinothalamic system. Sensory signals are transmitted to the brain more slowly and with less precision by this system than by the dorsal column system (facing page) (after Netter).

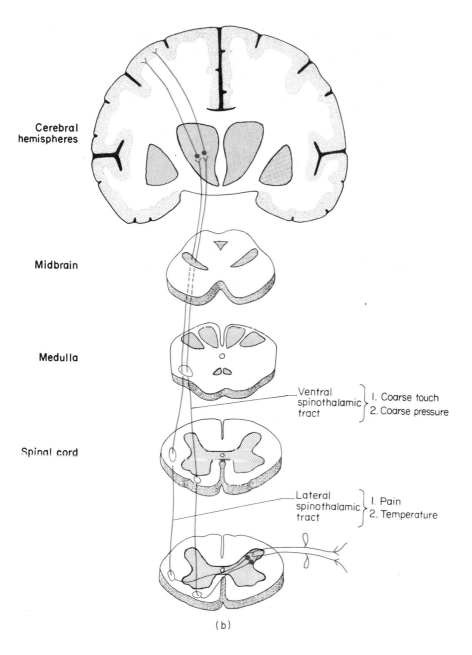

Cerebral
hemispheres

Midbrain

Medulla

Spinal cord

Ventral
spinothalamic
tract
}
1. Coarse touch
2. Coarse pressure

Lateral
spinothalamic
tract
}
1. Pain
2. Temperature

(b)

spinal cord in the dorsal roots and branch in the dorsal horn. Some branches are short and form synapses locally while the longest branch travels up the spinal cord to the medulla in the dorsal columns. On each side of the medulla there are two sensory nuclei, the dorsal column nuclei (the gracile and cuneate nuclei). The gracile nuclei receive sensory fibres from the hindlimbs and lower body, while the cuneate nuclei receive sensory fibres from the forelimbs and upper body. Within these nuclei sensory fibres of the dorsal column system form synaptic connections with second-order sensory neurones (those neurones which are second in the chain joining sensory receptors to the brain). The axons of the second-order neurones then cross the midline to the opposite side of the medulla and travel up to synapse in the thalamus. These fibres then pass to the somatic sensory area of the cerebral cortex (Fig. 9.35a). The actual part of the somatic sensory cortex to which nerve fibres travel depends upon the part of the body those fibres represent. Thus, signals from the leg arrive in that part of the somatic sensory cortex near the midline (Fig. 9.35b). In this way there is a spatial representation of the body across the somatic sensory cortex. The size of the cortical area devoted to each part of the body, however, is not proportional to the physical size of that part, but is related to the density of sensory receptors; therefore the fingers and lips, which have a high degree of discrimination for tactile stimuli, have a large area of cortical representation in comparison with the trunk, which has relatively poor discrimination.

Sensory-nerve fibres carrying less precise information to the somatic sensory cortex via the spinothalamic system enter the spinal cord through the dorsal roots, then travel up or down the spinal cord for several segments before forming synaptic connections with second-order neurones. Fibres of the pathway then cross the midline and travel up the spinal cord, through the medulla, finally terminating in the thalamus. Here they synapse with neurones which convey the signals to the somatic sensory cortex. Fibres of the spinothalamic system are generally smaller in diameter and hence have a lower conduction velocity than those of the dorsal column system. Each system does, however, consist of a chain of neurones which carries signals from sensory receptors to the somatic sensory cortex.

In the above description of the sensory pathways to the brain each successive synapse was considered merely as a relay by which sensory information is passed from one neurone to another; each synapse in the pathway does, however, modify the sensory signals in some way, so that the pattern of signals arriving at the somatic sensory cortex is not an exact replica of that in the sensory-nerve fibres entering the spinal cord. An example of the way in which signals may be processed by successive neurones in a sensory pathway is given with reference to the visual system (see p. 303).

Control of movement

Although motor acts are often classified as conscious or unconscious, the distinction between the two is often blurred. Thus, when learning a new skill, such as riding a bicycle, we must attempt to concentrate simultaneously on many different aspects of our often unsuccessful attempts to maintain equilibrium. In the light of painful mistakes we are able to focus voluntarily on the most faulty aspects of our performance. Once the skill has been learned we are able to keep balance without the need to concentrate on the details of the task and are able to devote our thoughts to other matters as we cycle along. In this way a given series of motor commands have passed from the conscious to the automatic domain of control.

At present we do not know exactly how the brain comes to a decision to perform a particular action, or where this decision is made; instead we are limited to descriptions of those pathways by which movements

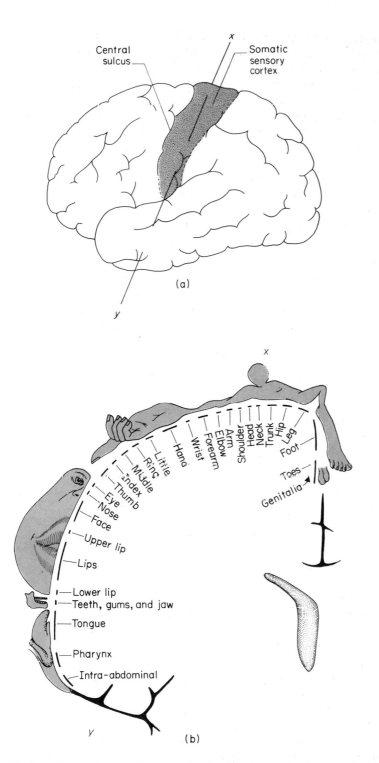

Fig. 9.35. Somatic sensory area of the cerebral cortex. (a) The somatic sensory cortex is situated just behind the central sulcus on a ridge known as the postcentral gyrus. (b) Signals from different parts of the body are sent to different regions of the somatic sensory cortex; body regions with the highest tactile discrimination have the largest areas of cortical representation. (After Penfield & Rasmussen.)

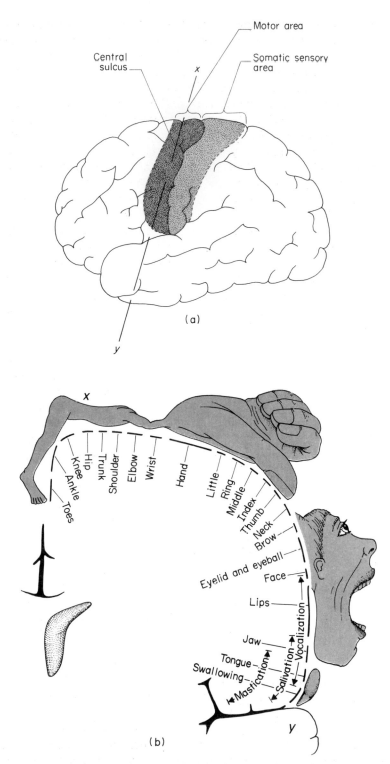

Fig. 9.36. Motor cortex. (a) This area occupies a strip of cortex just in front of the somatic sensory cortex. (b) Different parts of the body are controlled by different regions of the motor cortex; those muscles which may be controlled most precisely have the greatest area of cortical representation. (After Penfield & Rasmussen.)

are controlled once they have been initiated. It has been known for some time that localised muscular contractions can be produced by electrical stimulation of that area of the cerebral cortex known as the precentral gyrus. This lies just in front of the somatic sensory area and is known as the motor cortex (Fig. 9.36a). Different muscles can be activated by electrically stimulating different parts of the motor cortex. In this way it is possible to construct a map of the muscles controlled by each part of the motor cortex (Fig. 9.36b). As with the somatic sensory cortex, the cortical area representing a given part of the body is not directly related to its physical size. Instead muscles capable of delicate movements have a relatively large representation on the motor cortex, while muscles under comparatively poor nervous control have a small area of cortical representation.

There are two major systems by which signals are transmitted from the brain to the spinal cord to produce movement. These are known as the pyramidal (corticospinal) system and the extrapyramidal (extracorticospinal) system. Both systems send nerve fibres down the spinal cord, where they activate directly, or indirectly, both α- or γ-motor neurones. Neurones of the pyramidal tract have their cell bodies in the cerebral cortex. Their nerve fibres pass down into the white matter of the cerebrum where they all converge to pass through a narrow gap (the internal capsule) between brain centres. From the internal capsule, the nerve fibres travel on down through the brainstem, where most fibres cross to the opposite side of the midline at the caudal end of the medulla. The fibres then continue down the spinal cord and terminate in the spinal segment containing the motor neurones they are to control (Fig. 9.37). Thus, signals descending in the pyramidal tract are transmitted from the cerebral cortex to their destination in the spinal cord without intervening synapses.

The extrapyramidal system can be defined as any system other than the pyramidal system by which the brain sends motor commands to the spinal cord. The extrapyramidal system is anatomically more complex than the pyramidal system for two main reasons: first because of its heterogeneity and, second, because there are a number of synaptic relays in the system. The extrapyramidal system consists of the basal ganglia (a group of nerve centres deep within the cerebrum), several brainstem nuclei and the brainstem reticular formation (an area containing a complex network of small cells) (Fig. 9.38). Motor signals from the cerebral cortex are received and modified by this system, then sent down the spinal cord. The extrapyramidal system also receives information concerning such things as balance, body position and the visual world from sources other than the cerebral cortex. All this information is integrated and contributes to the final motor commands sent to the spinal cord by the extrapyramidal system.

Although, so far, the pyramidal and extrapyramidal systems have been considered separately, they are functionally interdependent. Thus, signals travel back from parts of the extrapyramidal system to the motor cortex, where they influence the activity of neurones in the pyramidal tract.

Because the extrapyramidal system receives sensory information from

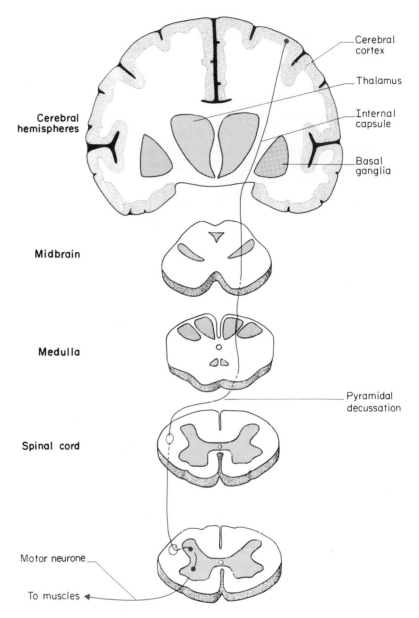

Fig. 9.37. Pyramidal system. Fibres of this system travel without synaptic interruption from the cerebral cortex to the spinal segments in which they influence motor-neurone activity. Most of the fibres cross the midline in the pyramidal decussation at the caudal end of the medulla (after Netter).

a number of sources and uses this to modify motor signals in the pyramidal system, it is extremely important for the maintenance of posture and balance and also for the production of smooth, slow movements. In view of this, it is perhaps not surprising that malfunction of the basal ganglia results in a number of motor disturbances such as the resting tremor which can often be observed in patients suffering from Parkinson's disease.

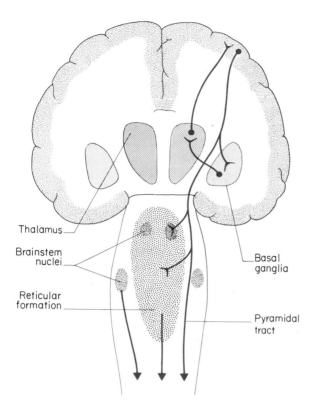

Fig. 9.38. Extrapyramidal system. This system is more heterogeneous than the pyramidal system; it carries signals from a number of different brain areas to spinal motor centres. These brain areas include the basal ganglia, brainstem nuclei and the brainstem reticular formation, all of which are influenced directly or indirectly by motor signals from the cerebral cortex.

The cerebellum has a large modifying influence on motor signals sent from the brain to the spinal cord via both the pyramidal and extra-pyramidal systems. The functions of the cerebellum are in some ways complementary to those of some components of the extrapyramidal system. Thus, the extrapyramidal system predominates in the production of smooth, slow movements, while the cerebellum is of primary importance in the generation of coordinated rapid movements.

Like the extrapyramidal system, the cerebellum receives signals related to body position and movement from a variety of sources. Thus, information about head movement (from the vestibular apparatus) and limb movement (mainly from muscle spindles and Golgi tendon organs) and signals from the visual and auditory systems converge upon the same region of the cerebellar cortex, so that, like the somatic sensory cortex, a map can be made to show where each part of the body is represented on it. Although the cerebellum receives such a wide range of information, it plays no part in the conscious perception of sensory stimuli. The cerebellum is itself also unable to initiate movements. For these reasons electrical stimulation of the cerebellum leads neither to sensations nor to movements; instead, the cerebellum uses incoming sensory information to compute the position in space of different parts of the body. In addition to sensory signals the cerebellum also receives signals from the motor cortex and from elements of the extrapyramidal system reflecting their output to spinal motor centres. By combining sensory and motor signals the cerebellum can predict the movements and positions of different parts of the body which will result from motor commands leaving the brain via the pyramidal and extrapyramidal systems. In this way the cerebellum can compare an intended movement with that which is actually being produced. The cerebellum also has connections with both the motor cortex and parts of the extrapyramidal system through which it can modify their subsequent output. Thus, if for example the cerebellum computes that a limb is moving too fast for a particular action, it will send messages to the relevant motor systems, inhibiting the overactive muscles and causing excitation of antagonists. In this way the limb movement stops at the right moment without appreciable overshoot. The interactions of the cerebellum with the two motor systems are shown diagrammatically in Fig. 9.39.

It appears, in fact, that the motor systems normally send commands to the spinal motor centres which would, in themselves, produce exaggerated movements; the moderating influence of the cerebellum, however, allows the production of smooth, accurate movements. Thus,

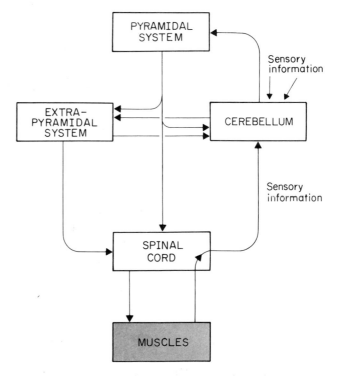

Fig. 9.39. Interactions of the cerebellum with the two motor systems. The cerebellum receives information about the motor commands travelling to the spinal cord via both the pyramidal and extrapyramidal systems; it also receives sensory information from a number of different sources, including muscles. In the light of all this information the cerebellum sends corrective signals back to the motor systems to modulate their subsequent output (modified from Guyton).

impairment of cerebellar function results in movements which are jerky, less precise and which often overshoot the desired final position; this overshoot can lead to pendular limb movements, which result from successive unsuccessful attempts to bring a limb to rest at a given final position. Cerebellar damage can thus be associated with 'intention tremor', in which muscular tremor develops when movement is attempted, but is absent at rest (Fig. 9.40). Like other movement disorders which result from cerebellar malfunction, 'intention tremor' might be expected if commands from the motor systems are not modulated by the cerebellum. 'Intention tremor' contrasts markedly with that seen after damage to the basal ganglia (see p. 272) in which tremor occurs at rest and is reduced during muscular activity. Loss of cerebellar corrective functions can also result in an inability to coordinate movements involving several different joints. Instead, movements of each joint are produced separately. This is perhaps not surprising if the patient

Fig. 9.40. Effect of cerebellar malfunction on movement. Damage to the cerebellum can disturb its corrective influence on motor commands; as a result movements become poorly controlled. Thus, when such a patient attempts to move his finger from his nose to a particular object the movement is irregular and becomes progressively more jerky as the object is approached (modified from Ruch).

has difficulty even executing and terminating movements about a single joint.

One of the major sources of sensory input to the cerebellum is the vestibular apparatus, which provides information about the orientation and movement of the head in space. This information alone cannot be used by the cerebellum to compute the position of the whole body; the cerebellum also receives signals from joint receptors in the neck, which indicate the angle of the neck. By analysing the signals from both the vestibular apparatus and the neck-joint receptors the cerebellum is able to compute the position of the whole body in space. In the light of this it is not surprising to find that cerebellar damage can be associated with a loss of balance.

We have already seen that the cerebellum receives a wide range of sensory information. Under normal circumstances the information from all the different sources should correspond, and all will contribute to the signals sent from the cerebellum to the motor centres. As conditions change, however, the relative importance of the sensory signals from different sources can vary. Thus, in the dark the cerebellum receives no visual information. The cerebellum of an astronaut in a space capsule, on the other hand, receives no information about head position from the vestibular apparatus, since there is no gravitational force to act upon the macular hair cells. Instead the cerebellum of the astronaut has to rely heavily on visual information.

In summary, movements are controlled by the two major motor systems which have their origins in the brain. The activity of both systems can be strongly influenced by the inflow of sensory information from a number of sources, some of which can act directly through the extrapyramidal system and some of which acts via the cerebellum to maintain balance and to produce coordinated, smooth and precise movements. At the present time, though, we have little insight into the way in which the 'decision' to produce movements is made in the brain.

The reticular formation

So far we have discussed the routes by which relatively specific information is transmitted from one part of the nervous system to another. We are all aware that the overall level of activity in the nervous system can alter under different conditions; thus,at some times we may be very alert and excited and could not sleep even if we wished, while at others it is hard to remain awake. These alterations in the degree of arousal are a result of variations in the level of activity in the area of the brainstem core known as the reticular formation. This consists of an extremely complex network of small interconnected neurones; it is not homogeneous but contains a number of nuclei with quite specific functions in addition to more diffuse areas. Thus, the reticular formation has a number of discrete functions in addition to that of setting the level of consciousness. We have already seen that some nuclei embedded in the reticular formation form part of the extrapyramidal system and are concerned with the maintenance of posture and balance. In addition, other areas of the reticular formation are responsible for, among other functions, the regulation of respiration, heart rate and blood pressure.

The reticular formation has widespread inputs, including the cerebrum, cerebellum, spinal cord and other regions of the brainstem; it has widespread outputs and is able to influence the activity of many areas from which it receives signals. In this way the reticular formation samples much of the activity in the CNS and can use this information to modulate overall nervous activity (Fig. 9.41).

Reticular activating system

Stimulation of one particular area of the reticular formation can cause an individual to become aroused or alert. This part of the reticular formation is known as the reticular activating system (RAS). The reticular activating system projects diffusely to most of the cerebral cortex, largely via the thalamus. In this way activation of the reticular activating system in turn alerts the cerebral cortex and makes it more receptive to incoming signals.

Although the reticular activating system is directly responsible for the production of the alert state, it is itself under the influence of two other regions of the reticular formation, both of which lower the overall level of arousal. Both these centres act together to produce sleep, although the role of each is different. (See below for a discussion of sleep.)

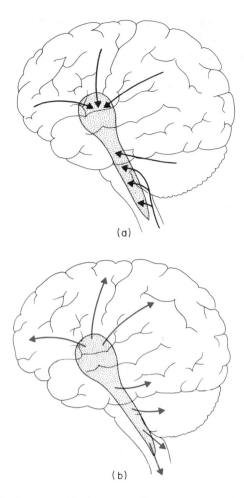

Fig. 9.41. Reticular formation. (a) The reticular formation is a complex network of small nerve cells located at the core of the brain. It receives signals from widespread parts of the CNS. (b) The reticular formation, in turn, distributes signals to many parts of the CNS. The connections of the reticular formation enable it to sample and modulate the overall activity of the CNS. (After Vander *et al.*)

Electroencephalogram
Electrodes placed on the scalp can be used to record electrical activity of the brain. Such a recording is known as an electroencephalogram (EEG). The EEG does not reveal the electrical activity of single nerve cells, but shows electrical changes which result from the activity of large populations of nerve cells in the cerebral cortex. The nature of the activity recorded depends upon the level of consciousness of the individual. Figure 9.42 shows some of the types of wave which can be observed. If a subject is awake but relaxed with his eyes closed his EEG pattern shows regular waves with a frequency of 8–12 Hz; this is known as the α-rhythm and apparently results from synchronised activity of many cortical neurones. If the subject opens his eyes and becomes alert, the α-rhythm is replaced by irregular low amplitude activity which results from desynchronisation of neuronal activity. The actual changes which

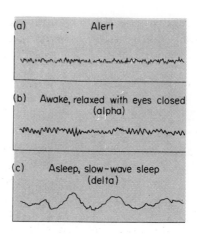

Fig. 9.42. Electroencephalogram (EGG). Electrical activity of the brain can be recorded by electrodes placed on the scalp; such recordings reflect activity in large numbers of cortical nerve cells and are known as electroencephalograms. (a) Recordings made from an alert subject show irregular low-amplitude electrical activity. (b) If the subject relaxes with his eyes closed, regular waves appear with a frequency of 8–12 Hz. This type of electrical activity is known as the α-rhythm. (c) When a subject first falls asleep the α-rhythm is replaced by larger, slower δ-waves. (After Vander *et al.*)

take place at the cellular level to produce the different types of EEG pattern are unclear. The EEG can be used clinically, however, to determine the position of localised brain abnormalities. Normally the shape and timing of the EEG patterns recorded from corresponding points on each cerebral hesmisphere are similar. A tumour or other localised lesion on one hemisphere would upset this bilateral symmetry in the EEG, allowing its position to be determined. The EEG can also be used in the diagnosis of the various forms of epilepsy. Epilepsy results from excessive activity of whole or part of the CNS and can be divided into partial and generalised forms. Partial epilepsy usually results from local brain damage. The damaged area causes very rapid discharge of neighbouring neurones. During an epileptic attack this abnormal discharge may remain localised or may spread to other brain areas. Generalised epilepsy involves larger areas of the brain and can be subdivided into grand mal and petit mal epilepsy. Grand mal is associated with convulsions which may last from several seconds to several minutes. Petit mal epilepsy is closely related to grand mal, and may similarly be associated with convulsions; these are, however, much shorter in duration. During both grand mal and petit mal episodes the EEG has a characteristic form.

Sleep

Although we spend a considerable proportion of our lives asleep, at present the physiological reason why we need sleep is not understood. It is, however, clear that the brain does not switch off during sleep.

Sleep can be divided into two types: slow-wave sleep and paradoxical sleep. Initially, on falling asleep, the α-rhythm is replaced by slower, larger (delta) δ waves (Fig. 9.42) and the person is said to be in slow-wave sleep. At intervals slow-wave sleep is interrupted by periods of paradoxical sleep, in which the EEG pattern resembles that of an alert individual (Fig. 9.42). In spite of this, a subject is less easily aroused during paradoxical sleep than during slow-wave sleep; paradoxical sleep is associated with a reduction in the tone of limb and body muscles, but can be accompanied by rapid movements of the eyes. For this reason paradoxical sleep is also known as 'rapid eye movement sleep' (REM sleep). Dreaming usually takes place during REM sleep. Periods of REM sleep normally last about 20 minutes and constitute approximately 20 percent of the total time spent asleep.

We have already said that the level of consciousness depends upon the interplay between three regions of the reticular formation. It is thought that slow-wave sleep is produced by a centre in the reticular formation of the medulla. It does this by depressing the activity of the reticular activating system (RAS). This depression apparently results from the inhibitory action on the recticular activating system of 5-hydroxytryptamine produced by cells of the sleep centre in the medulla. While the centre in the medulla inhibits the RAS it simultaneously activates the second sleep centre, which is located in the pons. This pontine sleep centre is responsible for the generation of paradoxical sleep. Sleep cycles probably result from fluctuations in the levels of 5-hydroxytryptamine in the reticular formation.

Emotion—the limbic system

Both emotion and motivation are entities which can be only poorly defined neurophysiologically; it seems that both are controlled by a part of the brain previously known as the rhinencephalon (=nose brain).

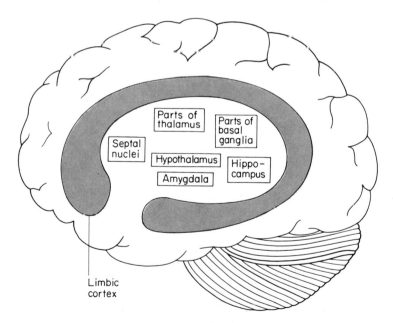

Fig. 9.43. Limbic system. This system is concerned with a number of functions including the perception of smell, production of feeding behaviour and the control of some biological rhythms, fear and rage and motivation. It consists of a ring of cortical tissue encircling a number of underlying structures all of which interact by means of a complex series of interconnections (modified from Guyton).

Originally this part of the brain was thought to be primarily concerned with the sense of smell; it is now known to have several other important functions. For this reason the term 'rhinencephalon' has been replaced by 'limbic system' to describe these structures. The limbic system is heterogeneous and consists of a ring of primitive cortical tissue on each side of the midline encircling a number of subcortical structures, including parts of the hypothalamus and thalamus, the amygdaloid nuclei, the septal nuclei and the hippocampus (Fig. 9.43). These components of the limbic system have relatively complex interconnections. Much of the output of the limbic system is channelled through the hypothalamus, which, therefore, has a central role in limbic function.

The limbic system is important in a number of different brain functions. These include analysis of olfactory signals, generation of some feeding behaviour, and the control of several biological rhythms, the emotions of fear and rage and the motivational state of the animal. For this reason these activities may become enhanced or depressed by damage to different parts of the limbic system.

Normally there is a balance between rage and placidity, so that an animal will only become enraged if severely provoked; lesions in certain parts of the limbic system, however, can turn a previously placid animal into one which will show rage with minimal provocation. A similar transformation in behaviour can also be produced by electrical stimulation of other regions of the limbic system. Enragement produced in this way not only results in the behavioural changes directly associated with rage, but also the characteristic effects of changes in activity of the autonomic nervous system (see p. 283), such as an increase in heart rate, blood pressure and respiration. Thus, artificially produced rage is indistinguishable from rage observed in normal animals. Fear and rage are related emotions, and so it is perhaps not surprising that experimental manipulations of limbic function can enhance or suppress fear as they can rage.

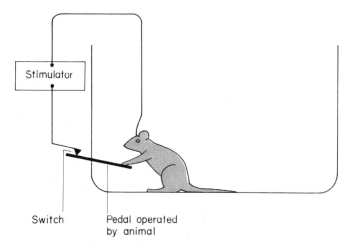

Fig. 9.44. Apparatus used to study reward and punishment. An electrode chronically implanted in the brain of the experimental animal is connected via a pedal switch to a stimulating device. The animal is able to operate the pedal and so stimulate its own brain; if stimulation has a rewarding effect, the animal will repeatedly press the pedal.

Experiments have been performed to discover whether electrical stimulation of particular brain areas has a rewarding or punishing effect.
To do this electrodes are chronically implanted into a selected brain area
of an animal. The electrode is then wired-up so that, by pressing a lever,
the animal can deliver stimulating pulses to its own brain (Fig. 9.44). If
the effect of stimulation is unpleasant, the animal will learn to avoid
pressing the bar; if the effect of electrical stimulation is rewarding, the
animal will press the bar repetitively. In fact, animals find stimulation of
some brain areas so rewarding that they will ignore food to continue
bar pressing and will only stop when completely exhausted. Among the
areas producing the highest rates of bar pressing are the hypothalamic
centres controlling thirst, food intake and sexual behaviour. The effectiveness of some areas as centres for reward are dependent on the
motivational state of the animal. For example, with certain electrode
placements, bar pressing may be more frequent when the animal is
hungry.

Learning and memory

Learning is hard to define precisely; a general definition is that learning
is a change in the response to a given stimulus that occurs in the light
of experience. Although our understanding of learning and memory is
increasing, much of the available information is of a behavioural nature
and we still have little information about the cellular mechanisms underlying these phenomena. While those studying learning have classified
different types of learning situation it is likely that, in each case, learned
information is stored by similar mechanisms.

Memory can be divided into two phases: short-term memory and
long-term memory. When an animal learns something this information
first of all enters short-term memory, where it will remain for a matter
of minutes to hours. (The precise duration of short-term memory depends upon the species of the animal studied and upon the experimental
methods used.) While information is being stored in short-term memory,
it is susceptible to disruption by a number of agents including electro-
convulsive shock (strong electric shocks applied to the head), low
temperature, coma and deep anaesthesia. Thus, any of these treatments
may produce a state known as retrograde amnesia, in which the
memory of recent events is disrupted, leaving earlier memories unaffected. Since more remote memories are resistant to disruption, it has
been concluded that the mechanism by which information is stored in
short-term memory differs from that for long-term memory. Because
short-term memory is disrupted relatively easily by procedures which
may be expected to have a profound effect on the electrical activity of
the brain, it has been suggested that information is stored in short-term
memory as reverberating electrical activity in the brain. As information
passes into long-term memory, on the other hand, it is stored in a more
durable form.

Presumably the formation of memories must lead to an alteration in
the transmission of electrical signals through parts of the brain. This
change in transmission could result either from a modification in the
effectiveness of previously existing synapses or from the formation of

new synaptic contacts. So far it is not possible to state which, if either, of these two possibilities is correct. If either of these alternatives is correct, they would involve either structural or enzymic changes (for example, altered transmitter substance metabolism). Because proteins are the enzymic and structural elements of living tissue, both RNA and proteins themselves have been studied as candidates for the storage site of long-term memory. Several different experimental approaches have been used to study the problem. These include the administration of chemical agents which block protein synthesis and measurement of incorporation of radioactive precursors into brain proteins or RNA. Such experiments have provided evidence to suggest that protein synthesis is involved in the formation of long-term memory, although the precise role of proteins in long-term memory still remains unclear.

Cerebral cortex

So far we have discussed those areas of the cerebral cortex which are involved in somatic sensations, motor control and those cortical regions which constitute part of the limbic system. The cortical areas concerned with the analysis of visual and auditory information are discussed in Chapter 10. In addition to these areas there are three 'association' areas. One occupies part of the frontal lobe, a second occupies part of the temporal lobe, while the third occupies part of both the parietal and the occipital lobes (Fig. 9.45).

One of the most remarkable characteristics of the human brain is its ability to deal with complex language. The ability to comprehend and

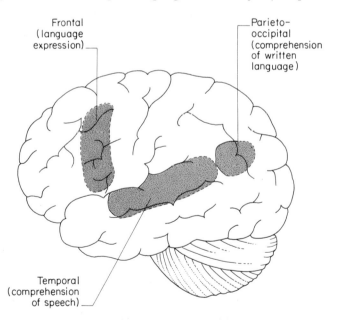

Frontal
(language
expression)

Parieto-
occipital
(comprehension
of written
language)

Temporal
(comprehension
of speech)

Fig. 9.45. Cortical areas concerned with language. One area in the frontal lobe is responsible for the motor aspects of language production; a second area in the temporal lobe is responsible for the comprehension of spoken language; while a third area on the boundary between the parietal and occiptal lobes is responsible for comprehension of written language. Damage to any one of these areas produces characteristic language deficits.

produce language is located in parts of the cerebral cortex; disorders known as aphasias have provided a great insight into the way in which the brain deals with language. Aphasias are language deficits which result not from deafness or blindness nor, on the motor side, from paralysis of respiratory or oral muscles, but instead from an inability directly related to language. Aphasias may be classified as sensory or motor. In sensory aphasias there is a deficit in the ability to understand language. This deficiency may be limited to comprehension of written or of spoken language. The cortical areas responsible for comprehension of written and spoken language are shown in Fig. 9.45. Motor aphasias, like sensory aphasias, can involve a deficit either in written or in spoken language; a patient with a motor aphasia can comprehend both spoken and written language but has difficulty expressing language. The areas responsible for the motor aspects of language are located in the frontal lobes just anterior to the motor areas (Fig. 9.45).

Although, in a child, both cerebral hemispheres have an equal potential to develop language, language functions in most adults are largely limited to the left hemisphere. For this reason the left hemisphere has been termed the 'dominant hemisphere'. It has become clear, however, that although the right hemisphere has only limited language capability it has other capabilities absent in the left hemisphere; thus, it seems that the cerebral hemispheres are specialised in such a way that the left hemisphere takes over most of the language functions, while the right hemisphere predominates in functions such as the analysis of spatial relationships in the external environment and of musical forms. In normal individuals information can be exchanged between the two hemispheres via the corpus callosum, a large tract of nerve fibres joining the two cerebral hemispheres. This allows the two hemispheres with complementary specialisations to cooperate with each other.

It is hard to define precisely the role of the anterior pole of the frontal lobes. Damage to these lobes has little effect on intelligence, although it can cause some change in character and impair memory. One of the most striking effects of damage to this area is loss of anxiety which would appear in normal individuals under certain conditions. For this reason operations have been performed in which connections are severed between the anterior pole of the frontal lobes and the remainder of the brain (prefrontal leucotomy) of patients suffering from chronic acute states of anxiety.

THE AUTONOMIC NERVOUS SYSTEM

Control of the internal environment

The autonomic nervous system innervates smooth muscle, cardiac muscle and glands and, together with the endocrine system, is largely responsible for the control of the internal environment of the body. In this way it controls blood pressure, activity of the gut, sweating, urine output, body temperature and a number of other functions. Almost all control of the autonomic nervous system takes place below the level of consciousness.

The activity of the autonomic nervous system, like that of the motor supply to skeletal muscles, is controlled both by reflex activity and through the influence of higher centres. The most important centres for this control are the hypothalamus and regions of the brainstem.

The autonomic nervous system is divided into two divisions with essentially antagonistic actions: the sympathetic nervous system and the parasympathetic nervous system. Unlike the nerves running to skeletal muscles, fibres of the autonomic nervous system synapse after leaving the spinal cord and before reaching their effector organs. Nerve

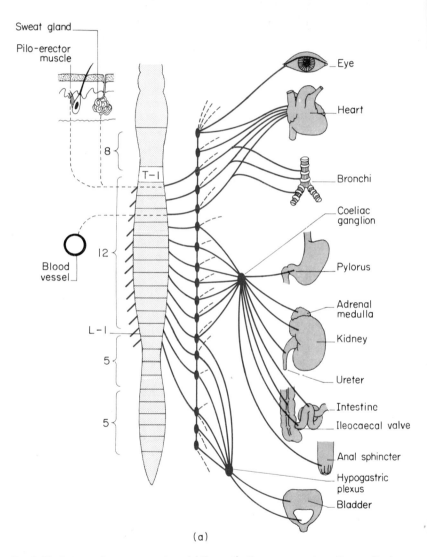

(a)

Fig. 9.46. Autonomic nervous system. (a) Sympathetic nervous system. Preganglionic nerve fibres leave the thoracic and lumbar segments of the spinal cord and run to sympathetic ganglia, where they form synaptic contacts with postganglionic neurones; the postganglionic neurones innervate effector organs. (b) Parasympathetic nervous system. Preganglionic nerve fibres leave the central nervous system in the cranial and sacral nerves; they form synaptic connections with short postganglionic neurones in or near their effector organs (facing page). (After Guyton.)

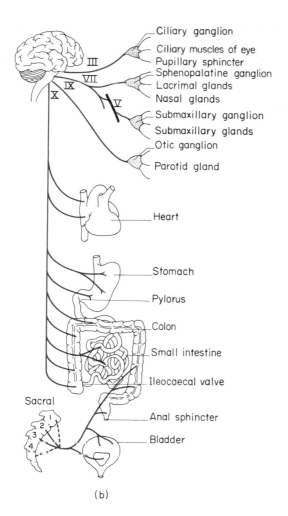

Ciliary ganglion
Ciliary muscles of eye
Pupillary sphincter
Sphenopalatine ganglion
Lacrimal glands
Nasal glands
Submaxillary ganglion
Submaxillary glands
Otic ganglion
Parotid gland
Heart
Stomach
Pylorus
Colon
Small intestine
Ileocaecal valve
Anal sphincter
Bladder

Sacral

(b)

fibres of the sympathetic nervous system originate in the thoracic and lumbar segments of the spinal cord (Fig. 9.46a). They exit in the spinal nerves and most form synapses in a paired chain of ganglia, the sympathetic ganglia, which run parallel to the spinal cord. The nerve fibres running from the spinal cord to these ganglia are known as preganglionic fibres, while fibres leaving the ganglia bound for their effector organs are known as postganglionic nerve fibres. Not all the preganglionic fibres of the sympathetic nervous system synapse in the ganglia of the sympathetic chain. Instead some travel to the more peripheral coeliac and mesenteric ganglia or to the adrenal medulla. The coeliac and mesenteric ganglia function in the same way as ganglia of the sympathetic chain. The adrenal medulla will be discussed later in this section.

Preganglionic nerve fibres of the parasympathetic nervous system leave the CNS at two levels, forming the cranial and sacral outflows (Fig. 9.46b). Unlike the sympathetic ganglia, those of the parasympathetic nervous system are not situated close to the spinal cord. Instead the ganglia are normally in or near the effector organ. For this reason preganglionic nerve fibres of the parasympathetic nervous system are longer than those of the sympathetic nervous system. In both the sympathetic and parasympathetic nervous systems, the transmitter substance released from the terminals of the preganglionic nerve fibres is acetylcholine (ACh); the transmitter substance released from postganglionic nerve fibres onto effector organs differs in the two divisions of the autonomic nervous system. At parasympathetic neuro-effector junctions ACh is the transmitter substance, while noradrenaline* is released at junctions of the sympathetic nervous system (Fig. 9.47).

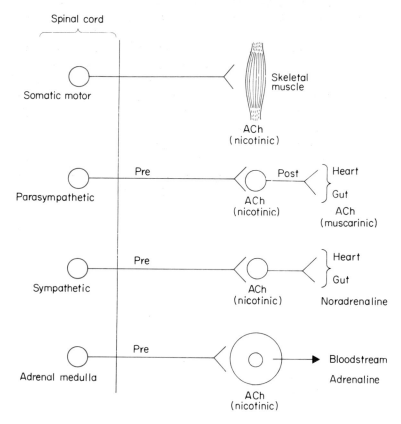

Fig. 9.47. Comparison of autonomic and somatic nervous systems showing the transmitter substances involved (functional not anatomical). All autonomic preganglionic nerve fibres release acetylcholine from their terminals; postganglionic neurones of the parasympathetic nervous system also release acetycholine (ACh), while most postganglionic neurones of the sympathetic nervous system release noradrenaline. The cells of the adrenal medulla are similar to postganglionic sympathetic neurones, but have no axons and secrete adrenaline instead of noradrenaline.

The adrenal medulla resembles a sympathetic ganglion in that it is innervated by sympathetic preganglionic nerve fibres. It has, however, no postganglionic nerve fibres; instead cells, which are innervated by preganglionic nerve fibres, when activated release adrenaline into the bloodstream. Thus, the adrenal medulla appears to be a modified sym-

* This is an over-simplification since ACh is released from a small number of postganglionic sympathetic nerve fibres, while the adrenal medulla produces and releases adrenaline and noradrenaline.

pathetic ganglion which has taken on an endocrine function. The structures of ACh, noradrenaline and adrenaline are given in Fig. 9.5.

The sensory receptor which provides the afferent limb of local autonomic reflexes may be associated with the viscera or somatic structures. Their sensory nerve fibres enter the spinal cord and influence the activity of preganglionic autonomic neurones, thus initiating reflex control of the appropriate visceral structures; Fig. 9.48 shows a typical sympathetic reflex arc. Sympathetic and parasympathetic activity may be modified independently by such reflexes, but a high level of activity in one division of the autonomic nervous system is usually associated with low activity in the other.

Most visceral structures are innervated by both divisions of the autonomic nervous system and in many instances parasympathetic and sympathetic effects oppose each other. Thus, activity in the sympathetic supply to the heart causes an increase in the heart rate, while activity in the parasympathetic supply decreases the heart rate (p. 127). In this way opposing innervations from the two divisions of the autonomic nervous system allows rapid and precise adjustments in the activity of visceral organs. The actions of the sympathetic and parasympathetic nervous systems on different visceral organs are complex and varied and so are not listed here individually; in general activity in the sympathetic nervous system tends to prepare the body for active responses to adverse conditions in the external environment, while the parasympathetic nervous system is mainly responsible for the control of internal functions

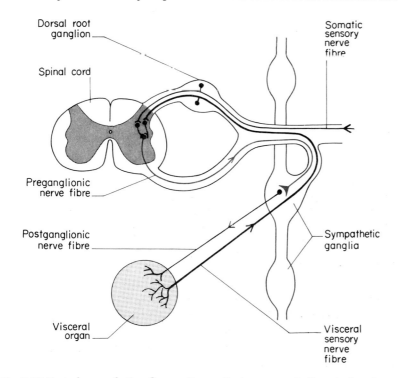

Fig. 9.48. Typical sympathetic reflex arc. Preganglionic neurones in the spinal cord can be excited by either somatic or visceral sensory nerve fibres; these can thus produce reflex changes in effector organs.

under quiescent conditions. Thus, under conditions which produce fear or anger (preparation for fight or flight), the sympathetic nervous system will become active and prepare the body by, among other things, diverting blood from the gut to skeletal muscles and by increasing the pumping action of the heart. On the other hand, parasympathetic activity decreases the pumping activity of the heart and enhances digestive function. In this way interplay between the two divisions of the autonomic nervous system maintains the internal environment without need for conscious control.

Nerve centres controlling the internal environment
Although a great deal of autonomic function is integrated locally by visceral reflexes, much of the activity of the autonomic nervous system is controlled from the brain. Thus, centres in the medulla are responsible for the autonomic control of heart rate, blood pressure and respiration. Details of these systems are given in the sections on cardiovascular and respiratory physiology.

The hypothalamus
The hypothalamus is situated at the base of the brain just rostral to the brainstem. Below it and connected to it is the pituitary gland (Fig. 9.49). The hypothalamus is an extremely important brain centre in the control of several aspects of the internal environment of the body. Thus, it controls body temperature, various forms of appetitive behaviour and the release of a number of hormones from the pituitary gland.

Stimulation of the hypothalamus may produce a large number of alterations in the activity of the autonomic nervous system; for example, it can produce marked changes in blood pressure and heart rate. The extent, however, to which the hypothalamus is directly involved in the continuous precise regulation of autonomic function is not known. The effects on autonomic function instead may be related to the part played by the hypothalamus in the control of emotions by the limbic system (see p. 279). Thus, hypothalamic modulation of autonomic function may merely represent one facet of the many physiological changes which may accompany an alteration in emotional state, rather than reflecting a continuous fine regulation of autonomic activity.

Hunger. Food intake is apparently controlled by two hypothalamic centres. There is a feeding centre which, when stimulated, causes an animal to search out food and eat. Destruction of this centre can result in a complete loss of appetite. Electrical stimulation of a second centre known as the 'satiety centre' inhibits eating, while its destruction causes animals to develop a voracious appetite and rapidly become obese. It is thought that the feeding centre is normally active but, after sufficient food intake, is inhibited by the satiety centre.

One hypothesis to explain the control of food intake is that some cells in the satiety centre effectively monitor the plasma glucose level. The satiety centre is activated by a rise in the level of blood glucose. As the glucose level falls, so the satiety centre becomes less active, releasing the feeding centre from inhibition and initiating a feeding drive (p. 175).

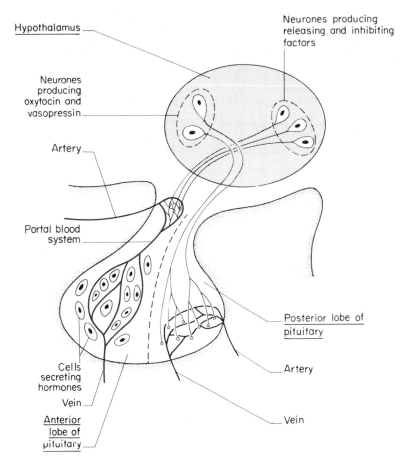

Hypothalamus

Neurones producing
releasing and inhibiting
factors

Neurones
producing
oxytocin and
vasopressin

Artery

Portal blood
system

Cells
secreting
hormones

Vein

Anterior
lobe of
pituitary

Posterior lobe of
pituitary

Artery

Vein

Fig. 9.49. Hypothalamus and pituitary gland. The hypothalamus is located at the base of
the brain; below it and connected to it is the pituitary gland. The hypothalamus
influences the endocrine activity of the anterior pituitary gland by secreting factors into
the portal blood system linking the hypothalamus and the anterior pituitary gland; these
factors either elevate or depress hormone release from the anterior pituitary gland.
Neurosecretory neurones with cell bodies in the hypothalamus send their axons into the
posterior lobe of the pituitary gland; here they release either oxytocin or vasopressin into
the blood.

Water balance. The hypothalamus maintains water balance by controlling both water intake and loss. It controls water intake by generating
the sensation of thirst when the osmotic pressure of the body fluids rises.
This is detected by cells known as osmoreceptors which are located
within the hypothalamus. These cells are stimulated when the osmolarity of the plasma rises (pp. 341, 415).

The hypothalamus controls the release of two hormones from the
posterior pituitary gland. Neurones with cell bodies in the hypothalamus
send their axons into the posterior pituitary, where, on stimulation, they
release their hormone into the blood. One of these hormones is oxytocin,
which acts upon the female breast and uterine myometrium (see Ch. 12,
pp. 395 & 396). The second hormone is vasopressin (also known as
antidiuretic hormone or ADH) which alters the permeability of the distal
tubules and collecting ducts of the kidney (see Ch. 8, p. 216). The

secretion of vasopressin, like the sensation of thirst, is mainly controlled by osmoreceptors within the hypothalamus. Thus, when the plasma osmotic pressure rises the secretion of vasopressin increases and enhances water retention by the kidney. The secretion of vasopressin is apparently influenced by blood volume as well as osmotic pressure (p. 341). Stretch receptors in the walls of large veins probably provide the hypothalamus with information about blood volume, since these vessels are the major reservoirs of blood. In this way a fall in blood volume would cause a rise in vasopressin secretion.

Body temperature. The hypothalamus plays an important role in the control of body temperature in mammals (see Ch. 15, p. 426); to do so it receives information both from peripheral temperature receptors and temperature receptors actually within the hypothalamus. Using information from these receptors the hypothalamus initiates reflex responses to maintain a constant body temperature. Thus, in response to cold the hypothalamus initiates such responses as shivering and peripheral vasoconstriction, while high temperatures produce responses such as sweating and peripheral vasodilatation.

Secretion of anterior pituitary hormones. The anterior pituitary gland has been termed the 'master endocrine gland' since four of the six hormones it produces control the secretory activity of other endocrine glands (see Chs. 11 (p. 335) and 12 for details of these systems); the secretory activity of the anterior pituitary is itself controlled by hypothalamic factors released into a blood portal system linking the hypothalamus with the anterior pituitary. In this way the hypothalamus indirectly controls some aspects of bodily and mammary growth (via growth hormone and prolactin, respectively), and the secretory activity of the thyroid gland, the adrenal cortex and the gonads.

From the foregoing sections it is apparent that the hypothalamus has a number of distinct functions in the control of the internal environment of the body. Since it is also involved in the emotional functions of the limbic system, it is perhaps not surprising that emotions can influence many aspects of bodily function. Thus, for example, it is well known that the menstrual cycle and fertility in the human female are particularly susceptible to emotional factors.

Chapter 10
Special Senses

VISION

Vision is the subjective sensation which is initiated by light falling on the eye. Although the eye is indispensable in seeing, its behaviour is by no means sufficient for vision; the nervous system to which it is connected is also necessary. For example the visual system has to make sense of a world in which everyday objects are normally distorted by perspective; the image of a man at 100 metres is smaller than one standing beside us, yet we know they are about the same size. The eye's input, together with the other sensory inputs, interacts with the current and stored information in the central nervous system to give the sensation of vision.

The eye is built on a similar plan to a camera, which in essence focuses images of the external world by means of a lens on to a film at its back; later the film is processed to give a picture. In the eye light is focused by the cornea (in terrestial animals) and lens on to the retina, the light-sensitive layer. This image is 'processed' initially by the nerve cells in the retina—which can be considered as part of the nervous system stationed in the periphery—and then by the visual areas of the nervous system to give the pattern we 'see' (Fig. 10.1). In each the amount of light entering the structure may be controlled, and the light may be focused by adjustments to the lens.

Light refraction

In order for vision to be effective, images of objects in the external world need to be brought to a focus on the retina. Before a ray of light can reach the retina it has to pass through the cornea, the aqueous humour, the lens and the vitreous humour. Once in the retina it has to traverse various structures before it reaches the light-sensitive region (see below).

The three characteristic features of the retinal image are: (1) it is reversed; (2) it is sharp and well defined (if it is accurately focused); and (3) its size depends on the angle which the object subtends to the eye. In normal young eyes all objects situated more than about 60 metres away are in focus at rest, nearer objects may be focused by increasing the thickness of the lens. Light is bent or refracted out of its original direction when it passes from one medium into another with a different refractive index; so, rays of light entering the eye are refracted at the cornea and at the lens before forming an image on the retina. In man the refraction which occurs at the air–corneal surface is greater than that at the lens, although of course the lens refraction can be varied whereas that at the cornea cannot.

The power of a lens is expressed by its focal length, which may be

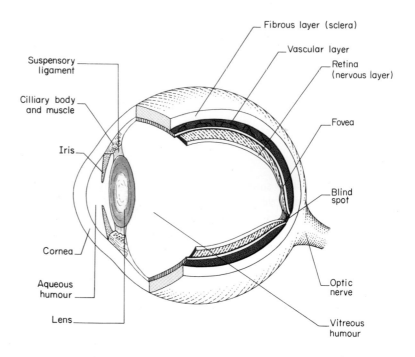

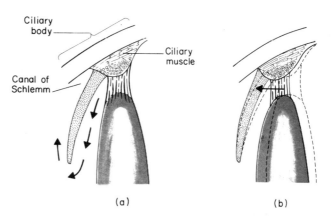

(a)   (b)

Fig. 10.1. General structure of the human eye.

Cornea—composed of collagen fibres with fixed cells between. Some of the oxygen supply obtained directly from the air. Richly supplied with free nerve endings.

Iris—a pigmented screen containing a circular sphincteric muscle and radially arranged dilator muscle fibres. The sphincteric muscle is supplied by parasympathetic fibres from the third nerve nucleus. The dilator muscle has a sympathetic nerve supply from T1 and T2.

Lens—composed of ribbon-like fibres arranged in concentric laminae. The front of the lens is less curved than the back. The strong capsule is attached to the ciliary body via suspensory ligaments. Tension in the lens is partly maintained by the pressure of the aqueous humour on the suspensory ligaments. When the ciliary muscles contract (inset b) the insertions of these ligaments are moved forward and the tension lessened, the front surface of the lens then bulges forward.

The front part of the eye is filled with aqueous humour and is divided into anterior and posterior chambers by the iris (which forms the pupil). The aqueous humour has a low viscosity and is similar in composition to plasma but contains hyaluronic acid which is kept depolymerised by hyaluronidase. It is formed actively in the ciliary process, flows from the posterior to the anterior chamber, and is reabsorbed into the canals of Schlemm (Fig. 10.1 inset a). It is produced—by an active process—at a rate of about 2 ml/min and is under a pressure of 25 mmHg. The vitreous humour, which fills the back part of the eye, is a transparent jelly-like substance with numerous collagen fibres in it.

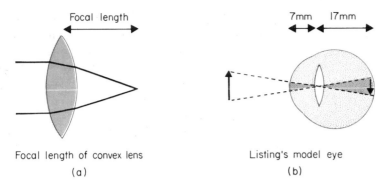

Focal length of convex lens

(a)

Listing's model eye

(b)

Fig. 10.2. (a) Focal length of a convex lens; (b) Listing's model eye.
  The image subtended by an object may be calculated by using the model eye. If the object is 1 metre in size and situated 10 metres in front of the eye the image is then $1/10 \times 17$ mm in size (by the method of similar triangles).

measured by allowing parallel rays of light to come to a focus (Fig. 10.2). The fatter the lens the stronger it is and so the shorter the focal length. Another way of expressing the power of a lens is to use dioptres, where a dioptre is the reciprocal of the focal length in metres; thus a lens of focal length of 2 m has a power of $\frac{1}{2}$ dioptre. This measure has the advantage that powers and dioptres are directly related and are additive. It is found experimentally that the normal human eye has a total power of about 59 dioptres, the lens contributing 16 and the cornea about 43.

  To reduce the complexity of the optics of the human eye the simplified or reduced model proposed by Listing is useful. This model has a 'lens' containing all the power of the eye situated 17 mm in front of the retina, which is about $\frac{1}{4}$ of the distance back from the cornea. This model has the correct power ($1000/17 - 59$ dioptres) and may be used for calculating the size of an image on the retina (Fig. 10.2b), and the angle subtended by an object, the visual angle.

Visual acuity
This is the appreciation of fine detail in a visual image. It would be expected that the limit to this would be set by the size of the visual receptors in the retina, which are about 1 μm in diameter in the central part of the eye. Under good conditions objects forming retinal images of

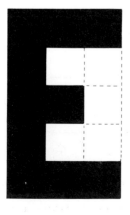

Fig. 10.3. Construction of Snellen's test type. Each square subtends an angle of 1 min of arc at the eye when viewed at the distance marked below the letter. A subject who can read the line marked '6-metres' at 6 metres has normal [6/6] vision; if he can only read the 60-metre letters at 6 metres then he has defective [6/60] vision.

about 2 μm across (which corresponds to a visual angle of about 30 seconds) can be distinguished, so that the eye's performance approaches the limits set by its structure. Actually the visual acuity measured by using an object of two lines is greater than that using two points (10 compared to 30 s) presumably because the image of the lines falls over several receptors on the retina rather than on one receptor. This is the reason for using lines on vernier scales on instruments, rather than points. Clinically, visual acuity is commonly measured by Snellen's test types (Fig. 10.3). The visual acuity varies markedly over the retina from the fine acuity at the macula discussed above, to very poor acuity at the periphery. This is partly a matter of receptor size and density, and partly the way the retina and visual cortex is organised.

Optical defects of the eye

Although the eye, regarded as an optical instrument, has several defects its overall performance is extremely good. The main kinds of fault found are those arising from the curvature of the refracting surfaces. If the cornea and lens were simple spherical lenses there would be considerable spherical aberration due to light passing through their edges. In fact, both the cornea and lens are flatter near the edges to reduce this defect and in addition the pupil usually cuts off the light at the edges and so improves the quality of the image. The fact that the front and back surfaces of the lens have different curvatures also helps to reduce aberration. Clinically eyes may be too long (myopia) or too short (hypermetropia) for the focal length of the refractive system and then additional lenses need to be supplied for correction (Fig. 10.4). Commonly also the

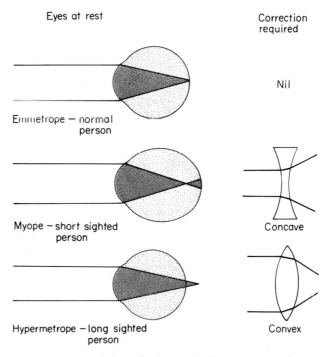

Fig. 10.4. Common refractive defects. The short-sighted person can see objects near him, the long-sighted person cannot without correcting lens.

curvature of the eye is not the same in the horizontal and vertical planes, so that both images cannot be in focus at the same time; this is astigmatism.

Accommodation

Young eyes can usually focus objects from the horizon (say > 60 metres) to a few centimetres away. This is done by increasing the curvature of the lens (mainly the anterior surface) by a contraction of the ciliary muscle which slackens the suspensory ligament of the lens and allows it to bulge forward (Fig. 10.1 inset b). This can, with difficulty, be shown by observing reflections from the surfaces of the lens—traditionally by a candle—in a darkened room. The front curvature of the lens changes from a radius of 12 mm to one of 6 mm.

During accommodation the axes of the eyes come together so that the images in both eyes fall on corresponding parts of the retinae. The pupils also constrict to increase the depth of field of the image and so improve the images on the retinae.

Amplitude of accommodation

The extra refractive power required to look at close objects can be calculated by considering the power of a lens which would be required to make such rays parallel; so an object 10 cm (0·1 m) away needs a

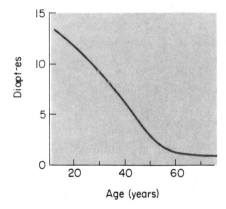

Fig. 10.5. Change in accommodation with increasing age.

lens of $1/0·1 = 10$ dioptres for this purpose. With increasing age the power of accommodation decreases steadily from an initial value of about 15 dioptres in young children to 1 dioptre at 50 years and over (Fig. 10.5). This is due to decreasing elasticity of the lens, which can no longer bulge as much as it did in youth.

Visual fields

The images on the retina are inverted and transposed from side to side, so that the lower temporal field projects its image to the upper nasal retina. If the eyes were out on stalks the fields would be very large, but as they are relatively sunken into the head the fields are restricted by the nose and brows (Fig. 10.6). (For those wearing spectacles, the fields

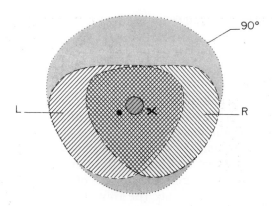

Fig. 10.6. Visual fields of left and right eyes to white light superimposed, showing area of overlap and blind spots due to the optic nerves (dot, left; cross, right). The fovea centralis is in the centre.

are further greatly restricted.) The fields from both eyes are largely superimposed so that binocular vision is possible. For this purpose the images need to be very accurately lined up on corresponding parts of the retinae.

When moving one's head, as in walking, the eyes are moved in a way appropriate to keep the image on the same part of the retina (stabilise the image). This can be done up to angular velocities of about 30°/s. This stabilisation is not complete, however; normally there are small oscillations of the eyes so that 'stationary' images are continually moving from one set of receptors to another. If an image is truly stabilised then it becomes invisible after a time, presumably due to 'adaptation' of the receptors.

The structure and functions of the retina

The structure of the retina
The light-sensitive part of the retina is made up of a carpet of photosensitive elements, the photoreceptors. These form a mosaic, whose fineness of grain and size varies across the retina; this is one factor determining the visual acuity. The neural part of retina consists essentially of three conducting cells lying end to end—the receptors, the bipolar and the ganglion cells—which form an 'input–output' pathway; and cross-connected cells—the horizontal and amacrine cells—which form a horizontal pathway (Fig. 10.7).

The input–output pathway is responsible for passing information from the receptors to the brain along the optic nerve, while the horizontal pathway allows for modification of this transmission process. In addition there are numerous other cross-connections in the retina, leading to a very complex structure with considerable local processing powers.

There are several other features of the retinal structure which need to be considered.
1   The photoreceptors are of two classes—the rods and the cones. The cones are responsible for colour vision, work in daylight (photopic) and

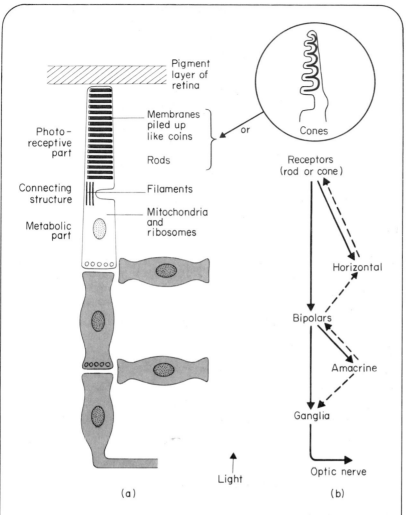

Fig. 10.7. Structure (a) and functional connections (b) of retinal layers in human eye. Visual pigment areas shown in red. In the central region of our eyes each cone is connected to one optic nerve as shown; in other parts of the retina several receptors converge on one bipolar cell and both cone and rod pathways converge on to the same ganglion cell. About 130×10^6 receptors terminate on 10^6 nerve fibres in each eye. At the fovea centralis (macula) the bipolar and ganglion cells and optic nerves are displaced to the side to allow the light to fall on to the cones directly. The receptors are packed closely together and are connected to each other by gap junctions. Light shone continuously on to a receptor cell activates the bipolar cell in its direct path but inhibits surrounding bipolar cells via the horizontal cells. Flashing lights have similar actions on ganglion cells via the amacrine cells. (After Werblin.)

number about 7×10^6. The rods are for monochromatic vision, work in twilight (scotopic) and number about 125×10^6. The world is grey in the twilight.

2   The distribution and density of these receptors across the eye has a characteristic pattern in our eyes, with rods mainly at the periphery and cones centrally (Fig. 10.8).

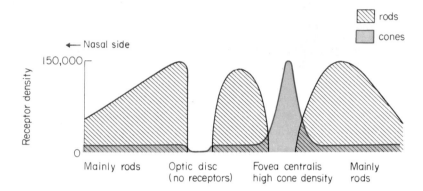

Fig. 10.8. Section across human retina in the horizontal plane showing density and type of receptor. Cones are shown in red.

3   The ratio of the receptors/nerves also changes across the retina. The overall numbers are that about 130,000,000 receptors converge on to 1,000,000 nerve fibres, a convergence of 130 : 1. At the fovea, however, the ratio is one cone to one nerve fibre, whereas at the periphery it is 300 : 1 for the rods. This also helps to explain the poor visual acuity at the periphery compared to the centre.

4   The discs in the photoreceptors are continually being turned over, with new discs inserted at the connecting structure and removed at the outer end by the pigment layer of the retina.

Photoreception

The structure of the outer segment of the photoreceptors, which consists of layers of photopigment, is well suited to capture light hitting it. When the pigment molecules (rhodopsin in the rods) are hit by photons they are raised to a higher energy level; this causes them to change their shape and in doing so they increase c-GMP (Fig. 10.9). This c-GMP acts as a transmitter to increase the membrane potential and so start the signal on its way to the nervous system. At any particular light level a dynamic equilibrium of pigment concentration, transmitter release and membrane potential is established.

Recently, cones containing red-sensitive (erythrolabe), green-sensitive (chlorolabe) or blue-sensitive (cyanolabe) pigments have been found in various animal and human retinae; these promise to explain the action of cones in a way similar to that for rods.

The visibility curve of the eye and colour vision

The eye is only sensitive to light over the range of about 400–700 nm. The exact range depends on the intensity of the light. The eye is also

(a)

Photons hit pigment, raises energy level

↓

Rhodopsin changes from cis to trans form, in several steps

↓

c-GMP released at a particular step

↓

c-GMP reduces Na permeability and so current at the cell membrane

↓

Membrane potential increases

(b)

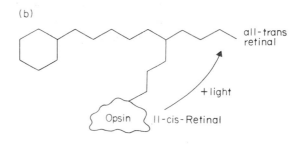

all-trans retinal

+light

Opsin   II-cis-Retinal

(c)

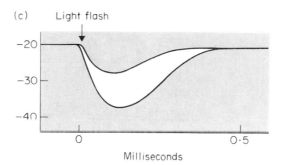

Fig. 10.9. (a) Sequence of events between light hitting a rod or cone and the change in membrane potential of the receptor. The cone response is faster than the rod one, because the visual pigment is embedded in the cell membrane so the transmitter has less far to travel.
(b) Change in rhodopsin following photon absorption; metabolic energy is required to reverse this process.
(c) The receptor potential (mV) of a cone to light flashes of low and high intensity. Rhodopsin consists of opsin, a glycoprotein, plus a chromophore, 11-cis-retinal derived from vit Al.

selective (non-linear) in not responding equally to equal amounts of energy in all parts of the visible spectrum.

The rods have a single peak of sensitivity at about 500 nm whereas cones show three peaks. It was originally suggested by Young, and later by Helmholtz, that colour vision could be explained by the presence of three different kinds of receptors and that all colours could be perceived as mixtures of these. This suggestion was based partly on finding people with various specific colour blindnesses (for example, to red or green) who could be imagined to lack the appropriate receptors; this appears to have been confirmed by recent research which has established the presence of three kinds of cones each containing one of these different pigments. There is, however, a considerable degree of overlap between the different spectra of the three cone classes. So a given wavelength will stimulate two or even three receptor types and so produce a unique ratio of responses (Fig. 10.10). Using the patterns on these three kinds of cones we can appreciate some 150 different colours (for example, in matching wool samples), a wavelength separation of only 3 nm.

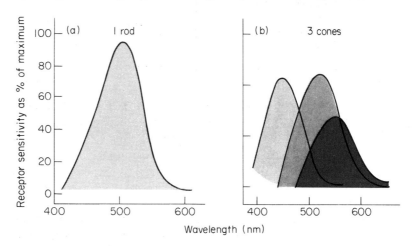

Fig. 10.10. The way in which the sensitivities of the rods and cones of human eyes varies with wavelength of applied light. Each is expressed as a percentage of the maximum for that species. The rods are actually some 10,000 times more sensitive than the cones. The peaks of the cone responses correspond to wavelengths of 445 (blue), 535 (green) and 570 (called 'red' for historical reasons though actually yellow) nm (after Schmidt and Thews).

The energy range of the eye and control of sensitivity of the retina

It has been estimated that only one quantum of light needs to be absorbed by each of about 10 rods for vision to be elicited; although, in practice, more light needs to fall on the cornea to give vision, because of the losses in the system. The eye, like the ear, approaches the theoretical minimum set by our physical construction. This means that the eye can work at extremely low light intensities, and yet it is also able to work at high light intensities; it can operate over a range from starlight to full sunshine, a range of intensity of some one hundred thousand million (1×10^{11}, Table 10.1). Within this vast range we see objects with high contrast; for example we can read 'black' print against a 'white' background in sunlight or in moonlight and yet the 'blackness' of the print in sunlight is brighter than the 'whiteness' of the background in moonlight. Part of the problem of understanding this, physiologically, is that our nerve fibres can only operate over a range of about one hundred, say an impulse frequency of 1–100 Hertz; how then can they signal changes of a hundred thousand million?

Table 10.1 Range of light intensities over which the eye can operate, with an indication of the mechanisms used. The eye does not operate instantaneously but summates signals over about 1/10 of a second in daylight. Vision also persists for this time after the stimulus is removed; this explains why we do not see the flicker present on a cinema screen or TV picture.

Millilamberts	Descriptive condition	Mechanism used
1×10^{-7}	Threshold after adaptation	Rod vision—can see stars
1×10^{-3}	White paper under moonless sky	(slightly to side of fovea)
1×10^{-2}	White paper under moonlight	Transition zone
1	Read newspaper with difficulty	
10	Comfortable reading	
100		Cone vision
1×10^{4}	White paper in sunlight	
1×10^{8}	Carbon arc	
1×10^{9}	Sun	Eye damage
1×10^{10}	Atom bomb (first 3 msec)	

It is clear that some method of compression must be used. This could be achieved either by: (1) each receptor operating over the whole range of intensity so that large changes of light intensity are required to stimulate the receptor, but this would decrease sensitivity to small changes in light intensity (i.e. give a poor contrast): or (2) for different receptors to be used for different parts of the range, but this would lead to poor visual acuity as many different receptors would be required and so the density of each kind would be low. The retina, in fact, performs operations which incorporate the best features of both systems in that it can signal with high contrast over a broad range of light intensities without sacrificing acuity. As we shall see, it does this by varying the sensitivity of the receptors to match the prevailing light conditions. They cannot, however, work well at high and low light intensities at the same time.

A clue to the mechanism by which this is done is given by the common experience that our vision changes on entering a dark room from a brightly lit one. Initially little can be seen but gradually the darkness

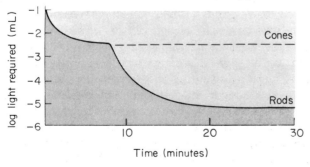

Time (minutes)

Fig. 10.11 The time course of dark-adaptation of human eyes in moving from light to dark conditions. The first smaller and faster part is due to cone adaptation, the second slower and larger part is due to rod adaptation. Adaptation is associated with an increase in the amount of unbleached visual pigment but is not directly related to it. The slow rate of adaptation is well matched to the gradual dimming of the ambient light during the evening twilight.

becomes less, so that more and more objects can be distinguished (Fig. 10.11). The explanation of this phenomenon, which involves both cones and rods, is not yet entirely clear but may be explained by some recent experiments. In these it has been shown that any receptor cell can only respond over a range of about a thousand units of illumination, but that its sensitivity can be varied so that it can respond to various light levels. The sensitivity of the rods can be varied to work from absolute threshold upwards by a factor of 10,000,000 (1×10^7, Fig. 10.12); the cones from

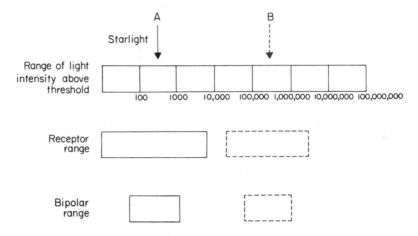

Fig. 10.12. Mechanism of dark adaptation in rods. The sensitivity of the receptors is adjusted to correspond to the mean level of light falling on the local portion of the eye; that of the bipolars is similarly adjusted to that of the receptor range. We show settings for low (A) and high (B) light levels (shown in red). The cause of these shifts in sensitivity of receptor and bipolar cells is probably due to a mean change in photochemical concentration of the surrounding area, signalled through horizontal cells (lateral inhibition).

Cones show similar mechanisms, but respond to levels of illumination to the right of that shown, i.e. from 100,000,000 up to full sunlight. Generally when rods are active cones are inactive and vice versa.

there up to bright sunlight, a further range of 10,000 (Table 10.1). The receptor sensitivity is thus adjusted to the prevailing light level; thereafter the receptor responds to light variations over a range of about 1000 around this level. The bipolar cells in turn work over a tenfold range within the thousandfold range of its receptor cell. In this way the retina operates with high contrast over a wide range of light inputs.

A disadvantage of this system is also part of one's common experience, i.e. that an area of similar light intensity appears to be light against a dark background but dark against a light background. This serves to emphasise the subjective nature of our senses.

The main, perhaps only, function of the ganglion cells is to pass on information to the optic nerve, although some of them seem to operate with the amacrine cells to retain a high sensitivity to small moving images within the visual field against the vast changes in contrast caused by blinking and eye movements.

The visual pathway

Impulses leaving the retina in the million or so nerve fibres of the optic nerve travel to the visual cortex, which is situated at the posterior pole of the brain in the occipital lobe (Fig. 10.13). Information from the left visual field (or right half of each retina) passes to the right lateral geniculate nucleus of the thalamus and vice versa. Each lateral geniculate nucleus sends fibres to the ipsilateral visual cortex. At the visual

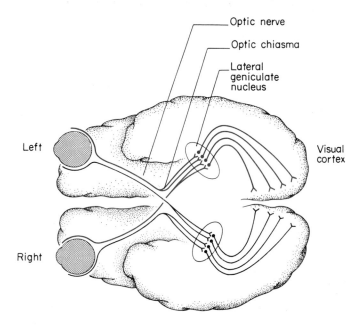

Fig. 10.13. Visual pathway. At the optic chiasma nerve fibres from the nasal half of each retina cross the midline to terminate in the contralateral (i.e. opposite) lateral geniculate nucleus; nerve fibres from the temporal half of each retina remain on the ipsilateral (i.e. same) side of the brain. Information from the visual world is passed to the visual cortex which lies at the posterior pole of the brain (after Kuffler & Nicholls).

cortex the area devoted to the macula is greater than that to the rest of the field (for example, 1 mm of macular retina = 16 mm of cortex, but 1 mm of peripheral field = 0·3 mm of cortex); this reflects the marked difference in the visual acuity between the macula and the periphery.

Some nerve fibres carrying visual information branch from the main visual pathway and transmit signals to the midbrain; here they may be involved in the control of eye movements and pupillary reflexes.

Visual information processing

It is part of our common experience that our visual system is more sensitive to change than to the steady state, and that we readily recognise objects by their outline only (e.g. all cartoons depend on this property of our visual system). These observations are explained in this section by adaptation and edge detectors.

In recent years a great deal has been learned about the way in which the visual system processes information received by the eyes. Here we shall show that at the various levels of organisation of the visual system (retina, optic nerve, geniculate body, visual cortex) the cells respond to visual patterns which are more specialised than those required at each of the 'lower' levels. It is also thought that other sensory systems (for example, auditory) work in a similar way.

Optic nerve. Single optic nerve fibres receive information from circular areas of the retina; the size of these areas is smallest at the fovea (0·1 mm) and largest at the periphery (2 mm), as would be expected from the ratio of the number of receptors per nerve fibre (p. 298). Optic nerve fibres can be divided into two types depending on their response to a circular spot of light falling on the centre of their receptive fields. One type is excited by a central spot of light ('on' centre) and is inhibited by a spot of light falling on to the periphery of its receptive field (Fig. 10.14a). The second type of optic nerve fibre ('off' centre) has the converse arrangement; they are inhibited by a central spot of light and excited by a spot of light at the periphery of the receptive field. This arrangement of receptive fields in which the centre and surround have antagonistic effects is an example of lateral inhibition (see p. 248) and means that diffuse illumination covering the whole receptive field is a relatively ineffective stimulus for optic nerve fibres. Since the only connection between the retina and the brain is the optic nerve, the higher cortical cells and hence the brain are not very sensitive to general levels of illumination, but are more concerned with contrasts in illumination between one part of the visual field and another.

Lateral geniculate nucleus. The receptive fields of cells in the lateral geniculate nucleus of the thalamus are very similar to those of optic nerve fibres, having antagonistic centres and surrounds.

Visual cortex. The left visual cortex receives information from the right visual field of each eye and vice versa. Unlike the neurones of the lateral geniculate nucleus, about 80 percent of the neurones in the visual cortex receive signals from corresponding areas of each eye; in other words, each neurone can be excited through either eye by a stimulus in a particular part of the visual field.

The receptive fields of cortical neurones are rather different from those of optic nerve fibres or lateral geniculate neurones; instead of being circular they are elongate, and these cells respond best to light or dark bars rather than to round spots. Cortical neurones can be classified into simple, complex and hypercomplex cells according to the complexity of the stimuli to which they respond optimally.

Simple cells respond to light or dark bars and edges with a particular orientation; their receptive fields show well-marked areas of excitation or inhibition (Figs 10.14b, c) reminiscent of those in the optic nerve and lateral geniculate nuclei. Cells such as those in Fig. 10.14b would be able to detect light bars; those like that in Fig. 10.14c could act as edge detectors.

Simple cells (like other cortical cells) are extremely sensitive to stimulus orientation. Thus neurones with receptive fields similar to those in

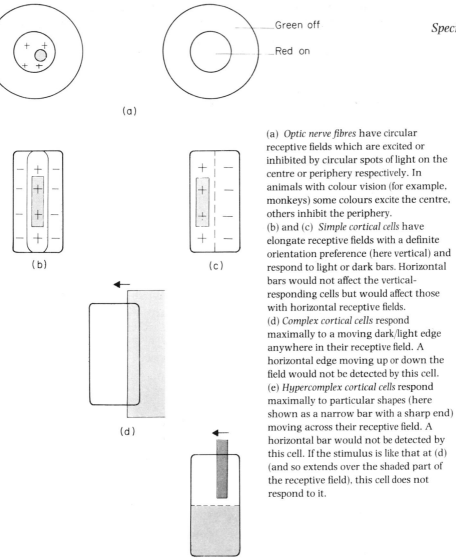

(a) *Optic nerve fibres* have circular receptive fields which are excited or inhibited by circular spots of light on the centre or periphery respectively. In animals with colour vision (for example, monkeys) some colours excite the centre, others inhibit the periphery.

(b) and (c) *Simple cortical cells* have elongate receptive fields with a definite orientation preference (here vertical) and respond to light or dark bars. Horizontal bars would not affect the vertical-responding cells but would affect those with horizontal receptive fields.

(d) *Complex cortical cells* respond maximally to a moving dark/light edge anywhere in their receptive field. A horizontal edge moving up or down the field would not be detected by this cell.

(e) *Hypercomplex cortical cells* respond maximally to particular shapes (here shown as a narrow bar with a sharp end) moving across their receptive field. A horizontal bar would not be detected by this cell. If the stimulus is like that at (d) (and so extends over the shaded part of the receptive field), this cell does not respond to it.

Fig. 10.14. Receptive fields of cells in the optic nerve and visual cortex emphasising the hierarchy of organisation in the system. In each case the stimulus which gives excitation of the cell is shown by the stippled area. In a, b and c a stimulus in the + area gives excitation while one in the − area gives inhibition. Uniform illumination of the whole field neither excites nor inhibits these cells (i.e. the resting rate of firing is unchanged).

Figs 10.14b and c respond optimally to vertical stimuli. As the stimulus is rotated then these cells would respond more and more weakly until with bars at right angles to the optimum direction of detection the stimulus would have no effect.

Complex cells (Fig. 10.14d) only respond to stimuli of a particular orientation anywhere in their receptive field and are not sensitive to their position within that field. They respond both to a stationary dark–light boundary and to one moving across their field.

Hypercomplex cells (Fig. 10.14e) respond only to moving stimuli of a particular orientation, as do complex cells. They also, however, respond best to particular shapes; thus some cells respond to (and hence detect) narrow strips (Fig. 10.14e) which end within their receptive field, while others detect corners or angles.

From the above description, we can see that successive neurones in the visual system require progressively more specific visual stimuli for them to become excited; in fact, there is evidence for cortical neurones which are even more selective than the hypercomplex cells described here. It is likely that this type of processing allows us to recognise intricate features of the visual environment.

Columnar organisation of the visual cortex

Neurones within the visual cortex are largely elongate with their long axes perpendicular to the cortical surface (see pyramidal cell Fig. 9.2). Thus neurones make synaptic connections over large distances through the thickness of the cortex, while their lateral connections are relatively local. As a result of this, cells which are connected together and have similar functions are arranged in columns perpendicular to the cortical surface. Thus, if a micro-electrode is driven through the cortex perpendicular to its surface, most of the cortical neurones are found to respond to bars of light of a similar orientation; if the micro-electrode penetrates obliquely through the cortex, then cells with differing orientation preference are encountered (Fig. 10.15). These observations illustrate two important principles of cortical organisation. First, that the cortex is arranged in columns perpendicular to the outer surface and, second, that neurones with similar functions lie close together.

The light reflex

Light shone into the eye causes a constriction of the pupil and also of the other pupil. The pathway involved is from the retina via the superior colliculus, the III nerve nuclear relay, the ciliary ganglion and ciliary nerve.

External eye structures

The human eye is protected by the bony orbit and the eyelids. In addition to sweat and sebaceous glands associated with the eye-lashes, the eyelids have tarsal glands which secrete an oily fluid that covers the edges of the lids and stops tear fluid overflowing.

The tear fluid is secreted mainly by the lacrimal glands which lie in

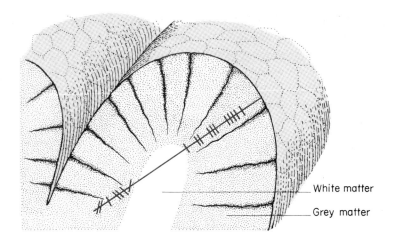

White matter

Grey matter

Fig. 10.15. Columnar organisation of the visual cortex. A recording micro-electrode driven through the visual cortex perpendicular to its surface will initially sample the activity of cortical cells which respond to a similar stimulus orientation (indicated by the short lines along the electrode track).

If the electrode travels obliquely through the cortex it will encounter cortical cells of differing stimulus orientation preference. This occurs because the electrode samples activity of neurones in different cortical columns (after Hubel).

the upper and outer part of the orbit. It moistens and lubricates the eyeball and eyelid and, as it contains a bactericidal enzyme called lysozyme, sterilises it to some extent. Blinking, which occurs about twenty times a minute and lasts some 300 milliseconds, probably helps to renew the fluid over the cornea and waft foreign bodies to the inner edge of the orbit. The tear fluid drains through the tear duct into the nose.

HEARING

This is the sense by which animals can perceive sound. All of the higher vertebrates, and many of the lower animals, have a specialised organ, the ear, by which this is done. In its simplest form hearing allows the different sexes to find each other but its more general function is in acting as a 'distance receptor' for warnings about impending danger. This primitive form of communication is highly developed in birds and mammals.

The structure and functions of the different parts of the ear

The ear is composed of outer, middle and inner parts (Fig. 10.16). The outer ear receives aerial waves of sound, the middle ear conducts them inwards to the inner ear which, by means of the receptor cells, changes them into nerve impulses sent through the auditory nerve to brain centres, finally leading to the perception of sound.

Fishes have a primitive inner ear which receives water-borne sound waves; amphibia, living in air, have in addition a middle ear to receive and conduct airborne sound waves; mammals have an outer ear which

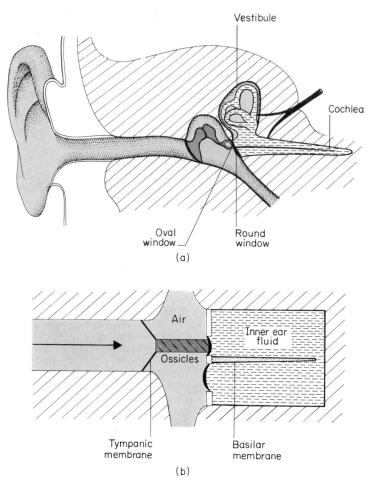

Fig. 10.16. Diagrams of the human ear showing: (a) the anatomical structure and (b) the schematic structure; both show the inner ear 'straightened out'.

The tympanum picks up the air waves of sound, transmits them through the chain of three small bones—the ossicles—to the oval window of the cochlea where they are analysed and changed into nerve impulses. Diagram (b) shows how the ossicles act as a piston against the fluid of the inner ear, amplifying the pressure (after Békésay).

probably both extends the range of frequencies and in some animals gives directional properties to their hearing.

The outer ear consists of the visible parts of the ear and a short tube—the external auditory meatus—leading to the middle ear. In man the outer ears serve no auditory purpose for they are too small to deflect the sounds we commonly use. To modify the pathways of waves the reflecting surface needs to be of similar or larger dimensions than the wavelength; as the wavelengths common in speech are about 30 cm (1 ft) our small ears can have little effect. This phenomenon can be seen at any harbour; a breakwater deflects the waves of the sea but a flagpole or small buoy does not. In animals with bigger ears who use higher frequencies, for example, dogs, the outer ear probably has a useful effect, particularly if they can be moved. The eardrum reproduces the vibration of the air molecules which hit it and stops vibrating almost immediately when the sound wave stops.

The middle ear is an irregular space beyond the eardrum which contains the three small bones—the ossicles—providing a mechanical linkage between the large, delicate eardrum and the smaller, stronger oval window. It is filled with air and connected to the throat by the eustachian tube. This allows pressure equalisation to occur when opened during yawning or swallowing.

The ossicles (malleus, incus and stapes) are suspended in air and linked together by ligaments in such a way that they transmit sound waves onwards from one to the next, but can slide sideways on each other so that they do not transmit vibrations arriving at right angles to them. This is thought to be important in minimising the vibrations from our own vocal cords acting on our inner ear.

The function of the middle ear is the efficient transmission of sound energy. It does this by having a large lightweight collector—the tympanic membrane or eardrum—connected by a mechanical lever to a smaller membrane in contact with fluid; there is, therefore, a diminution in movement but an increase in pressure between the eardrum and the oval window. The efficiency of transfer is about 30 percent, a figure much greater than the 0·1 percent which remains when the ossicles are removed. With no ossicles the whole head needs to be vibrated for hearing to occur. In engineering terms the middle ear is a matching device transferring energy from a region of low impedance to one of high impedance.

The inner ear occupies a complex and irregular space in the temporal bone together with the vestibule and the semicircular canals. The cochlea, which is responsible for hearing, is a tapered canal some 3·5 cm ($1\frac{1}{2}$in) long, twisted into a spiral of $2\frac{1}{2}$ turns. Its name comes from a snail's shell, which it resembles.

The cochlea is divided by a longitudinal partition into two canals; the upper one is the vestibular, which is connected to the vestibule (see organs of balance later), and the lower, the tympanic (Figs 10.16 and 10.17). These canals are filled with the perilymph, which is rather like cerebrospinal fluid, but may not be connected to it (Table 10.2). At the outer end of these canals both are connected to the middle ear, by the oval and round windows respectively, the former being covered by the

Table 10.2. Composition of perilymph and endolymph. The endolymph is rather like intracellular fluid but has a large +ve potential (+80 mV) with respect to the others, which have little potential with respect to extracellular fluid.

ECF		Perilymph	CSF	Endolymph
Na$^+$		150	152	16
K$^+$	mmol/l	5	4	145
Cl$^-$		121	122	107
Protein (mg/100 ml)		50	21	15

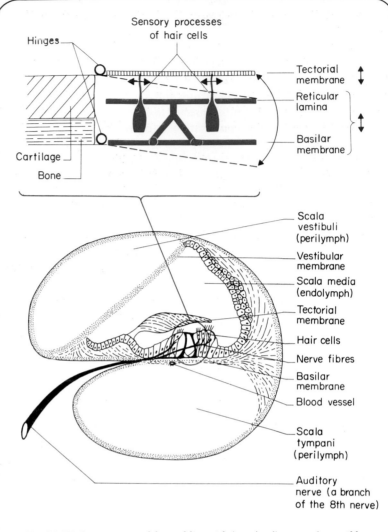

Fig. 10.17. Cross-section of the cochlea, with (inset) a diagram of a possible way in which the hair cells of the organ of Corti are stimulated by movement of the basement membrane. In this scheme it is supposed that the basement membrane, reticular lamina and hair cells move together and the tectorial membrane moves separately. The processes of the hair cells receive a shearing action from this movement which stimulates these cells.

foot of the stapes. At the narrow inner end the canals are connected together by a small opening, the helicotrema, which keeps the 'resting' pressures in both canals the same.

The partition between these canals (Fig. 10.17) is composed of the fibrous basement membrane on which lies the organ of Corti, containing the hair cells which are the auditory receptors. Nerve fibres start around these hair cells and run axially towards the centre pillar of the cochlea to the ganglion of the auditory nerve. This sensory part of the cochlea is bathed in a separate fluid, the endolymph (Table 10.2), which is kept from mixing with the perilymph by the thin vestibular or Reissner's membrane.

When a sound is applied to the ear the stapes moves, the pressure in the vestibular canal changes, the basement membrane and associated structures are deflected, the pressure in the tympanic canal changes and finally the round window moves to equalise the pressure between the cochlea and the middle ear. The movement of the basement membrane and organ of Corti is complex, the current view is shown in the inset to Fig. 10.17. One analogy is to imagine two parallel doors hinged at the same ends and then moved together. The resulting shearing action between them distorts the hairs of the receptor cells and so stimulates them. The action potentials then set up in this way pass via the spiral ganglion of the cochlear nerve to the brain to give the sensation of hearing.

Electrical recording from regions around the organ of Corti picks up a potential—the cochlear microphonic—whose amplitude varies with movement of the basement membrane. This potential very faithfully follows the frequency and amplitude of applied sound and indeed can be used to drive an audio amplifier. In addition to this varying voltage of some 5 mV maximum value, the cochlear duct has a steady resting potential of about + 80 mV with respect to the other parts of the cochlea. These potentials are not understood but at present it is thought that the cochlear microphonic potential may be a generator potential (see p. 244) which produces the action potentials in the cochlear nerve, although under some conditions the two can become dissociated.

Sound and its analysis

Sound is the sensation produced when rapid vibrations of air strike the eardrum. If these variations have a uniformity of pattern then they are called periodic and are heard as tones; if there is no uniformity they are called aperiodic and are heard as a noise. The periodic sounds may vary in four different ways:

1 Frequency is the rapidity of recurrence of the pattern; it is measured in Hertz (cycles per second).

2 The amplitude or intensity is the energy of the sound. It may be measured in absolute terms of pressure at the eardrum as dynes/cm², but is more usefully measured on a relative scale. The scale used is based on the ratio of the test sound to a standard sound. The standard sound chosen is the minimum which can be heard, under optimum conditions, for a note of 1000 Hz by a child. The log of this ratio is taken because hearing, like other senses, is proportional to the log of the intensity. The unit so obtained is called a bel; in practice decibels (= 1/10 bel)

$$\text{Decibel (dB)} = 10 \log \frac{\text{test sound}}{\text{standard sound}},$$

are used as being of a more convenient size. In most practical systems the instrument used to measure such sounds is constructed to have a frequency response as similar to the ear as possible (see below). The loudness of some typical sounds is shown in Fig. 10.18a.

(a)

Decibels	Relative amplitude	Typical sound
0	1	Threshold of hearing 0·0002 dyne/cm² Whisper at 1 metre
40	10,000	Normal conversation at 4 metres
80	100,000,000	Very heavy street traffic Range of sound at a disco
120	1×10^{12}	Pneumatic drill Ear pain
160	1×10^{16}	Near a jet plane

(b)

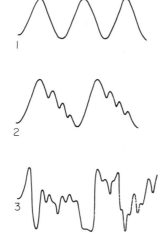

Fig. 10.18. (a) Typical sounds on a decibel and relative scale. (b) 1, Sine waves; 2, more complex compound wave; 3, noise.

3   The wave form is the nature of the pattern. The simplest sinusoidal wave (Fig. 10.18b) is difficult to attain but may be produced by striking a tuning fork very gently. Usually compound waves are produced and heard; the simplest of these is when a sinusoidal wave has overtones added to it. All compound waves may be regarded as the sum of a series of simple waves (Fourier's law) and may be so analysed.

4   Phase refers to the timing of corresponding points in waves when two or more waves occur together.

Phenomena of hearing

Pitch. The pitch of a tone depends on the frequency of the applied sound. Low frequencies give a deep tone, high frequencies a shrill tone. In children the normal range of frequencies is between 15 and 20,000 Hz. With increasing age the upper frequency decreases until at 50 years the upper limit is about 10,000 Hz. Within the middle range of frequencies

(1000–3000 Hz) we can commonly discriminate notes whose frequency differs by only 2–6 Hz.

Loudness. The loudness of a sound depends on its energy level and on its frequency. If, for example, a note of 100 and 1000 Hz of equal sound intensity are listened to, then the 1000-Hz note appears louder. This is because the sensitivity of the ear varies with frequency, being highest at about 3000 Hz and falling off to either side (Fig. 10.19). Figs 10.18a

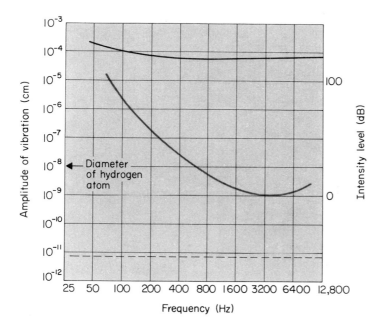

Fig. 10.19. Sensitivity of the ear. The red curve indicates the amplitude of vibration of the tympanic membrane at the threshold of hearing. The lower dashed line shows the corresponding amplitudes of vibration of the basilar membrane at the threshold of hearing; the upper black line shows it at the pain threshold.

The audibility curve is largely determined by the energy losses in the middle and inner ears; the low frequencies are decreased by the elasticity of the system and the high frequencies by the mass (after Von Békésy).

and 10.19 also show that the ear operates over an extremely wide range of sound intensity, some 1×10^{12} times. At the threshold of audibility for the middle frequencies the effective pressures and movements are exceedingly minute—displacements of the eardrum of the dimensions of a hydrogen atom are effective (10^{-8} cm). We can almost hear the random movements of the atoms. This extreme sensitivity is due partly to the amplification given by the arrangement of the middle ear and partly by that of the organ of Corti, in which pressure on the basilar membrane is transmitted into a shearing force on the sensory cells.

'It is interesting that the ear is just insensitive enough to low frequencies to avoid the disturbing effect of the noises produced by muscles, bodily movements etc.' (Von Békésy). With greater sensitivity at low frequencies we might hear our heads vibrating as we walked.

Timbre. The human eye is unable to distinguish between identical colours produced by different mixtures of the primary colours. The ear, on the other hand, can dissect a sound into its component parts such as its fundamental, its overtones, the amplitude of these, phase changes, general noises due, for example, to different musical instruments. This means that two sounds may have a similar frequency structure but differ in other respects and sound different. These sounds are said to have a different timbre. Herein lies part of the value of hearing.

Distortion. Because the ear does not behave linearly to sounds of different frequency it is said to show frequency distortion. The ear also shows non-linear distortion, in that for pure tones above a certain low amplitude it introduces harmonics, and also produces combination notes as the 'beating' of two simple notes.

Speaking and listening. We can speak without deafening ourselves, so that we can listen while speaking. Part of the explanation for this is in the layout of the vocal cords and middle ear. The vocal cords are symmetrical, which means that they transmit a minimum of vibration to the head; they also vibrate in an up-and-down direction, a direction to which the ossicles are not very sensitive. So the ossicles have the dual function of increasing air conduction from the tympanic membrane but decreasing bone conduction from our larynx. Lower animals have a variety of mechanisms for achieving the same end, for example, the frog has a large eustachian tube which opens when it opens its mouth wide preparatory to vibrating its tongue to croak; its own 'croak' therefore hits both sides of the eardrum simultaneously, thus cancelling out.

Sound localisation

It is common experience that we can localise the origin of sound in our environment. The auditory cues for sound localisation depend on having two ears, and vary depending on the frequency and nature of the sound. A note of 1000 cycles has a wavelength of about 30 cm, a length somewhat greater than that between the ears of most people. So, for simple tones of frequencies lower than this, different parts of the same wave strike the two ears simultaneously therefore there is a clear phase difference between the note striking one ear and the other. For high notes this is confusing as the wavelength is much shorter than the size of the head. For simple high-frequency notes the head is an obstacle to their passage, i.e. casts a shadow—so that the intensity is greater in one ear than the other (at 15,000 Hz the difference is 20 dB for a sound facing the ear). For noises, intensity, time and timbre all play a part (the latter because of differential effects on the component parts), so that localisation is relatively easy (Fig. 10.20). Sounds originating on an extension of the midline of our head cannot easily be localised without exploratory movements of the head.

Sound localisation in normal environments is more difficult than in the laboratory because sound waves often travel along complicated paths to the ears. In normal life localisation often depends on cues which are visual as well as aural and on the knowledge of the environment; by inspection we see what is likely to have produced the sound.

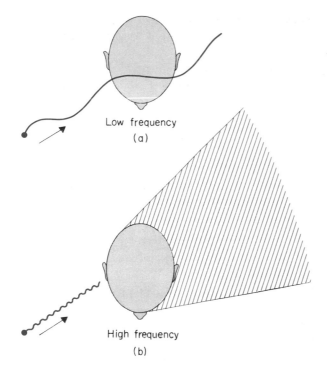

Fig. 10.20. Sound localisation. At low frequencies (a) there is a phase difference between the waves reaching each ear with little deflection of the sound; at high frequencies (b) the sound is deflected by the head so casting a 'shadow' and leading to differences in intensity between the ears.

Theories of hearing

There has been a good deal of controversy about the way in which the ear distinguishes between sounds of different frequency. The current view is that frequencies are distinguished by virtue of the place on the basement membrane which vibrates at that frequency and enhanced by brain mechanisms which 'sharpen up' this signal. At very low frequencies (below 100 Hz) it is possible that the rate of firing of auditory nerves is in time with the sound applied. As notes become louder this is signalled, as in other sensory systems, by an increase in frequency of action potentials in the auditory nerve and by recruiting more nerves.

Frequency discrimination in the cochlea

Pure tones applied to the ear lead to the formation of a series of waves in the vestibular canal of the cochlea. These waves start at the oval window and as they travel down the canal their height increases to a maximum and then drops off rapidly towards the end.

At the point of maximum deflection there is also a maximum change in phase of the wave. These two effects—maximum deflection and phase change—occur at a characteristic place on the basement membrane for each applied frequency; high frequencies near the middle ear and low frequencies away from the middle ear (Fig. 10.21). This deflection of the basement membrane then sets up action potentials in the auditory nerve, which are 'recognised' as being due to these applied frequencies.

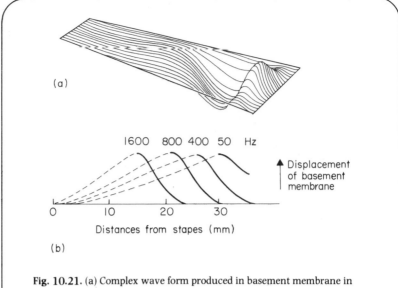

(a)

1600   800 400 50   Hz

Displacement
of basement
membrane

0   10   20   30

Distances from stapes (mm)

(b)

Fig. 10.21. (a) Complex wave form produced in basement membrane in response to sound waves reaching the ear. (b) The deflections produced in the basement membrane of the cochlea by pure notes of various frequencies applied to the ear. Low notes cause deflection of most of the membrane whereas high notes do not; this is because of the asymmetry of the wave form (after Ganong).

The physics of these wave motions are complex and not completely understood. A rough approximation is to consider a long whip whose handle is moved up and down at various frequencies: at very low frequencies the handle and the whole of the whip moves up and down in synchrony and so the greatest movement is at the tip; at a higher frequency the handle and the nearest part of the whip move in synchrony and the tip waggles in a complex way. Eventually at very high frequencies the 'break point' for synchronous movement is very near the handle and the rest of the whip waggles in a complex way. The 'break point' corresponds to the characteristic place for each frequency.

Action potentials in the auditory nerves

These cochlear phenomena give rise to complex patterns of action potentials in the auditory nerves. They can best be understood by considering the response of a single auditory axon arising from the part of the basement membrane corresponding to a note of say 6000 Hz. Imagine that notes of various frequency are now applied to the ear. If the intensity of these notes is very low then only the hair cells corresponding to the peak position of the cochlear waves will be stimulated; so that stimulation at 6000 Hz only will produce action potentials in the nerve (Fig. 10.22). As the intensity of the applied note increases the cochlear waves produced by lower tones will also stimulate the 6000-Hz hair cells, so that action potentials will also be produced by these. In this way increasing intensity of sound will produce stimulation at lower and lower applied frequencies.

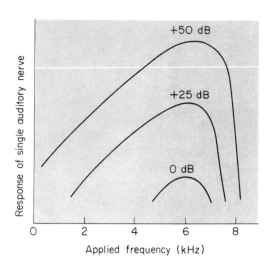

Fig. 10.22. Diagram of response of a single auditory nerve to three intensities of applied tones. The response is measured as impulses in a single axon arising from a hair cell at the 6-kHz region of the basement membrane. At low intensity (O dB) only frequencies around the 'natural' frequency of the fibre are stimulated; at higher intensities stimulation also occurs at lower applied frequencies. Note the asymmetry of the response to lower and higher frequencies.

Central auditory pathways and brain mechanisms

Axons pass from the cochlear nucleus via a variety of pathways to the interior colliculus, where auditory reflexes arise, and are then relayed via the medial geniculate body in the thalamus to the auditory areas of the cortex (in the temporal lobe). At these higher areas the neurones generally receive inputs from both sides. There are widespread association areas adjacent to the primary auditory area.

Until recently it was thought that the basement membrane showed a 'blunt' frequency response to an applied tone; it is now thought that this was an artefact produced during experiments and that the primary cochlear response to applied tones is very sharply localised. There is evidence that this process continues as one ascends the auditory pathway to the cortex, so that good discrimination is obtained eventually. The most likely explanation for this, latter, phenomenon is the surround inhibition discussed on p. 248.

The peripheral speech apparatus

Speech consists of two parameters (Fig. 10.23), pitch and articulation. The pitch (i.e. frequency) is produced by the vocal cords which are ligaments situated in the larynx in the throat; these vibrate in the airstream coming up from the lungs. The pitch is normally higher in women than men and can be varied consciously. The whole mechanism is called phonation.

Articulation is the mechanism by which various sound qualities are

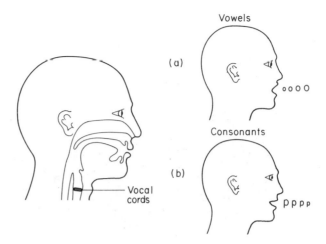

Fig. 10.23. The vocal cords vibrate at different rates to give the tones. (a) The tongue divides up the open mouth into two resonating cavities to produce the various vowels, (b) Consonants are more complex to produce; some are produced in the nose (n, m, ng), some with (ex)*plosive* bursts of air (p, b, t, d, k, and g) and some with turbulent air flow (v, f, s, z the *fricatives*). (After the *Observer*, 1982.)

imposed on the basic pitch, using the resonances in the mouth and other hollow spaces in the head. In speaking, sequences of characteristic sounds are produced; called phonemes. In whispering the vocal cords do not oscillate, only articulation occurs.

In speech the glottis is closed, the chest contracted until a pressure of 4–6 cm H_2O (40–60 Pa) is built up, and then the air is released in puffs. These puffs cause the vocal cords to oscillate; this oscillation is then modulated by the air spaces in the mouth and sinuses. The main frequencies used in speech correspond to the range of maximum sensitivity of the ear, i.e. 600–2500 Hz (Fig. 10.9).

THE ORGANS OF BALANCE

The vestibular apparatus is a complex structure consisting of three semicircular canals and two spaces which lie in the bone beside the inner ear near to the cochlea (Fig. 10.24). They give the body information about (1) the direction of gravity, and (2) the movement of the head in space.

The basic sensory receptors consist of a cell with a hair process which protrudes into either gelatinous material (the cupula in the semicircular canals) or into a dense calcium complex (the macula or otolith organ in the saccule and utricle). If the hair is bent in one direction the rate of firing of the receptor cell is increased, if in the other direction it is decreased.

Gravity detection
The otolith organs of the utricle and saccule are mainly responsible for signalling the position of the head in the gravitational field. With the head held normally there is little activity from the receptors in the utricle

(a)

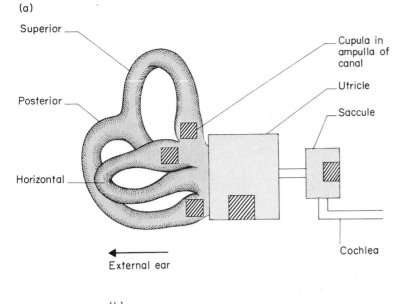

Superior

Posterior

Horizontal

Cupula in
ampulla of
canal

Utricle

Saccule

Cochlea

External ear

(b)

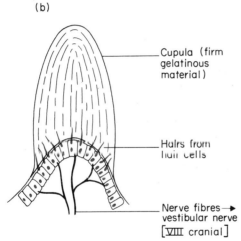

Cupula (firm
gelatinous
material)

Hairs from
hair cells

Nerve fibres →
vestibular nerve
[VIII cranial]

(c)

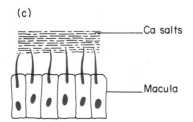

Ca salts

Macula

Fig. 10.24. (a) Diagram of the membranous labyrinth. The entire space is closed and filled
with endolymph. The cross-hatching show the basic forms of sense organ in
(b) semicircular canals, and in (c) utricle and saccule.

and that from the receptors in the saccules on each side of the head are equal. If the head is tilted sideways one saccule hangs down and gives an increased output, the other points upwards and gives a lesser output. If the head is bent backwards and forwards the utricles start to signal as they are now both hanging down (Fig. 10.25).

Other gravity information probably comes from joint and pressure receptors in the rest of the body, for example soles of the feet. Visual cues also play a large part. A room needs to be tilted quite far from the horizontal before we are aware that the walls are not vertical, for we prefer to pay attention to visual rather than other cues.

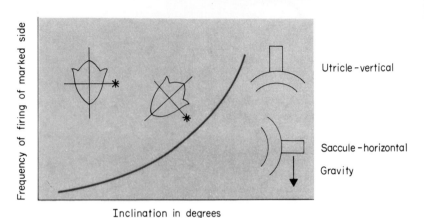

Fig. 10.25. Diagram of the orientation in space of the saccule and utricle and of the effect of tilt on the output of a saccule receptor.

Detection of movement in space
Linear movement in space is detected by the otolith organs of the saccule and utricle, for their high calcium content makes them denser than the endolymph bathing them and therefore they respond more to acceleration than the fluid around them.

Rotational movement is detected by the receptors of the semicircular canals. The plane and direction of the rotation determines which of the canals is most affected and in which direction. Once acceleration has stopped the cupula, and so the impulse traffic, returns to normal with a $T_{\frac{1}{2}}$ of about 10 seconds (Fig. 10.26).

Space travel
The effective absence of gravity in space travel leads to various vestibular effects as well as other physiological consequences. It has been suggested that these vestibular effects might be overcome by the creation of an artificial gravity; created, for example, by having a dumb-bell-shaped spacecraft rotating continuously. This kind of scheme would not work, for the following reason. Normal gravitational waves are effectively parallel because we are small compared to the scale of the earth; in such a spacecraft, however, the centrifugal force produced would not be nearly parallel on our scale because of the much smaller craft. This leads

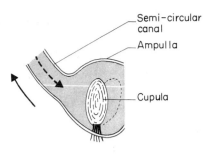

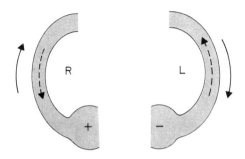

Fig. 10.26. Action of the horizontal semicircular canals to rotation—inset shows that the cupula, situated in the ampulla or swelling of the canal just before it enters the utricle, nearly fills the whole canal.

When the canal accelerates in one direction the fluid lags behind and so causes movement of the cupula in the opposite direction; with constant acceleration the fluid (and so the cupula) returns to its normal position; deceleration produces the opposite effects. Bending the cupula in one direction (towards the utricle) usually causes stimulation, in the other direction, inhibition.

to complex signals from the vestibular apparatus resulting in great feelings of nausea.

SMELL AND TASTE

Smell and taste are chemical senses, in that they enable us and other animals to distinguish between molecules of different kinds in our environment. Their importance varies with the species; it is probably true that they are less important in us than in lower animals.

Smell
Smell is a distance sense, distinguishing desirable and enjoyable substances from those which are unattractive and potentially dangerous. The distance over which it operates varies with the substance and the conditions.

Receptors and their stimulation
The olfactory receptors are situated in the upper part of the nose; their axons converge on second neurones in the olfactory bulb, whose axons then cross the midline to the other bulb and also go to a diffuse olfactory

region of the brain via the first pair of cranial nerves (Fig. 10.27). The function of these very complex anatomical connections, which have been discovered recently, is not clear. The receptors are covered by a layer which is secreted by interspersed supporting cells.

These receptors respond only to substances which are dissolved in the mucus covering them and thus able to come in contact with their membranes. Before a substance can be smelt it must be (a) volatile, i.e. in the air, and (b) able to dissolve in water in the mucus and then (c)

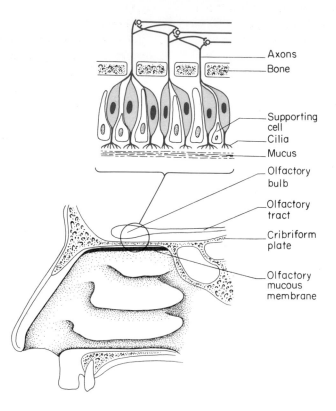

Fig. 10.27. Diagram of olfactory area with inset showing receptor cells and arrangement of nerve fibres (in red). The olfactory area is about 5 cm² (1 in²) in man and contains 10–20,000 axons. The air breathed in bypasses the olfactory area apart from eddies or unless we 'take a sniff'.

There is considerable convergence of receptor cell axons on to groups of cells in the olfactory bulb (the olfactory glomeruli), with fibres crossing the midline from one bulb to the other. The arrangement of connections is much more complex than shown.

possibly be fat soluble to react with the receptor membrane. Probably odoriferous molecules react with receptors as a substrate–receptor complex; this then sets up a generator potential and so gives rise to action potentials in the nerves to the olfactory bulb. The mechanism of this reaction is not known. It is known that most odoriferous substances have from 3 to 20 carbon atoms in them.

The sensitivity of smell is remarkable in that concentrations of some 1×10^{-12} molar in air may be detected; though it must be remembered that this still represents a molecular concentration of 1×10^{11} molecules per litre of air. Changes in concentration need to be quite large (30 percent) before a difference is detected. We can discriminate between some 3000 different odours including those made artificially and so never met before. The mechanism for this is unclear. Our language contains few words distinguishing between olfactory qualities; instead we use the name of the object to describe the smell as well as the object (clove, rose, onion, rubber, etc). Attempts have been made to classify smells as floral, peppermint, pungent, etc. with molecular 'shapes' for each, but this has not proved very useful.

It has been suggested that a common chemical sense exists as well as the specific sense of smell, for example, the response to onions and formaldehyde leads to nasal, respiratory and eye responses as well as those of smell. The response is mediated by the naked endings of trigeminal pain fibres in the olfactory mucosa.

We all know that continuous exposure to agreeable or disagreeable odours leads to a decrease and eventually unawareness of their presence. This response is partly due to adaptation in the olfactory receptor and partly to central changes; it is generally specific for that odour.

Taste

Our sense of taste is not as distinct from other senses, particularly smell, as might commonly be thought, for example, the 'taste' of our food is compounded of the sensation of taste, of heat, cold, texture and especially of smell. This latter can be shown by eating with the nose tightly held or while suffering from a 'cold'. Unlike smell, there are words for the qualities of taste which are separate from the objects tasted. In fact it is only deemed necessary to have four elementary taste qualities: sweet, salty, sour and bitter. Most tastes are combinations of these. Taste qualities interact, as when vinegar is added to pickles, or sugar reduces the bitterness of coffee; this appears to occur centrally, rather than at the receptors.

Receptors and their stimulation

The taste receptors are collected in groups of about 10 in the taste buds, situated on the edges and upper surfaces of the tongue and walls of the posterior pharynx. These occur singly or in clusters, being particularly associated with the fungiform and vallate papillae (Fig. 10.28). Centrally the sensation of taste has no separate representation but is represented on the facial sensory area.

Substances in solution enter the taste bud through its pore and are presumed to combine with receptors on the taste cells (this combination must be weak for it is readily reversible by dilution with water). The generator potentials set up seem to respond to all the four modalities of taste mentioned above though maximally to one.

Modalities of taste

Taste buds which are maximally sensitive to sweet, sour, bitter or salt substances tend to be distributed over the tongue as shown in Fig. 10.29. In eating we make use of this, for example small boys roll sweets across the tip of their tongues, men gulp beer to the back of their mouths.

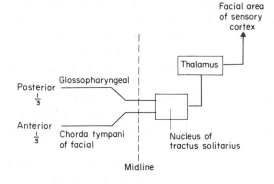

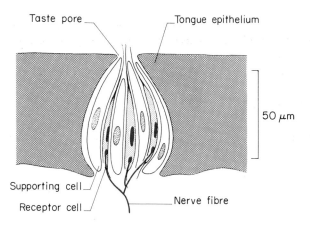

Fig. 10.28. Diagram of taste bud with inset showing the innervation of the tongue (in red). The receptor cells are formed at the edge of the taste bud, migrate inwards and die in about 4 days; these are not true nerve cells but are innervated by nerve fibres. These nerve fibres innervate about five receptor cells with multiple endings. There are about 10,000 taste buds in man.

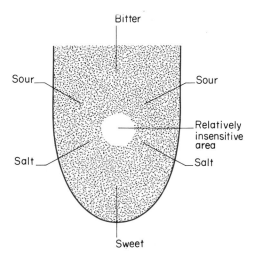

Fig. 10.29. Distribution of the modalities of taste on the tongue.

The thresholds for stimulation vary with the modality tasted; it also decreases with increase of the area of tongue involved and is greatest at 30–40°C. There are considerable individual differences in threshold. The ability to taste phenylthiocarbamide (PTC) is inherited as an autosomal recessive trait. Thirty percent of Caucasians cannot taste it.

Sour tastes are due to the hydrogen ion concentration of acids. The reference is HCl, whose threshold is about 0·1 mmol/l.

Salty tastes are due to the anions of inorganic salts, particularly the halogens (Cl, Br). The reference is NaCl which has a threshold of about 20 mmol/l.

Bitter tastes are often organic compounds, for example morphine, caffeine. Quinine sulphate is the reference, with a threshold of about 8 μmol/l.

Sweet substances are also usually organic, for example sucrose, maltose, lactose, glucose though lead and beryllium and some proteins also taste sweet. Sucrose has a threshold of about 10 mmol/l but saccharin has one of 23 μmol/l so we can 'sweeten' without adding calories to the food.

Adaptation to a given taste gives a decrease in the sensitivity to that taste and often a change in sensitivity to other substances. Acids cause a general depression; adaptation to salt leads to increased sensitivity to sour, sweet and bitter substances.

Chapter 11
Endocrinology and
Metabolism

It is convenient to deal with the endocrine system separately from the nervous system. There is, however, no sharp distinction between them: both are control systems which involve interactions between cells; both use chemical substances as mediators between cells; both cause actions at a distance. On the whole, the endocrine system deals with events more slowly than the nervous system and has effects which are more generalised.

If hormones were 'visible' then the concentration of them in the body would be seen to vary from moment to moment throughout life. Each cell is thus bathed with very many local and general hormones all the time; the resulting action depends on the interaction of many of them at any one moment (Table 11.1). This complexity makes for difficulty in description because some simplifying scheme must be adopted. The one chosen here is to describe the hormones separately but to devote sections to their integrative action. The first section applies to most of the hormones, and ought to be referred to from time to time.

GENERAL FEATURES OF HORMONES

Hormones are carriers of information to cells rather than substrates for cell enzymes to work on; this explains their low concentration in the blood. Hormones are usually part of a regulatory system (see p. 10); as such two categories can be recognised:

1 Where the hormone is used to regulate a later step in the sequence, its concentration may vary widely. The control of plasma glucose concentration by insulin (Fig. 11.34) is a good example; others are the catecholamines, aldosterone and ADH etc.

2 Where the hormone concentration itself is controlled then its concentration is very constant. Thyroxine is the best example of this (Fig. 11.17). In these cases a constant level of the hormone is required for the orderly performance of various functions. In some cases the rates

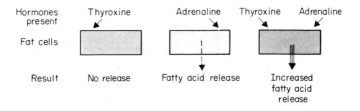

Fig. 11.1. Specific action of one hormone depends on the 'permissive' presence of another.

Table 11.1. The principal endocrine glands of the body and their most important secretions (from Passmore & Robson).

Glands	Secretions	Chemical nature
Anterior pituitary or adenohypophysis	Growth hormone (GH)	Protein, mol. wt 21,500
	Adrenocorticotrophic hormone (ACTH)	Polypeptide, mol. wt 4500
	Thyroid stimulating hormone (TSH)	Glycoprotein, mol wt 33,000
	Luteinising hormone (LH) ⎫ gonatrophic	Glycoprotein, mol. wt 33,000
	Follicle stimulating hormone (FSH) ⎭	Glycoprotein
	Prolactin	Protein, mol. wt 200,000 (sheep)
Posterior pituitary or neurohypophysis	Vasopressin or antidiuretic hormone (ADH)	Polypeptide of 9 amino acids
	Oxytocin	Polypeptide of 9 amino acids
Thyroid	Thyroxine (T4)	Iodinated amino acid
	Triiodothyronine (T3)	Iodinated amino acid
	Calcitonin	Polypeptide, mol. wt 3600
Parathyroids	Parathyroid hormone	Polypeptide, mol. wt 8500
Adrenals		
Cortex	Aldosterone	Steroid
	Cortisol	Steroid
	Corticosterone and many other steroids, including androgens and oestrogens in small amounts	Steroid
Medulla	Adrenaline (USA epinephrine)	Catecholamine
	Noradrenaline (USA norepinephrine)	Catecholamine
Kidneys	Renin	Protein, mol. wt 65,000
	Erythrogenin	Glycoprotein
Testes	Testosterone and other androgens	Steroid
Ovaries	Oestradiol and other oestrogens	Steroid
	Progesterone	Steroid
Placenta	Chorionic gonatrophin (human CG)	Glycoprotein, mol. wt 30,000
	Oestrogens	Steroid
	Progesterone	Steroid
Pancreas	Insulin	Polypeptide, mol. wt 5800
	Glucagon	Polypeptide, a single chain, mol. wt 3500
Gastrointestinal tract		
Stomach	Gastrin	Polypeptides, mol. wts 2000–7000
Duodenum	Secretin	Polypeptide, mol. wt about 2700
Small intestine	Cholecystokinin–pancreozymin (CPZ)	Polypeptide, mol. wt about 2700

of these functions are proportional to the concentration of the hormone: in others the hormone just needs to be there; this is called the permissive action of the hormone.

The actions of hormones are multiple and complex, to help you get a general picture of these it is useful to consider their functions from two aspects. Firstly in terms of their actions on the body. These are:

1 To enable and promote the development of physical, sexual and mental characteristics;

2 To enable and promote the adjustment of performance levels, e.g. some physiological adaptations are lost in the absence of certain hormones;

3 To keep certain physiological parameters constant, i.e. a homeostatic function.

Secondly hormone function can be considered from an 'anatomical' way:

(a) Hormones that act directly on the target organ (the 'effector' hormones), examples are the sex hormones.

(b) Hormones that control the synthesis and release of effector hormones (the 'tropic' hormones), an example is thyrotropic hormone. A special subgroup in this category are those hormones from the hypothalamus which control the production and release of the hormones of the anterior pituitary (the 'releasing' or 'release-inhibiting' hormones). This group is the main link between the CNS and the endocrine system.

Hormones are usually secreted directly into the circulation (via the extracellular space) in extremely low concentrations and are recognised by specific cells, which then respond in characteristic fashion. Some hormones are secreted as precursors and then formed in other organs or in the circulation. Typical hormone concentrations in the blood range from 10^{-6} to 10^{-12} mol/l compared to (about) 10^{-1} mol/l for Na to 10^{-3} mol/l for amino acids. Measurement of hormone concentrations is, therefore, difficult and has only recently been achieved for most of them. Although these hormone concentrations are low, 10^{-12} mol/l still means that 1 ml of plasma contains a hundred million molecules of a hormone.

Some hormones are secreted so near their target organs that they can reach them by diffusion without using the circulation; the cells producing these are known as the *paracrine* cells in contrast to the *endocrine* cells, which produce the blood-borne hormones. It has recently become clear that the same hormone is used in different parts of the body as local transmitters, without that at one site interfering with another site. This particularly applies to the newly discovered peptides. A simple analogy is to think of a person talking to his friends in one part of a room, while colleagues hold a conversation (in the same language) in another part; the result depends on the volume of the sound.

Blood transport

Many hormones circulate in plasma largely bound to plasma proteins; free hormone is often quite small compared to the total in plasma and is in equilibrium with the bound fraction:

$$\text{free hormone} + \text{protein} \rightleftharpoons \text{hormone–protein complex.}$$

Only the free hormone can react with the target cells.

Hormones are carried in the plasma in this way for a variety of reasons: (1) it prevents small molecules from being excreted by the kidneys; (2) it enables molecules which are relatively insoluble in water to be carried; (3) it enables a 'store' of hormone to exist in the blood which buffers the supply of hormone to the tissues.

The plasma concentration of any hormone will depend upon its rate of secretion into and its rate of removal out of the plasma (Fig. 11.2). Normally, for any hormone, some secretion is always occurring and inactivation by the liver or kidneys is always taking place. The plasma concentration may therefore be altered by changing either or both of

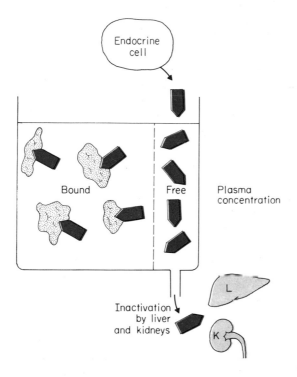

Fig. 11.2. The plasma concentration of a hormone is the resultant between the rate of production and the rate of inactivation. The hormone may be free in the plasma or bound to binding proteins; if bound the total amount in the blood is often largely determined by the amount of binding protein present. The amount excreted is often used as an indication of blood concentration.
1   Several hormones—for example, GH, steroids—are secreted in bursts rather than continuously as indicated; the reasons for, or advantages of, this are unclear at present.
2   A very small amount of hormone is inactivated by specific receptors in the tissues.

these rates. Usually the rate of inactivation is not changed (except that higher plasma concentrations lead to greater amounts inactivated); control is therefore exercised by changing the rate of secretion of the hormone. Glands usually have only a small amount of preformed hormone in them and so if the demand for secretion increases, although some may be discharged immediately, most must be made from precursors.

Hormone action on target cells

The primary event is a specific interaction between the hormone and a 'receptor' in the responsive cell; the greater the hormone concentration the greater the number of receptors occupied (Fig. 11.3). Cells which do not recognise certain hormones contain no 'receptors' for these hormones. The present view is that there are two main ways in which cells recognise hormones.

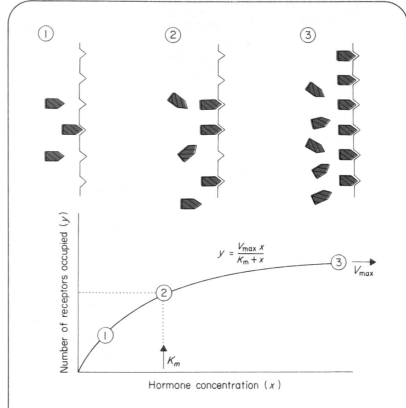

Fig. 11.3. Binding of a hormone to receptors, showing Michaelis–Menten kinetics. At low hormone concentration only a few sites are occupied; at high concentration virtually all are occupied. The characteristics of the system are defined by V_{max} (3), the maximum number bound, and K_m, the concentration of hormone at which half the sites are occupied at any one time (2). The higher the affinity of binding the greater the molecules bound for any substrate concentration (i.e. K_m decreases).

At the cell surface. Certain hormones react with specific receptors at the cell membrane and are then thought to cause the release of a second messenger within the cell—cyclic AMP (cAMP), cyclic GMP (cGMP) or Ca^{2+}—which then activates a specific kinase in the cell; this, in turn, causes the physiological changes specific to that cell. Hormones acting in this way are the catecholamines and peptide hormones. It may be significant that these hormones are large, are not fat-soluble and would not otherwise gain easy access to the cell interior (Figs 11.4 and 11.5).

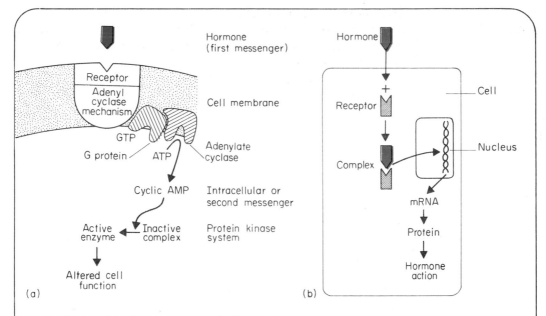

(a)

(b)

Fig. 11.4. Models of hormone action. The flow of information in the activation of (a) adenylate cyclase by the binding of a hormone to its specific receptor; (b) a cytoplasmic receptor by the intracellular binding of a hormone.

The G protein plays a critical role in both activation and deactivation of adenylate cyclase; the subsequent steps shown greatly amplifies the hormone signal. The 'cytoplasmic' receptors may actually be in the nucleus.

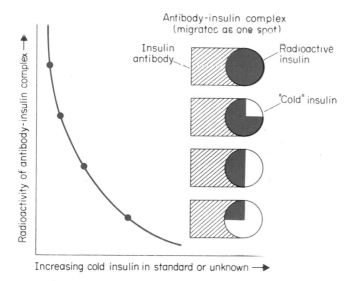

Fig. 11.5. Principle of saturation analysis for hormones. This involves the displacement of labelled (hot) hormone from a specific binding protein by unlabelled (cold) hormone. As increasing amounts of unlabelled hormone are added then proportionate displacement of labelled hormone occurs and so the radioactivity decreases. The system is calibrated with known amounts of pure unlabelled hormone. The figure shows a specific binding protein produced as an antibody to a peptide hormone ('immunoassay'). For many hormones the specific binding protein may be a transport or receptor protein ('protein-binding assay'). The system illustrated uses the binding protein bound to a surface, more commonly it is free in suspension and is then isolated chemically (after Tepperman).

This kind of receptor is sometimes known as a fixed receptor. Several different hormones act through the adenyl cyclase system and yet their actions are separate. How is this done? Each receptor only recognises its 'own' hormone and so is only triggered by it; once cAMP is formed it probably activates a specific kinase which then has various cellular functions. In the cell the adenyl cyclase mechanism links receptors at the cell surface with various parts of the biochemical machinery within the cell; just as in the whole body the action potential mechanism links different skin receptors to various parts of the cns.

Within the cell. Steroid and thyroid hormones are small and fat-soluble and probably enter most cells by diffusion. In responsive cells they react with a specific receptor protein in the cytoplasm. This complex then enters the nucleus and initiates a series of events which leads to an increase in the rate of transcription from specific genes. There is then an increase in the steady state concentration of mRNA made from these transcripts. The proteins made from these mRNAs give the physiological response. This kind of receptor is sometimes known as a mobile receptor.

Mechanism of hormone action

Hormones regulate existing functions in cells rather than initiate new functions; their action is thus on the rates of existing processes.

1 *Changing enzyme activity.* Many processes in the body are reversible because the forward and reverse reactions occur through different pathways:

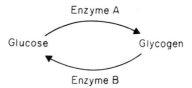

thus glucose is stored as glycogen through the action of synthetase (enzyme A) and produced from glycogen via phosphorylase (enzyme B). Hormonal control is exercised by altering the activity of these enzymes in the pathways; insulin acts on enzyme A and glucagon on enzyme B and so, by their relative concentrations, determine the overall direction of the process.

2 The activity of enzymes may be augmented by increasing the concentration of the active form of the enzyme; this is done by making more of the enzyme or by changing the shape of the enzyme from an inactive (or less active) form to a more active form (as in 1 above). The general rule is that hormones which produce 'intracellular messengers' produce structural changes in pre-existing enzymes while hormones which combine with the nucleus produce more of the enzyme.

3 *Changing membrane transport.* Many hormones act by increasing or decreasing the transport of a substance into target cells. Examples are: the rate of glucose entry into most cells is increased by insulin and decreased by other hormones; water and ion movements in the kidney

are influenced by various hormones. Such changes in membrane transport or movement then cause secondary effects in the cells. Thus an increase in glucose into cells leads to a higher intracellular concentration of glucose and so (by mass action) causes a greater formation of glycogen. As knowledge about the detailed working of hormones accumulates these two classes may merge, e.g. increased transport of sodium following aldosterone treatment may be due to the synthesis of more transport proteins. In this case changing membrane transport is only a special case of increased enzyme activity.

Recent methods of measuring the very low concentrations of hormones found in body fluids (Fig. 11.5) have greatly increased our knowledge of endocrine actions.

THE HYPOTHALAMUS AND PITUITARY

The pituitary is a small (0·5 g) endocrine gland situated above the roof of the mouth and just below part of the brain (the hypothalamus). It is a compound gland derived partly from a downgrowth of nervous tissue and partly from an upgrowth of the mouth (Rathke's pouch). Clinical investigators found that patients with disorders of the pituitary showed any of a range of diseases; for example they might remain as dwarfs or grow into giants; they might have an increased or decreased sexual development; their metabolism might be increased or decreased; they might pass large quantities of urine. Some of these changes are typical of other endocrine glands and it is now clear that the pituitary acts largely by controlling other endocrine glands, rather than acting directly.

The anterior pituitary and hypothalamus are controlled both by blood-borne hormones from target organs and instructions from the central nervous system (Fig. 11.6). The outcome of this interaction determines the amount and type of secretion from it.

Harris showed that removal of the pituitary gave the typical deficiency signs. If the pituitary was replaced elsewhere in the body, or put back into its normal site but insulated from the brain by waxed paper, function was not restored; direct blood vessel connection between the hypothalamus and pituitary was necessary for restoration of normal function. The explanation for this is that hormones are secreted by the hypothalamus and carried by the blood to the pituitary where they influence the secretion of other hormones. This arrangement probably also prevents the hypothalamic transmitters from reaching peripheral tissues in any large concentration, and therefore allows peripheral tissues to use the same transmitters (e.g. somatostatin, LHRH etc).

The anterior pituitary secretes six main hormones. To comprehend their actions they may be thought of in two different ways: (a) whether they act directly on tissue target cells or on other endocrine cells, or (b) whether their general function in the body is on reproduction or on growth, development and metabolism. Fig. 11.7 shows the six hormones and indicates how each fits into these two classifications. Some hormones do not fit a category too clearly thus prolactin causes development

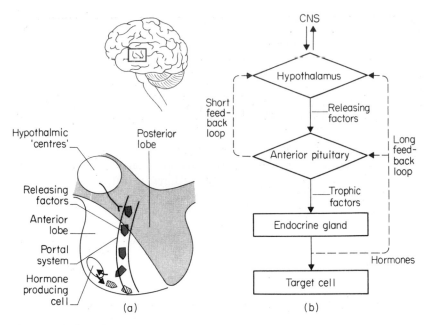

Fig. 11.6. Outline of (a) structure of anterior pituitary and hypothalamic connections, and (b) feedback pathways important in endocrine control.

The feedback from gland to pituitary (b) is negative, i.e. if the blood concentration of the hormone increases then the pituitary output of the releasing factor decreases and vice versa. This system is good for maintaining a constant level of plasma hormone but does not allow the level to change. One way of achieving this is by varying the sensitivity or response of the pituitary to the plasma hormones; this may be how the hypothalamic releasing factors work.

The portal system really consists of two sets of capillaries with joining vessels in between. The anatomy of the pituitary is also, alas, rather more complex than the diagram suggests. The anterior pituitary is made of glandular tissue; the posterior pituitary of nervous tissue.

of the breasts and is concerned with milk production during reproduction; follicle stimulating hormone and luteinising hormone act directly on the germ cells to cause them to develop, and are then involved in causing the gonads to secrete as an endocrine organ.

Recently a previously unknown group of regulatory polypeptides has been isolated from the hypothalamus and pituitary; these have a morphine-like action and so are often called the 'endogenous opiates'. They are derived from larger precursors. Many other peptides are currently being discovered (see also p. 143).

GROWTH HORMONE

As its name implies, the long-term actions of growth hormone (GH) are on growth: excess stimulates growth, lack reduces growth. If this occurs in young animals (more specifically when their bones can still grow, Fig. 11.20) then linear growth is affected and giants or dwarfs are produced. If an excess occurs in the adult then there is over-growth of the ends of the long bones, of the jaws and of the soft tissue in man; this is called acromegaly.

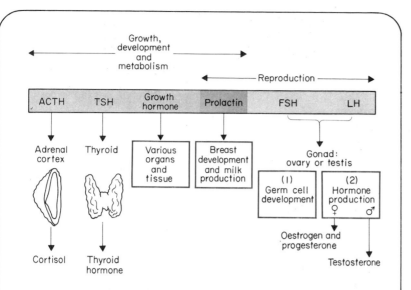

Fig. 11.7. Main pituitary hormones. These either have direct actions on target cells (boxes) or cause secretion of hormones by other endocrine glands (other shapes). The reproductive hormones have both roles (after Vander *et al.*).

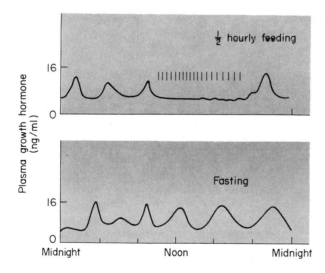

Fig. 11.8. Human growth hormone concentrations in the plasma of a normal subject under the different conditions shown. A single meal causes a 2-hour depression.

Production and structure. GH is produced and released by the acidophil cells in the anterior pituitary. Growth hormones from various species are different and so show species specificity; they are proteins with molecular weights in the range of 20,000–40,000. There is recent evidence that GH acts on growth via somatomedin, a substance found in serum and produced in the liver or kidney under the influence of growth hormone.

Blood concentration. The anterior pituitary contains about 1000 times more GH than the other hormones. This may be because GH has primary effects on body cells rather than acting as a controller of other endocrine glands. There is a total daily release of human GH of about 5 percent of the gland content. Serial measurements of plasma GH show marked bursts of activity, reaching 10 times the basal level (Fig. 11.8). The $T_{\frac{1}{2}}$ (Fig. 11.9) of circulating growth hormone is about 25 minutes; it is metabolised in the liver and kidneys.

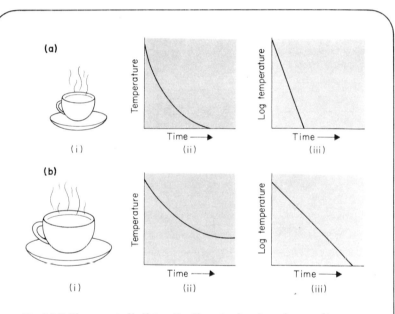

Fig. 11.9. *The concept of half-time.* $T_{\frac{1}{2}}$. The rate of cooling of a cup of tea depends on two factors:
(1) The temperature difference between the tea and the environment; as the temperature difference decreases then the rate decreases (a[ii], b[ii]).
(2) The rate of cooling also depends on fixed factors such as the size and insulating properties of the cup, presence of draughts, etc. Any one cup of tea will always cool at the same rate under the same conditions, but a different cup of tea will cool at a different rate; for example, a larger cup of tea will cool more slowly (because it has a larger volume for its surface area) (compare a and b). The rate of cooling will therefore be characteristic for each cup of tea.

   If the temperature is plotted on a log scale, then the curves become straight lines (a[iii], b[iii]). The slope of this straight line characterises the rate of cooling of the tea. A convenient measure of this is to take the time required for the temperature to fall to half; this is the $T_{\frac{1}{2}}$.

Specific effects. As GH is a polypeptide it would be expected to act at the cell membrane rather than in the cell. At present the picture is confused but it appears that GH does have a membrane action, though its subsequent effects are not through cAMP. Its cellular action may involve control of protein synthesis at a translational level; it also increases the transport of neutral and basic amino acids into cells.

The effect of GH may be considered on two time scales, corresponding to a long-term growth effect and a short-term effect on metabolism.

The long-term effects of GH determine the size of the animal when developing, and the maintenance of normal function once it has grown; they may be summarised as an increase in the protein content of the body with a decrease in the fat content. That is, GH tends to make the composition of the tissues appropriate for a growing organism; the incorporation of protein (anabolic action) is accompanied by reduced nitrogen excretion. Thus, these effects are to promote protein synthesis and hence cause a positive nitrogen and phosphorus balance to control:

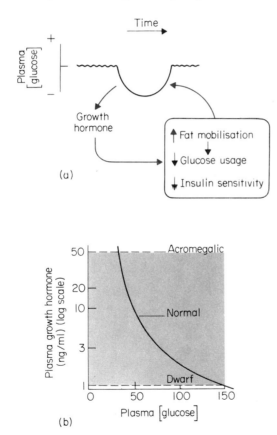

Fig. 11.10. (a) Short-term effects of a fall in glucose concentration on fat mobilisation and glucose usage. Rises in plasma glucose have opposite effects. (b) Effect of plasma glucose concentration on plasma GH concentrations in normal subjects. The upper edge of the red tint shows the situation in acromegalics, the lower edge that in dwarfs. In normal subjects a change in plasma glucose concentration causes changes in GH concentration; in dwarfs and acromegalics the plasma GH concentrations are fixed and unaffected by glucose concentration.

the size of muscle, cartilage and bone; the size and function of kidney; the bodily retention of the common ions Na$^+$, K$^+$, Cl$^-$, Mg^{2+} and Ca^{2+}, etc. This aspect of GH on developing tissues is discussed further on p. 346.

The short-term effects can best be understood and remembered as adjustment of starvation (Fig. 11.10). In this (and indeed between meals) the flow of food from the gut stops and body food stores are metabolised. The plasma glucose concentration drops, this switches on a release of GH which in turn changes energy usage from carbohydrates to fats. As a consequence the breakdown products of fats (ketones) appear in the blood and the plasma glucose rises. Conversely a rise in plasma glucose concentration decreases the production of GH and this eventually leads to a fall in the plasma glucose concentration. These inter-relationships between glucose and GH are particularly important in disease; it is, however, not altogether clear if changes in glucose concentration in normal people act in this way.

Control of GH release. The release of GH from the anterior pituitary depends on the interaction between hormones released by the hypothalamus and factors in the blood according to the general scheme shown in Fig. 11.11. The hypothalamus probably releases a growth-hormone-releasing hormone (GHRH) which acts to increase the release of GH. A variety of signals act on the hypothalamus to cause release of GHRH, for example 'stress' and sleep. As noted above, an important blood factor acting on GH release is the glucose concentration in the plasma, which bears a reciprocal relationship to plasma GH concentration (Fig. 11.10). The mechanism of action of glucose here is obscure, but is thought to involve the action of a metabolite of glucose on an intracellular messenger. Fig. 11.11 summarises some of the other factors acting on GH release.

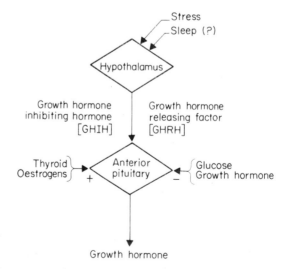

Fig. 11.11. Some stimuli acting on GH release from the anterior pituitary gland. It is not certain that these stimuli act where shown. GHIH is also known as somatostatin.

These hormones are concerned with the reproductive activities of the body, both directly or via other endocrine organs. The pituitary gonadotrophins are described in detail in the reproductive section, and are only outlined here; the actions of prolactin are dealt with in this section. In brief, in the male, follicle-stimulating hormone maintains the spermatogenic epithelium, and luteinising hormone causes the secretion of testosterone output from the testes; in the female, follicle stimulating hormone is responsible for the early growth of the ovarian follicles whereas luteinising hormone is responsible for their final maturation and oestrogen secretion and then is responsible for their conversion to the corpus luteum (the final stage in the cycle) and the secretion of progesterone.

Prolactin. This hormone consists of 198 amino acids; its plasma concentration is 2–15 μg/l, with no difference between the sexes. In pregnancy the plasma concentration rises to about 300 μg/l. It is responsible for growth of the mammary glands and milk production.

Control of prolactin release. The release of prolactin from the pituitary is normally inhibited by a hypothalamic factor (prolactin release inhibiting hormone—PRIH); control is exercised by varying this inhibition. Normally, in man, the prolactin release is increased by nipple stimulation or 'stresses' of various kinds, with a diurnal rhythm in which rises occur at night. During pregnancy the secretion increases, rising to a peak at the time of parturition; thereafter it falls to non-pregnant levels but is increased promptly by suckling, at least for the first 3 months of nursing.

The melanocyte stimulating hormone. (MSH) exists in two forms of 13 and 22 amino acids both related closely to ACTH. In mammals MSH is involved in skin pigmentation; in man its normal role is unclear.

HORMONES OF THE POSTERIOR PITUITARY GLAND

As in the anterior pituitary, this is a point of contact between neural and hormonal systems. In this case two hormones, antidiuretic hormone (ADH, or vasopressin*) and oxytocin, are made in the bodies of nerve cells in the hypothalamus and then—in combination with protein carrier molecules, the neurophysins—pass slowly down the axons of these cells to the nerve endings in the posterior pituitary gland (Fig. 11.12). Here they are stored and released, with the carrier molecules, into the bloodstream by exocytosis; the rate of their release is controlled by impulses which originate, like the hormones, in the hypothalamus and which travel down the same axons as the hormones did. The whole complex of hypothalamic nuclei, stalk and posterior pituitary can be

---

* ADH was originally called vasopressin because it constricts vessels in pharmacological doses.

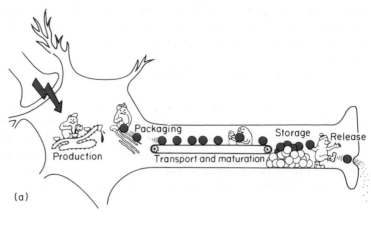

(a)

Fig. 11.12. Posterior pituitary. (a) Cartoon of the steps involved in producing hormones in secretory neurones. (b) Hormones made in the hypothalamus move down the tract to the endings in the posterior pituitary and are then released into the circulation. Protein carrier molecules (the neurophysins) pass down the tract with the hormones and are also released into the blood with them. The fibres from the supra-optic nuclei control ADH secretion, while those from the paraventricular nuclei control oxytocin secretion (a after Pickering).

considered as a functional unit, in that it makes, transports, releases and controls the release of these hormones; the advantage of this system is not clear.

The pre-synaptic membrane shown in Fig. 11.12(a) has a much larger surface area than shown, due to multiple branching of the end of the nerve.

Antidiuretic hormone. The main action of ADH is in the control of water excretion (see p. 216); it does this by increasing the water permeability of the collecting ducts of the kidneys. This leads to increased water reabsorption from the urine, which becomes more concentrated thus conserving water. The way this occurs is not yet clear except that it causes the release of cAMP in the tubular cells, which perhaps causes 'water channels' to be inserted into the apical surface of the cells.

Release of ADH. The secretion of ADH depends upon the osmolality of the extracellular space fluids; raising this by about one percent causes increased secretion of ADH, reducing it abolishes ADH secretion. The osmoreceptors for this effect are situated within tissues in the brain, supplied by the internal carotid artery; it is likely that some at least are situated in the hypothalamus; a possible mechanism is shown in Fig. 11.13.

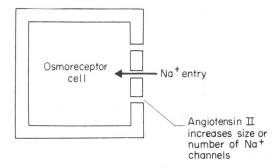

Fig. 11.13. Model of osmoreceptor action. Sodium entering the cell causes a depolarisation and increased rate of firing of action potentials. The amount of sodium entering increases as the sodium concentration of the CSF bathing the cell increases. Angiotensin II increases the sodium leak hence sensitising the system to osmotic changes.

In addition to this simple control system based on osmolality, ADH secretion also depends to some extent on the volume of the extracellular space; if this is reduced by 10–30 percent, for example by bleeding, then ADH secretion occurs no matter what the osmolality is. This mechanism does not play much part in day-to-day life, but is used in emergencies; probably the left atrial pressure receptors—which indirectly measure blood volume—are responsible. To replace extracellular fluid the body needs to replace sodium chloride and water. The renin–angiotensin mechanism (p. 355) retains sodium; the ADH mechanism retains water both directly and indirectly once the osmolality has increased (see also p. 216).

Alcohol produces its characteristic diuresis by inhibiting the release of ADH.

Oxytocin is involved in various aspects of reproduction. Its primary effect is to cause certain smooth muscles to contract; thus it causes contraction of uterine muscle during parturition and causes milk ejection during breast feeding (by an action on the myoepithelial cells lining the ducts). The signals for oxytocin release in these situations are stretching of the cervix and breast stimulation respectively.

THE THYROID GLAND

General effects. The thyroid gland makes and releases molecules which influence the growth, development and metabolic activity of all cells (Fig. 11.14). In man an increase in circulating thyroid hormone (TH) concentration increases the metabolic rate and causes an over-excited 'nervous' type of behaviour. A decrease slows the metabolic rate and causes sluggish mental activity. Reduced thyroid activity in childhood produces dwarfs who are mentally retarded, whereas reduced growth hormone secretion in childhood produces dwarfs of normal intelligence.

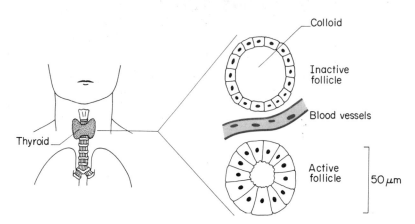

Fig. 11.14. Position of thyroid gland in the neck below larynx and across trachea, with schematic structure of follicles in active and inactive states. The blood supply is between the follicles and accounts for 1 percent of the cardiac output.

Synthesis of thyroid hormones. The cells in the thyroid follicles actively accumulate iodine and amino acids from the plasma. The amino acids are made into thyroglobulin which, together with the iodine, is secreted into the colloid within the follicle. The iodine reacts with part of the thyroglobulin in a series of reactions ending in the production of TH. The hormone is stored on the surface of the thyroglobulin until required. When required the hormone–protein complex is taken into the cells, the hormone (i.e. T_3 and T_4) is split off and then secreted into the plasma. Figure 11.15 shows a more detailed account.

Blood transport and concentration. Most of the TH in the plasma is associated with the plasma proteins and less than 1 percent is free (Fig. 11.16). This free TH is in equilibrium with the bound form and is the part responsible for the tissue actions of the hormones. The amount of binding protein may be changed in various states (increased in pregnancy) thus increasing the capacity to store TH. If the concentration of TH per unit of binding protein remains constant then the free TH will not be altered. A measurement of total blood TH (or of bound iodine) does not therefore reflect the thyroid status of the individual. Such a system of transport means that the TH supply to the tissues must be relatively constant, perhaps with minor and slow modifications imposed

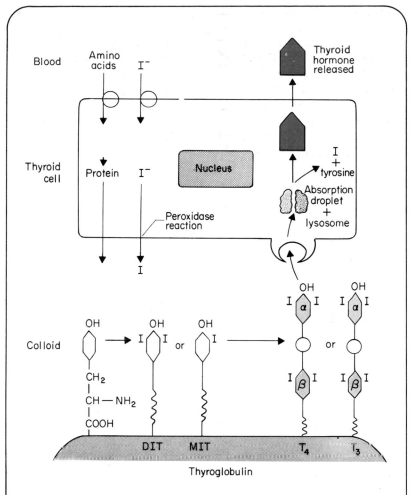

Fig. 11.15. Synthesis of thyroid hormones. For simplicity the side chain is only shown on the first step. Thyroglobulin is made by the follicle cells and exported to the colloid. Iodide (I^-) is actively taken into the cells and converted to iodine (I_2) by a peroxidase reaction at the inner surface of the cells. Release of free iodine causes iodination of the tyrosine molecules on the surface of the thyroglobulin molecule, forming MIT (monoiodotyrosine) or DIT (diiodotyrosine). These then couple spontaneously to form T_3 (triiodothyronine) or T_4 (tetraiodothyronine or thyroxine). When required colloid is taken back into the cells by pinocytosis, the thyroglobulin degraded and the thyroid hormone released into the blood. This degradation occurs by means of the enzymes in lysosomes. MIT and DIT are also liberated by this means and are further split into iodide and tyrosine, giving an iodine cycle.

by changes in the thyroid-binding globulin. This fits with the observation that replacement therapy is adequate if a constant level of TH is maintained in the plasma.

The binding proteins are of three kinds (1) thyroxine-binding globulin (TBG), (2) thyroxine-binding prealbumin (TBPA) and (3) albumin. T_4 binds to all these in the proportion of 60, 30 and 10 percent. T_3 binds

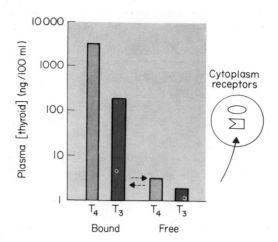

Fig. 11.16. Distribution of thyroid hormones in the plasma. The free T_4 and T_3 concentrations are about 3 and 1·5 ng/100 ml (40 and 20 pmol/l) respectively.

Note that: (1) bound thyroid hormones are given per unit of plasma not as per unit of binding protein; and (2) log scale on the ordinate.

rather weakly to TBG and albumin only. The free T_3 and T_4 is distributed throughout the extracellular space of the body whereas the bound hormone is confined to the blood spaces. This means that a larger fraction of T_3 than of T_4 is to be found in the extracellular space.

Metabolism and turnover of TH. Little TH is excreted by the kidneys, presumably because of the low free levels in the plasma. Some is excreted in the bile but most is metabolised by peripheral tissues (mainly muscle and liver) and the iodine recycled. As with other hormones the turnover of the hormone is related to its speed of action. Thus T_3 starts to act in about 5 hours and has a $T_{\frac{1}{2}}$ of turnover of 1 day, whereas T_4 takes about 2–3 days to act and has a $T_{\frac{1}{2}}$ of turnover of some 7 days. There is some evidence that T_4 may act as a prohormone for T_3 and it is possible that T_3 is the active hormone in the tissues. This fits with the observation that T_3 has a biological activity about five times greater than T_4.

Effects of TH. TH enters cells, probably binds to receptors inside them (Fig. 11.14) and goes to the nucleus. This is followed in about 5 hours by the production of new mRNA, then by new ribosomal RNA, then by new protein synthesis and an increase in oxygen consumption some 36 hours later. The TH concentration affects the rates at which various processes occur in the body, excess causing an increase and a deficit a decrease. Table 11.2 outlines some of the effects in man.

Control of thyroid function. The thyroid gland is controlled by the thyroid-stimulating hormone (TSH) released from the anterior pituitary gland (Fig. 11.17). It acts on the thyroid cell membrane producing an intracellular messenger, cAMP (Fig. 11.3); this in turn acts on all stages of TH production to change the thyroid cell from a state of torpor to one of great activity. Thus, there is increased amino-acid and iodine uptake,

Table 11.2. Thyroid effect at various levels of organisation. Most of these changes require prior synthesis of new proteins before they occur (after Tepperman).

Level of organisation	Hypothyroid	Hyperthyroid
Behaviour:	Sluggish mentally and physically	Active and restless mentally and physically
	Sensitive to cold	Sensitive to heat
Whole body:	Low BMR	High BMR
	Deficient growth	Negative N_2 balance
	Raised blood cholesterol	Lowered blood cholesterol
Organs:		
CVS	Cardiac output ↓, rate ↓	Cardiac output ↑, rate ↑
GI	Sluggish movements	Hyperactive movements
	Low glucose uptake	High glucose uptake
Muscle	Weakness, hypotonia	Weakness, tremors, brisk reflexes
Tissues:	QO_2 of liver, kidney, muscle, etc. ↓	QO_2 of same tissues ↑
	Normal QO_2 of brain, retina, testes	Normal QO_2 of brain, etc
	Adrenaline effects decreased	Adrenaline effects potentiated
Enzymes:	↓ oxidative enzymes	↑ oxidative enzymes

*Until recently the thyroid status of patients was assessed by measuring the O_2 consumption at rest; now it is assessed by measuring the uptake of radioactive iodine by the thyroid.

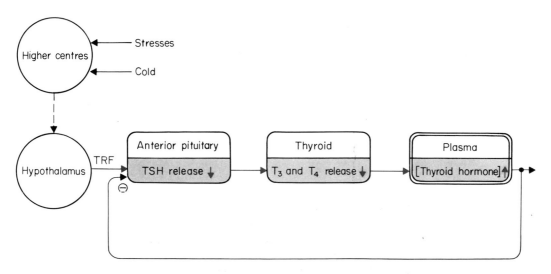

Fig. 11.17. The control of plasma thyroid concentration. A disturbance of plasma [thyroid hormone] (shown as an increase) is followed by the consequences shown (arrows in other compartments) which correct the disturbance. The plasma (and tissue) concentrations of T_3 and T_4 are kept within a narrow range.

increased iodination, increased pinocytosis, etc. There are typical morphological changes in the thyroid (Fig. 11.14). The end result is an increased TH secretion into the blood.

TSH production from the anterior pituitary is controlled by a negative-feedback system mediated by TH in the blood (but it appears that there is no TH effect on the hypothalamus). The anterior pituitary in turn is affected by a thyrotropic-releasing hormone (TRH) originating in the hypothalamus. The rate of release of TRH is affected by body temperature, food intake, etc.

THE CONTROL OF GROWTH

Animals start as single cells and by repeated divisions reach a size of many millions of cells. During this process cells differentiate and, by an orderly sequence of changes, become grouped into tissues and organs; this process involves an increase in the total protein and in the length and size of the animal, not just an increase in weight. With our limited knowledge of the way this process works it appears to be extraordinarily complex; here we can only indicate some of the more important factors.

Genetic and external factors. It is supposed that genetic factors determine the range of size which an animal can attain; the actual size within this range is determined by external factors. Of these the most important is the food supply. This must be adequate in protein, essential vitamins and other factors and in calories. If the calorie supply is too low then protein is burnt for energy; this deprives the animal of amino acids for protein synthesis.

Growth periods. Man shows two periods of rapid growth: the first of these is just after birth and is actually a continuation of the rapid growth which occurs in intra-uterine life; the second is during adolescence (Fig. 11.18). This curve of total growth conceals differences between various organs; thus we all know from observation that babies have big heads

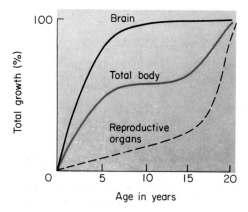

Fig. 11.18. The relative size of the whole body, brain and reproductive organs in man at various ages (after Vander *et al.*).

(containing big brains) early in life and that the reproductive organs do not grow to adult size until adolescence.

Hormonal influences. The main hormones affecting growth are growth hormone, thyroxine and the androgens (Fig. 11.19). Insulin is also necessary during the entire growth period.

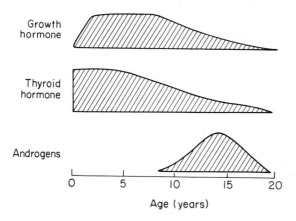

Fig. 11.19. The relative importance of various hormones in human growth at various ages (after Ganong).

Growth hormone is of major importance in controlling growth from birth to adolescence. It exerts this influence by its action on amino-acid transport into cells and by the synthesis of new protein. The most obvious target organs during growth are the long bones, growth of which controls the linear dimensions of the animal. The long bones grow by converting one edge of the cartilaginous plate at their ends to bone and simultaneously forming new cartilage at the other edge of the plate. The cells responsible for changing cartilage to bone are the osteoblasts. GH promotes both of these processes by stimulating protein synthesis at both sites. This effect is probably via somatomedin, released by the liver under the influence of GH (Fig. 11.20).

Thyroxine is essential in normal amounts for growth; excess does not produce overgrowth as with GH, but causes an increase in catabolism of protein and other nutrients. Thyroxine at normal concentrations thus has a permissive effect on the action of growth hormone on protein synthesis; in its absence amino acid uptake and protein synthesis are not much stimulated by GH. Thyroxine is particularly necessary for the rapid early growth of the nervous system; in its absence mental retardation occurs which cannot subsequently be corrected (cretinism). This defect may be due to a failure of myelination.

Androgens and oestrogens. Androgens from the adrenals in both sexes and testosterone from the gonads in males start to be secreted from about 10 years of age in humans, and increase progressively to a plateau in the next 5–10 years. In females oestrogen secretion from the ovaries increases in a similar way over this period, and causes development of the female reproductive organs, but has little effect on protein synthesis;

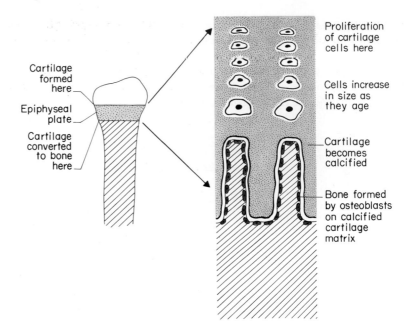

Fig. 11.20. Growth of a long bone showing conversion of cartilage to bone. End of long bone extended and then the shape of the shaft is altered by remodelling (after Passmore & Robson).

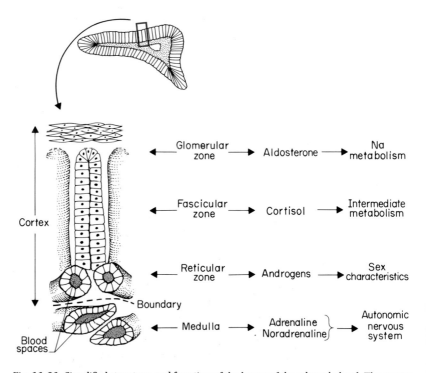

Fig. 11.21. Simplified structure and function of the layers of the adrenal gland. The cortex accounts for about 90 percent of the total size. The blood supply is communal; only the medulla has a nerve supply.

oestrogen probably also stimulates adrenal secretion of androgens. Androgens increase protein synthesis, causing both an increase in the size and development of the reproductive organs and of the general body size. They stimulate rapid growth of the epiphyseal plates of the long bones but also, eventually, stop it by causing complete conversion of the plate to bone. This accounts for the spurt and then stopping of growth in children and explains why eunuchs are often very tall; growth, though slower, goes on for longer.

ADRENAL GLAND

The adrenals are small 'cocked-hat' shaped glands closely applied to the upper poles of the kidneys. They are really compound glands, both in the sense that they are derived from two types of embryonic tissue, and in the fact that the outer part (the cortex) secretes steroid hormones whereas the inner part (medulla) secretes catecholamines. The main effects of the outer part are in the control of salt and organic metabolism. If the adrenals are removed death occurs in about a week, due mainly to sodium and water loss. If salt is given the animals live longer and effects due to cortisol deficiencies on organic metabolism become apparent. These general properties are shown in Fig. 11.21.

The adrenals are the glands, *par excellence*, of homeostasis. They are responsible for the relative freedom that higher animals have in a changing environment. Survival is possible in their absence but animals must then have: (1) regular food, (2) a high salt intake, (3) a narrow environmental temperature range.

ADRENAL CORTEX

The hormones produced by the adrenal cortex have numerous and diverse effects on the body, but their similar chemical structure gives them several common features which will be discussed first. Table 11.3

Table 11.3. Comparison of steroid and other hormones. The catecholamines behave like the peptide group, and thyroid hormone like the steroid group (after Catt).

Type of hormone	Peptide	Steroid
Derivation	Gut and related structures*	Mesodermal origin
Common features	Few—except for basic amino-acid composition—may be very complex	Molecules all very similar
Mode of action	Interact with cell-membrane receptors →intracellular messenger →?structural changes in pre-existing proteins	Affect nucleus (after combining with binding protein in the cytoplasm)→synthesis of new proteins
Form of circulation	Water-soluble and usually free in circulation	Usually sparingly water-soluble and circulate with binding proteins

*They may, however, have been derived ultimately from neuroectoderm.

lists the ways in which the steroids are similar to each other and contrasts their behaviour with the other large group of hormones, the polypeptides.

The characteristic structure of the steroid hormones contains four rings on one plane with the functional groups projecting on one or other side of the plane of the main structure (Fig. 11.22). Different steroids are produced by attaching different functional groups (—H, =O, —CH$_3$, —OH) to the basic ring structure. These groups project above (β) or below (α) the plane of the rings. These relatively minor changes alter

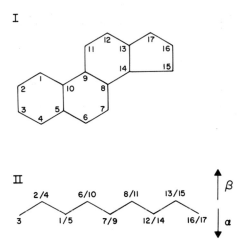

Fig. 11.22. The basic steroid structure consists of three cyclohexane rings and one cyclopentane ring. The carbon atoms are numbered as shown for identification of the functional groups. The main side chain is attached to atom 17, both in cholesterol and its derivatives. The four rings are not in a flat plane as conventionally shown in I, but have the approximate form shown in II, in which the observer is looking along the plane of the structure. Groups are either attached above (β) or below (α) the plane. The numbers on II refer to the molecule in that plane.

the steroids so that the reaction between the steroid and the appropriate receptor becomes very specific. They also change the specificity of the steroids to the transport proteins so that some are highly bound (cortisol) while others are hardly bound at all (aldosterone).

The steroids are all synthesised from cholesterol by a small number of steps. Therefore, if there is an error of metabolism in cholesterol then all are affected. There is practically no storage of steroids, so they are made as required. The steroids have a T$_{\frac{1}{2}}$ in the circulation of 20–100 minutes. They are inactivated in the liver by removal of the functional groups, then made water-soluble by conjugation with glucuronide and sulphate and excreted in the kidneys. The ring structure is excreted intact.

The various groups of adrenal hormones have largely different effects and will be discussed separately.

Glucocorticoids

As with most other hormones the actions of the glucocorticoids range over several layers of organisation from whole animals to membrane interaction. Again, as for most other hormones, no single or simple explanation of the effects can be given.

In man, sheep, dog, cat and cow the main hormone in this group is cortisol. In rodents it is corticosterone. In man most of the plasma cortisol is bound to a transport protein (transcortin). There is a circadian rhythm of plasma cortisol concentration which seems to be related to the sleep pattern and is mediated by the pituitary (Fig. 11.23). It may be due to an inherent rhythm in the brainstem.

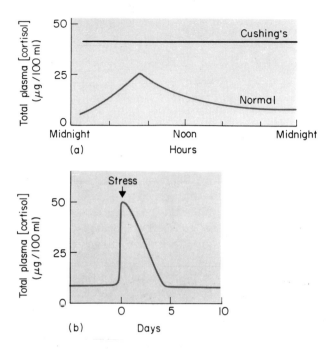

Fig. 11.23. Patterns of total plasma cortisol concentration: (a) throughout one day in the normal subject and a patient with Cushing's syndrome (a disease characterised by an abnormally high glucocorticoid production); and (b) the effect of stress on a normal subject.

Presumably the free cortisol concentration follows the total concentrations. In normal subjects these changes in cortisol are preceded by similar changes in ACTH. Recent evidence shows that this smooth curve is actually composed of irregular bursts of ACTH and cortisol secretion. The 'bursts' occur more frequently in the early morning giving the daily rhythm shown.

Specific effects. There is still uncertainty about the actions of cortisol at normal tissue concentrations, though there is wide acceptance of the view that cortisol exerts a 'permissive' action on metabolic and enzyme functions, thus permitting a variety of other hormones to exert their effect. In some cases these permissive effects are to induce new proteins which are then acted upon by other hormones.

The adrenals are involved in the body's response to 'stress', i.e. exposure to noxious or potentially noxious stimuli. In short-term sudden stress the immediate reaction is the release of catecholamines leading to the 'fight or flight response'; the longer-term reaction is the release of increased amounts of cortisol (Table 11.4). Many situations in everyday

Table 11.4. Psychosocial situations shown to be associated with increased plasma concentration or urinary excretion of adrenal cortical steroids (after Vander *et al.*).

Normal people

(a) *Acute situations*	(b) *Chronic life situations*	(c) *Experimental techniques*
1 Aircraft flight	1 Predictable personality— behaviour profile: aggressive, ambitious, time-urgency	1 'Stress' or 'shame' interview
2 Awaiting surgical operation		2 Many motion pictures
3 Final exams (college students	2 Discrepancy between levels of aspiration and achievement	
4 Novel situations		
5 Competitive athletics		
6 Anticipation of exposure to cold		
7 Decreased during weekends		
8 Many job experiences		

life lead to increased cortisol secretion and indeed the word 'stressful' is often used for these situations which do lead to this response. These effects of cortisol in stressful conditions act through their high blood concentrations; they can be understood and remembered as follows. Animals faced with such situations: (1) have no time to eat so the effects of cortisol are to supplant and conserve the energy derived from circulating glucose; (2) might require to repair damaged tissues and so have a need for higher free amino acid concentration in the body. These effects are accomplished by inhibiting glucose usage in peripheral tissues: by causing a flow of amino acids to the liver for new glucose synthesis (gluconeogenesis); by changing muscle metabolism from glucose to fatty acids and by mobilising fatty acids from adipose tissues. The response is rather like the 'fasting response', which occurs between meals (see p. 361), except that it leads to deposition of glycogen as a consequence of the gluconeogenesis. This action led to the name— glucocorticoids—used to describe this group of steroids.

High plasma cortisol concentrations—pharmacological effects. If cortisol concentrations are raised for prolonged periods—by certain diseases or during treatment of various disorders—then: (1) the generalised breakdown of protein which occurs in skin, muscle, bond and lymphoid tissue, gives wasting and/or weakness of these tissues and a negative nitrogen balance therefore develops. This increased protein breakdown will greatly increase the plasma glucose and glycogen stores; this leads to an increase in insulin secretion causing a greater conversion of glucose to fat and the obesity of Cushing's syndrome (Fig. 11.24); (2) suppression of the typical effects of injury (see 'triple response', p. 117) either when produced mechanically or by allergens. These injury effects are an increased fluid leakage, increased leucocyte appearance and eventually the laying down of fibrous tissue. This suppression is 'useful' or otherwise, depending on the circumstances; in allergies it is useful, in wound healing not. An attractive suggestion for this action of cortisol is that it stabilises the lysosomes in cells and so does not allow them to escape

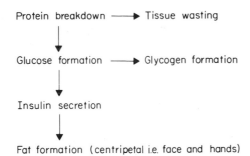

Fig. 11.24. Steps in the production of obesity caused by excess glucocorticoids.

from partially damaged cells, to prolong or increase the local damage; by this means the degradative enzymes present in the lysosomes can be contained. Perhaps associated with this action of cortisol is its effect on decreasing the circulating eosinophils.

Low plasma cortisol leads to:
1   Inability to cope with physically 'stressful' conditions, whether man-made, for example surgical operations, or the natural hazards of living.
2   Inability to handle a water load. This is not primarily a matter of altered kidney function but seems to be because of a maldistribution between the body fluid compartments.
3   There is a fall in blood pressure, and noradrenaline no longer has a constrictor effect on the cardiovascular system.
4   The blood glucose concentration falls, and liver glycogen decreases. This is partly because of the removal of the 'permissive' effect of cortisol on other hormones and partly by removal of its direct effect on gluco-neogenesis.

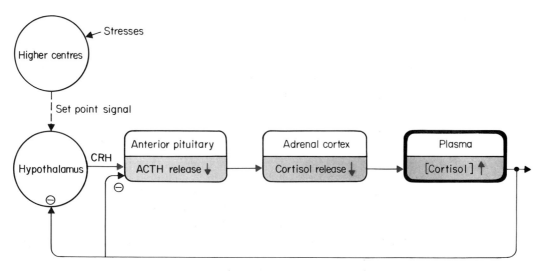

Fig. 11.25. The control of plasma cortisol concentration. A disturbance of plasma cortisol (shown as an increase) is followed by the consequences shown (arrows in other compartments) which correct the disturbance.

Control of cortisol secretion. ACTH (adrenocorticotrophic hormone) released from the cells of the anterior pituitary is responsible for maintaining normal adrenal structure and function, for the normal release of the glucocorticoid secretion (Fig. 11.25) and for exerting a permissive action in mineralocorticoid secretion (i.e. necessary, but not the main stimulus). ACTH release in turn is increased by corticotrophic-releasing hormone (CRH) and inhibited by plasma cortisol. CRH structure has not yet been worked out, but is probably a small (10 amino acids) polypeptide. ACTH structure for several species has been worked out. It consists of a polypeptide of 39 amino acids (Fig. 11.26) with an active part and a species-specific part.

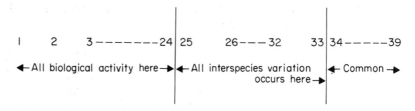

Fig. 11.26. Outline structure of ACTH. Each number represents an amino acid. Amino acids 1–13 are found in some species as melanocyte-stimulating hormone.

Mineralocorticoids

Aldosterone, the main hormone in this group, is secreted at a daily rate depending on the salt balance. There is a diurnal rhythm of plasma aldosterone with much the same pattern as that of cortisol (Fig. 11.23). This seems to be partly related to posture, in that lying flat inhibits it.

Specific effects. Aldosterone causes the kidneys to retain Na^+ and excrete K^+ and Cl^-, mainly by an action on the distal tubules. This leads to a secondary retention of water (mainly via antidiuretic hormone) and so an expansion of the extracellular volume results. Similar effects on Na^+, K^+ and H^+ handling occur in the salivary glands, sweat glands and intestine and perhaps in the proximal renal tubule.

The way in which aldosterone acts on target cells is not yet clear. It seems probable that the hormone–receptor complex enters the nucleus of appropriate cells and that new protein synthesis occurs, but it is likely this new protein is a 'permease' which increases the Na^+ leakiness of the cells leading to a secondary increase in sodium transport.

Control of plasma aldosterone concentration. The aldosterone concentration in the blood is related to sodium metabolism. If the total body sodium is high—by sodium loading—then the blood aldosterone concentration is low and sodium is excreted; if the body sodium drops—by sodium deprivation—then the blood aldosterone concentration is high and sodium is retained. To some extent potassium changes show opposite effects.

It is difficult to imagine a system which would measure total body sodium directly, but easier to imagine one which would measure some indirect effect of sodium. The current view is that the size of the extracellular fluid volume is used as an index of the sodium loading of the body. If the body is deprived of sodium chloride then the osmoreceptors (p. 341) cause an excretion of water to correct the osmotic pressure, hence the extracellular volume—and the blood volume—shrinks. There is thus a fall in the pressure in the arterioles perfusing the kidney and this leads via a series of hormones—the renin–angiotensin system (Figs 11.27 and 11.28)—to an increase of aldosterone release. An increase in body sodium works via the same system to reduce the plasma aldosterone and hence causes excretion of sodium. The apparatus which detects changes in perfusion pressure is the juxtaglomerular apparatus (Fig. 11.29), situated where the blood supply enters the glomerulus, with the distal convoluted tubule nearby. It can be thought of as a cuff around the afferent vessels to the glomerulus.

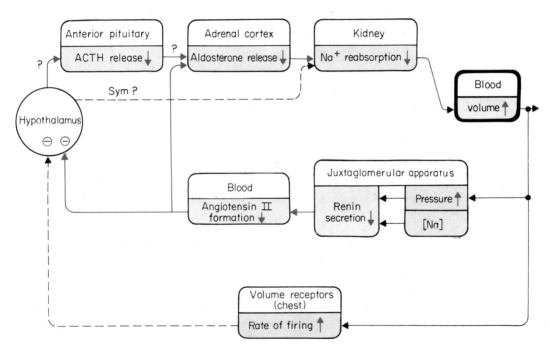

Fig. 11.27. The control of blood volume (simplified). A disturbance of blood volume (shown as an increase) is followed by the consequences shown (arrows in other compartments) which correct the disturbance.

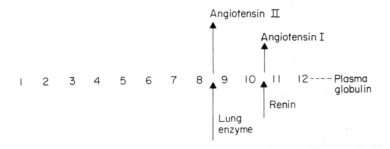

Fig. 11.28. Production of angiotensin II. Activation consists of chopping pieces off the larger plasma globulin molecule. Each number represents an amino acid.

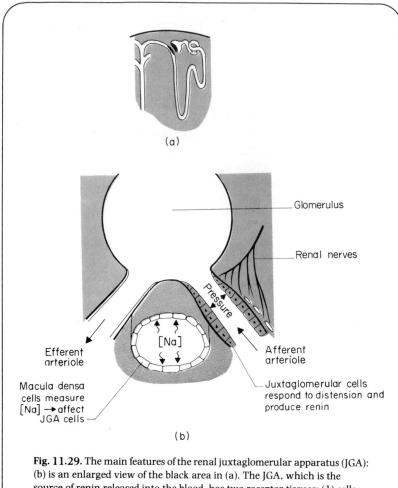

Fig. 11.29. The main features of the renal juxtaglomerular apparatus (JGA): (b) is an enlarged view of the black area in (a). The JGA, which is the source of renin released into the blood, has two receptor tissues; (1) cells arranged as a 'cuff' around the afferent arteriole which monitor vascular distension, and (2) the macula densa, which monitors the sodium concentration. Renin secretion is also regulated by activity in the renal nerves.

The situation is more complex than this in that the sodium concentration in the distal convoluted tubule is also measured by the juxtaglomerular cells and used to control aldosterone production and release. Plasma potassium probably also has a direct effect on aldosterone release by the adrenals. ACTH is probably permissive rather than controlling.

A bewildering array of other information is available on the renin–angiotensin system, most of which does not fit easily into a simple scheme. Three separate aspects which do will be mentioned:

1   Renin-like activity is present in many organs such as brain, salivary glands, uterus, arterial wall, etc. Release from these would presumably be triggered by low perfusion pressures and thus expand the blood volume as described.

2   Angiotensin II acts on the brain to cause increased drinking. This

action seems understandable in that if sodium is retained water also needs to be retained to expand the extracellular volume.

3 Angiotensin II is a very potent constrictor substance thus causing a raised blood pressure by a direct action on arterioles in addition to an indirect action via fluid retention.

Adrenal androgens. These have a keto group in the 17 position of the basic ring structure and are hence known collectively as the 17-keto-steroids. Half of the plasma androgens are bound to a sex-hormone-binding globulin (SHBG). Secretion is partly under the control of ACTH from the pituitary. The role of these androgens in normal man is unclear in that they are unable to maintain secondary sex characteristics in castrated men or animals; it has been suggested that they may be responsible for female libido. Their main clinical interest is that inborn errors of metabolism and tumours may lead to excessive production and 'virilisation' of females or young boys.

ORGANIC METABOLISM AND THE GLUCOSE REGULATING HORMONES

The hormones dealt with so far are those which have a variety of effects on many systems in the body. In the next two sections the emphasis is rather different, in that we are concerned with the control of particular functions by several hormones: the first of these functions is organic metabolism; the second is calcium metabolism; neither system is much affected by anterior pituitary secretions. The hormones involved are part of a regulatory pathway but are not themselves kept constant (p. 326).

Animals eat intermittently, with periods of fasting in between, yet their cells require a continuous supply of energy. The mechanisms which have evolved to cope with these alternating periods of feeding and fasting are hormonally controlled; the states themselves are often called absorptive and post-absorptive. We shall start by considering the events which occur during these states, and then describe the hormones which control these changes.

Interconversion of carbohydrates, fats and proteins. In order to understand the events during feasting and fasting it is essential to realise that body cells—particularly those of the liver—have a remarkable ability to convert one type of molecule into another. These conversions allow animals to survive and grow on a very wide range of foods, provided that certain essential items are present (see below and p. 168). It is found that not only are molecules of one sort converted to those of another following digestion, but that virtually all the molecules in the body are being broken down and rebuilt all the time. So the body is in a continual anabolic–catabolic state.

Figures 11.30 and 11.31 show the main pathways for interconversion of protein, carbohydrate and fat. The main difference between protein and the other two are: (1) protein cannot be stored to any extent—it is merely used for replacing protein molecules in cells—and

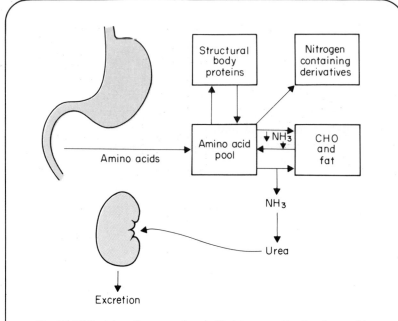

Fig. 11.30. Protein pathways and pools. Protein cannot be stored or used to supply energy directly (after Vander *et al.*).

1   An amino acid is changed to a carbohydrate by this reaction (deamination):

$$\underset{\text{enzymes}}{\text{R—CH—COOH}} \longrightarrow \text{R—C—COOH} + NH_3$$

$$\underset{\text{amino acid}}{\overset{|}{NH_2}} \qquad \underset{\text{carbohydrate} + \text{ammonia}}{\overset{\|}{NH_2}}$$

2   A carbohydrate is changed to a new amino acid by transfer of a nitrogen group from another amino acid (transamination):

$$\underset{\text{amino acid}}{\overset{|}{\text{R—CH—COOH}}} + \underset{\text{carbohydrate}}{\overset{\|}{\text{R}_1\text{—C—COOH}}} \overset{\text{enzymes}}{\longrightarrow} \underset{\text{carbohydrate}}{\overset{\|}{\text{R—C—COOH}}} + \underset{\text{amino acid}}{\overset{|}{\text{R}_1\text{—CH—COOH}}}$$

it cannot be metabolised directly but must first be converted to carbohydrate or fat; and (2) carbohydrate and fat can be stored as such and are the main fuel to supply the body with energy. The figures show the similarities of handling of all three molecules; we have omitted the minor losses which occur through the kidney and when cells are lost from the skin, etc.

The essential items which must be in a diet (p. 166) are:

1   A minimal amount of protein to replace the nitrogen which is excreted.

2   Certain essential amino acids, which cannot be made from others in the body. The best source of these is animal muscle, but an adequate mixture of plant protein can supply all our essential amino acids.

3   Certain vitamins and fatty acids are required, which again cannot be made in the body. Vitamins are used as cofactors or coenzymes for chemical reactions and not for energy.

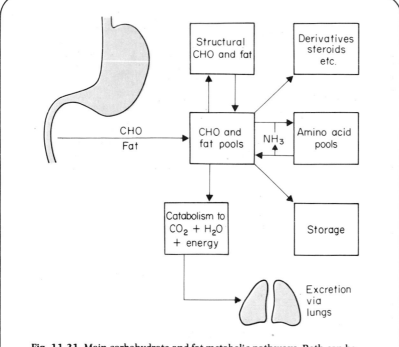

Fig. 11.31. Main carbohydrate and fat metabolic pathways. Both can be stored and used for energy directly. Much more fat than carbohydrate is stored, however (after Vander *et al.*).

Outline of absorptive and post-absorptive events

Food taken into the digestive tract starts a series of events leading to the production of the digestive enzymes which break the food down to a mixture of smaller molecules. These are then absorbed into the blood—either directly or via the lymph—causing a rise in the blood concentrations of glucose (+ galactose and fructose), amino acids and fats. During this period glucose is used as a major source of energy by the cells; most of the amino acids and fats, as well as the excess carbohydrate not used for energy, are stored as fat or glycogen. A small fraction of the ingested amino acids and fats is used to resynthesise the body proteins and structural fat which is being continuously degraded.

Between meals the tidal wave of newly ingested molecules decreases, so that the direction of flow of materials is now reversed; some cells—mainly those in the liver and fat tissue—now export rather than import molecules. The blood sugar is kept constant by breaking down the glycogen stores in liver and some other cells, and by the synthesis of new glucose; this is necessary because brain, renal medulla and erythrocytes have no other energy source. The energy supply of most other cells is switched from glucose to fatty acids, thus sparing glucose; most of the energy supply for the body now comes from fat. Protein and fat synthesis decreases, leading to a net breakdown of these materials.

These various adjustments are made by the hormones insulin, glucagon, GH and the catecholamines. Of these insulin is the hormone of absorption; it is released by the rise in blood glucose during a meal. Glucagon, GH and the catecholamines are the hormones of fasting; they are released by lowering the blood sugar. These four hormones are directly involved in the control of organic metabolism; others are, however, also involved in a permissive way, i.e. their presence is necessary for action to occur but they are not controlling. Thus, background concentrations of thyroid hormones and cortisol are necessary for the proper action of most of these hormones. In keeping with the close links between digestion and these metabolic processes, it is now clear that gut hormones, gastrin, secretin and pancreozymin augment the release of insulin and glucagon in preparation for the appearance of glucose and amino acids in the blood during a meal.

The absorptive state

Following a meal monosaccharides and amino acids enter the portal blood and go to the liver; the fats enter the lymph as chylomicrons

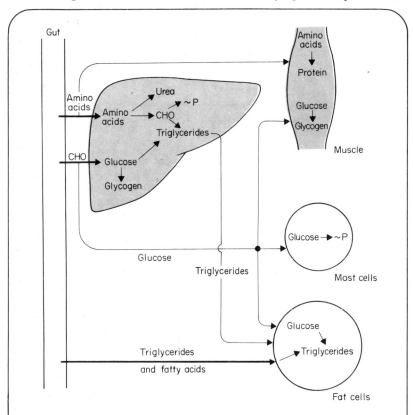

Fig. 11.32 Metabolic pathways of absorption. Absorbed carbohydrates (CHO) are converted to glucose in the liver and then used to form glycerol, triglycerides or glycogen. Most of the food entering the body is stored as fat and to a lesser extent as glycogen. Fat is a good way for mobile animals to store energy, for it is light (it contains little water) and has twice the energy yield of carbohydrate and protein per gram of material. Glucose is stored as glycogen for osmotic reasons (partly after Vander *et al.*).

containing triglycerides, and so enter the general circulation. The mono-saccharides consist of galactose, fructose etc., as well as glucose but can, for simplicity, be considered as glucose (Fig. 11.32). We shall consider separately the fate of each of these substances which enters the body following an average meal; say carbohydrate, fat and protein in the ratio of 65:17:18, respectively (p. 177).

Glucose. Much of the absorbed carbohydrate enters the liver cells; here it is stored as glycogen or converted to fat which is then stored partly in the liver but mainly elsewhere after transport in the blood. The glucose which does not enter the liver is used by almost all body cells as the main energy source; the remainder is stored as glycogen (mainly in skeletal muscle), and converted to fat in the adipose-tissue cells.

Amino acids. Much of the absorbed amino acids also enter the liver cells; here they are converted into carbohydrate (keto acids) by removal of the NH$_3$ which is converted to urea and excreted via the blood and kidneys. These keto acids enter the Krebs cycle to supply most of the liver's energy during the absorptive phase. They are also converted to fats which are then stored in the liver or elsewhere. The amino acids which are not taken up by the liver enter other cells (mainly muscle) only to replace that which is being continuously broken down; excess amino acids are not stored as protein but as carbohydrate and fat.

Triglycerides. Practically all absorbed fat enters the adipose-tissue cells, where it is stored. During the absorptive phase adipose-tissue cells receive triglycerides: (a) directly; (b) from those manufactured in the liver; and (c) those made in the fat cells from glucose. A little fat is metabolised by tissue cells to produce energy during this phase.

Post-absorptive or fasting state
The main problem during this state is that the plasma glucose concentration must be maintained (at 70 mg/100 ml, i.e. approx. 4 mmol/l)—because the brain, renal medulla and erythrocytes require glucose—and yet there are limited stores of glycogen available to do so (Fig. 11.33). The body solves this problem in three ways: (1) breaking down glycogen stored in the liver; (2) forming new glucose from fat and protein (gluconeogenesis); and (3) converting the metabolism of most cells to work on fat rather than glucose—thus sparing glucose.

Sources of plasma glucose
The glycogen in liver and muscle is broken down to supply glucose for the plasma. The liver glycogen can be broken down directly, but muscle lacks the necessary enzyme (glucose-6-phosphatase) to produce glucose from glycogen; instead pyruvate and lactate are produced in muscle, go to the liver and there are used to synthesize glucose. This source supplies glucose for about 8 hours at average body rates (100 cal/h).

Triglycerides are broken down to glycerol and fatty acids; the glycerol can be converted to glucose by the liver, and the fatty acids (which cannot be so converted) are used directly by the body cells as an energy source.

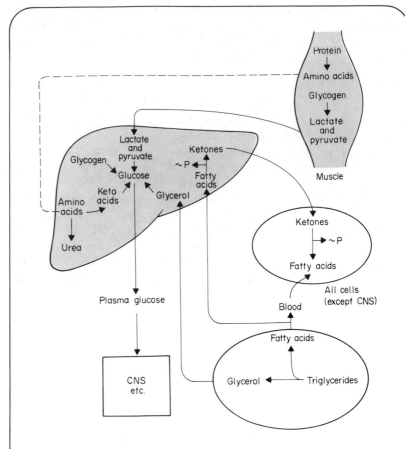

Fig. 11.33. Metabolic pathways during the post-absorptive state. Initially blood glucose is supplied from liver and muscle glycogen, later glycerol from body fat is used to make glucose. In prolonged starvation protein breakdown occurs. The main energy sources for cells are fatty acids, but glucose is used by the brain, renal medulla and erythrocytes (partly after Vander *et al.*).

During prolonged fasting—which can go on for about 60 days provided water is supplied—an additional source of glucose is the body proteins, mainly of muscle (see p. 170). A large fraction of this protein, which is apparently not absolutely essential for function, can be broken down to amino acids which are then converted to glucose. In normal life, however, protein is not used for this purpose to any great extent.

Glucose sparing and fat usage
Almost all body cells, including the liver but excluding the nervous system, are converted to using fat during fasting, thus sparing glucose for the use of the nervous system. The triglycerides of adipose tissue are catabolised to glycerol and fatty acids; these fatty acids circulate in the blood, are taken up by tissue cells and then used in their Krebs cycles. In the liver, however, these fatty acids are converted to ketone bodies (e.g. acetone), instead of entering the Krebs cycle, and then released into the

circulation to be used by tissue cells in their Krebs cycle. If the concentration of ketone bodies is high, as in prolonged starvation or untreated diabetes mellitus, then it gives a characteristic odour to the breath. Fig. 11.33 shows a summary of these events.

Insulin

Insulin is formed and stored (to a small extent) in the β-cells of the Islets of Langerhans in the pancreas (Fig. 11.36a). These make up 60–90 percent of the 1–2,000,000 islet cells. α-cells, secreting glucagon, and D-cells, which secrete gastrin, are present in small numbers. The pancreas is derived from the gut and here the general rule that gut hormones are polypeptides applies.

Insulin is composed of two peptide chains joined by disulphide bonds (Fig. 11.36b). It is actually made as a pro-insulin in the ribosomes of the rough endoplasmic reticulum of β-cells and stored by the Golgi apparatus as granules (Fig. 11.36a). These granules contain trypsin-like enzymes which split off the C-peptide, forming insulin. On receiving a signal to cause insulin release these granules fuse with the cell

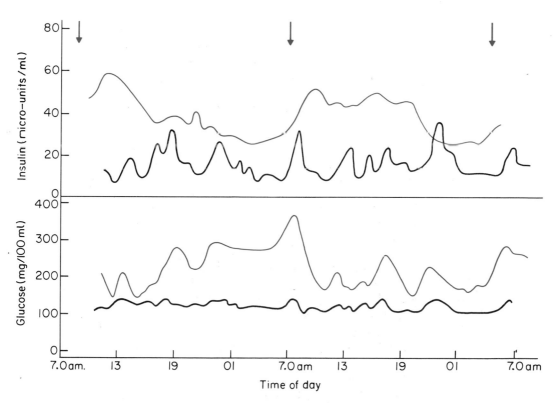

Fig. 11.34. Plasma glucose and insulin concentration in a normal subject over 24 hours (black lines), with results from a diabetic subject shown in red (the times of insulin injections shown by the red arrows). In the normal subject plasma glucose concentration is quite closely controlled in the diabetic it is not. (Conversion factor 18 mg/100 ml $\simeq$ 1 mmol/1) (after *Scientific American*).

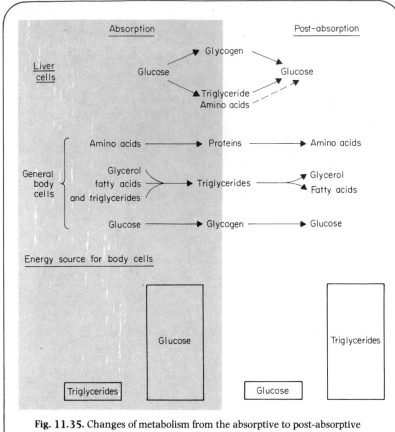

Fig. 11.35. Changes of metabolism from the absorptive to post-absorptive states (partly after Vander *et al.*).

membrane and release both insulin and the C-peptide (and perhaps some 5 percent pro-insulin).

In a normal subject the plasma glucose concentration does not vary much during the day but the insulin (and other hormone concentrations) do (Fig. 11.34). In the unusual situation of a glucose tolerance test (in which 50 g of glucose is drunk), the plasma glucose concentration does rise. The insulin is released into blood going to the liver and about half is inactivated by passage through the liver; this means that the liver, which is an important site of action of insulin, receives a higher insulin concentration than the rest of the body. Insulin probably exists in the plasma in two components: one in the same form as in the pancreas and the other one in a different form. The former acts on the tissues. Injected insulin disappears with a $T_{\frac{1}{2}}$ of 4 minutes, but becomes bound to the tissues and lasts there much longer.

Specific effects of insulin. Insulin is the main hormone which changes the body from the fasting to the absorptive phase; a large injection of it will reproduce most of the effects of absorption, and cause a profound fall in blood glucose—without, of course, uptake from the gut. It has two

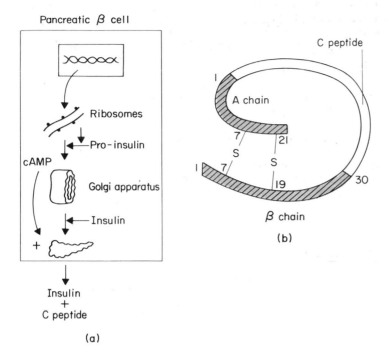

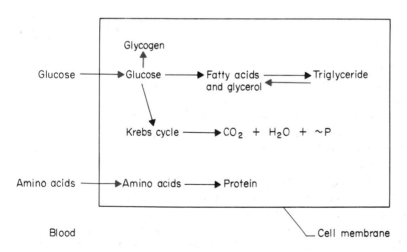

Fig. 11.36. (a) Production of insulin in pancreatic cells; (b) structure of pro-insulin. The size and structure of the C peptide varies with the species, that of insulin (shown shaded) shows only slight variations. (The numbers of amino acids are shown.)

Fig. 11.37. Main effects of insulin on metabolism (shown in red). Insulin increases glucose and amino-acid uptake by membrane effects (but not in the brain or liver); this leads to increased synthesis by mass action. It also has a direct effect on the glucose–glycogen pathway and inhibits triglyceride breakdown.

primary membrane effects (Fig. 11.37): it increases both the uptake of glucose and of amino acids into most cells, thus stimulating glycogen, fat and protein synthesis by an increase of glucose and amino-acid concentration in the cells. It also has direct stimulating effects on the rate-limiting steps in glycogen synthesis from glucose, causing increased synthesis. It decreases the breakdown of triglycerides, thus causing a net increase in triglyceride stores. Insulin, therefore, promotes glucose utilisation and decreases glucose synthesis.

In the absence of insulin these reactions are reversed. The uptake of glucose and of amino acids by most cells is reduced; this, plus the absence of the direct enzymic effects, causes breakdown of glycogen, fat and protein. The raised triglycerides in the plasma causes more to enter cells and increases the triglyceride usage for energy. The glucose uptake by brain and liver is not much affected because it is not dependent on insulin.

Control of insulin release. The rate of release of insulin, and hence its plasma concentration, is mainly controlled by the plasma glucose concentration (Fig. 11.38). Other factors interact with glucose: for example,

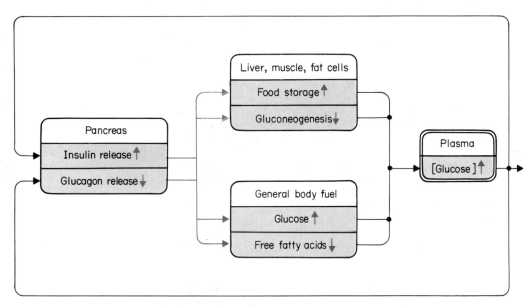

Fig. 11.38. Interactions of insulin and glucagon on the control of plasma glucose concentration. A disturbance of plasma [glucose] (shown as an increase) is followed by the consequences shown (arrows in other compartments) which correct the disturbance. Insulin and glucagon are controlling elements keeping plasma glucose constant. Their actions on cellular metabolism go hand in hand with their actions on plasma glucose.

metabolic fuels, several peptide hormones and some autonomic transmitters (Fig. 11.39a). When the β-cells are stimulated to release insulin they can do so immediately (< 1 min) from the storage granules, but then show a slower sustained release which requires new hormone synthesis. There is no effect of the anterior pituitary on insulin release.

The β-cells have an adenyl cyclase mechanism. An increase in cAMP

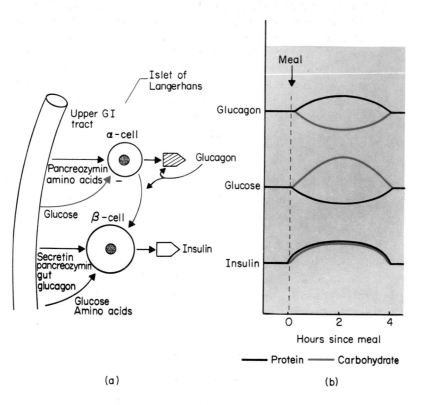

Fig. 11.39. (a) Factors affecting insulin and glucagon release—all are positive except for the red line (after Catt). (b) Effects on the plasma concentrations of glucagon, glucose and insulin of a protein and a carbohydrate meal. The ordinate is in arbitrary units.

Both glucose and amino acids by mouth give much larger insulin and glucagon responses than if given intravenously. This is thought to be due to the augmentation by the digestive enzymes shown. The amino acids which can be converted to glucose (glucogenic ones, for example glycine) give a bigger glucagon response than the others.

in the cells causes an increase in insulin secretion and production, and a decrease inhibits secretion. It has been suggested that glucose enters the β-cells, is metabolised to glucose-6-phosphate, forms ATP and stimulates cAMP production. This is an unusual way for an increase in cAMP to occur, usually membrane receptors are required.

The metabolic effects of glucagon, growth hormone and adrenaline
These three hormones have effects which, at an elementary level anyway, are opposite to those of insulin. They are, therefore, the hormones of fasting, whereas insulin is the hormone of feasting (Table 11.5). The precise effects of these hormones and the importance of each is not yet clear, but the overall effect is that they raise the plasma glucose and fatty-acid concentrations; this latter effect increases fatty-acid entry into

Table 11.5. Summary of the main effects of hormones which oppose those of insulin. The glucagon action on liver is on two time scales; a rapid one on glycogen breakdown and a slower one on gluconeogenesis.

Action	Result in plasma concentration	Glucagon	Adrenaline	Growth hormone
Glycogenolysis	↑ glucose	yes	yes	—
Gluconeogenesis	↑ glucose	yes	—	—
Fat metabolism	↑ fatty acids and glucose	—	—	yes

and usage by the body cells and thus, as a secondary consequence, spares glucose.

The main stimulus for their release is, as expected, a decrease in the plasma glucose concentration, but the pathway for this effect and the detailed interactions differ with each hormone. Glucagon is a small polypeptide released from the α-cells of the pancreas; a related form is released from the intestinal mucosa during glucose absorption. It has an action on the membrane of liver cells, causes the release of cAMP which then mediates its intracellular effects. The detailed factors acting on glucagon release are quite complex (Fig. 11.39), and, like insulin, release is influenced by gut hormones. The release of growth hormone by the pituitary and adrenaline by the adrenal medulla occurs when the hypothalamic glucose receptors are stimulated by low glucose.

Interactions of insulin and glucagon. The way in which insulin and glucagon normally interact is more complex than suggested in the above account, and cannot properly be described as simply as we have done. In addition to the states of feasting and fasting, the kind of food to be metabolised alters the hormones required to do it. If a carbohydrate meal is eaten, the blood is flooded with glucose and other sugars, insulin is secreted and the combination of this insulin and carbohydrate depresses glucagon production (Fig. 11.39). If a protein meal is eaten, the blood is flooded with amino acids; this causes a large rise in plasma insulin but now there is no corresponding rise in glucose uptake to keep up the plasma glucose. In this circumstance the plasma glucagon concentration also rises, thus keeping up the plasma glucose by glycogenolysis and gluconeogenesis in the liver. Possible mechanisms for this are suggested by Fig. 11.39. Normally, of course, meals consist of mixtures of carbohydrate, fat and protein so that the hormone response is a mixture of insulin and glucagon responses. It is therefore more useful to think in terms of the relative blood concentrations of both hormones in any given situation, rather than of each hormone separately. This can be done by considering the molar ratios of insulin and glucagon (I/G ratio).

It has been found that if the molar ratio of insulin to glucagon exceeds 5 then glucose is being stored in the liver, if it is less than 5 then glucose is being released by the liver. After a large carbohydrate meal the ratio may rise as high as 30:1; during starvation it may fall to 1:2 (Fig. 11.40). Thus, during energy storage the molar ratio of insulin to glucagon is high, favouring the deposition of glycogen, protein and fat;

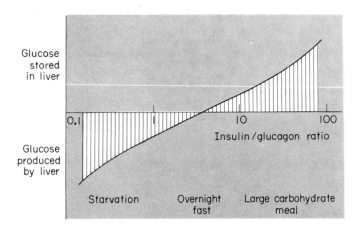

Fig. 11.40. Effect of insulin/glucagon ratios on glucose storage or production by the liver. The various conditions producing these ratios are indicated along the bottom of the graph. The insulin/glucagon ratio is on a log scale; the liver glucose is on an arbitrary scale.

This figure is based on the idea that insulin promotes the entry of materials into cells, while glucagon promotes the release of stored nutrients; thus the α- and β-cells of the pancreas act as a single functional unit controlling nutrient movement (see also Fig. 11.38) (after Ganong).

during energy mobilisation the molar ratio is low favouring glycogen breakdown, protein conversion to glucose and fat breakdown.

BONE AND THE CALCIUM REGULATING HORMONES

In this section we are concerned with the control of one substance—calcium—by the action of three hormones—vitamin D, parathormone and calcitonin—working at a variety of sites. We will tackle this by first considering the general properties and movements of calcium and then the action of these hormones on this.

Calcium

Calcium is widely distributed in the body and is essential for most bodily functions. Its biological roles are:

1 *Structural*, i.e. bone and teeth (contains 90 percent of total), membrane integrity.

2 *Electrical*—cardiac action potential.

3 *Regulatory*

(a) extracellular (mM range)—blood clotting

(b) intracellular (μM range)—muscle excitation–contraction coupling; exocytosis n–m junction, glands

Ca^{2+} on K^+ permeability

? cell–cell coupling

? oocyte activation

? non-muscle motility.

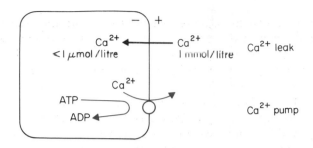

Fig. 11.41. Pump/leak movements of calcium across the cell membrane. The numerical values for free calcium are approximate.

The distribution of calcium between the extracellular fluid and the cells is regulated by calcium pumps at the cell membrane. Figure 11.41 shows this schematically, with approximate figures for free intracellular and extracellular calcium concentrations.

Phosphate is important for such substances as ATP, buffers, bone and teeth. It is present mostly in plasma as HPO_3^{2-} (85 percent), less as $H_2PO_4^-$ and organic phosphates with little bound. The actual concentration is some 5–6 mg/100 ml (1·8 mmol/l of phosphorus) in children and about half in adults. Most of the absorbed phosphate is in the inorganic form, or is converted to this during digestion.

Bone is a heterogeneous material, that is, it is made up of several different materials combined together. The advantages of such combinations are that they combine the virtues of each of the constituent materials. Everyday examples of such materials are reinforced concrete and glass-reinforced plastics. These consist essentially of fibres (steel or glass) which resist extension and a matrix (concrete or resin; a polymer) which binds the fibres together and withstands compression. Bone consists of a combination of three substances: *collagen fibres* bound together with a soft *mucopolysaccharide polymer*, hardened with deposition of *bone minerals* on the collagen fibres. Man-made structures and natural bone are therefore similar in construction; in both instances the orientation and density of fibres is arranged at stress points to give maximum strength.

The bone minerals are laid down as small flat crystals some 20 nm long by 3 nm thick, attached to the fibres in a regular way. Each crystal is surrounded by a watery shell which allows exchanges of the ions to occur. Because of their small size, the crystals have a high surface to volume ratio facilitating rapid exchanges.

Bone is a living structure; it contains three types of cell: (1) the osteoblasts which are found where new bone is being made, so are perhaps producing extracellular elements collagen and mucopolysaccharides; (2) the osteocytes whose function is unclear but may be descendants of osteoblasts; and (3), osteoclasts which are giant multinucleated cells present where bone is being resorbed, and which perhaps ingest parts of bone and break it down. These cells are polarised with villi and phagocytic processes absorbing bone at one side. It is probable that hormonal control of these cells occurs on two time scales: (1) a minute-to-minute modulation; and (2) a coordinated stimulation of the differentiation of one cell type or another so that the bone either undergoes proliferation or resorption. Ordinarily these processes of remodelling bone go on all the time.

The calcium in bone is laid down as a compound of Ca^{2+}, PO_4^- and OH^- called hydroxyapatite $[3Ca_3(PO_4)_2, Ca(OH)_2]$. Bone also contains Na^+, Mg^+ and K^+ combined in some unknown way; these contain 40, 60 and 5 percent, respectively, of the body's stores of these substances. At present it is not clear how this deposition of calcium crystals occurs; it is possible that the hydroxyapatite is precipitated directly or via an intermediate substance which then hydrolyses. It is known that the enzyme alkaline phosphatase, which is found in association with the osteoblasts, is important for this process.

Resorption of bone is also not understood. Perhaps osteoblasts cause a local increase in the concentration of citric and lactic acids with formation of a Ca–citric acid complex, and consequent movement of calcium out of the bone.

Calcium regulation

The plasma calcium concentration is normally kept within a range of ± 3 percent of its normal value of 10 mg/100 ml (2·5 mmol/l); approximately half of this exists as free calcium, the rest is bound. After intense exercise the plasma Ca^{2+} rises transiently by about 10 percent, in phase with the fall in pH. It takes part in three main exchanges and is regulated by three hormones. The exchanges are between the plasma and the gut, bone and kidney (Fig. 11.42). The hormones acting on these exchanges are vitamin D, parathyroid hormone (PH) and calcitonin. The first two act to increase the plasma calcium concentration, the last to decrease it. It seems, however, that considerable exchanges continue in the absence of these hormones, so that they should be considered as 'fine controls'. In the absence of these hormones the plasma calcium settles down at a lower than normal concentration (7 mg/100 ml, 2 nmol/l); the net hormone effect is therefore to raise plasma calcium.

Vitamin D is the term used to describe a group of closely related sterols produced by the action of ultraviolet light on various precursors (Fig. 11.43). The release of the active form 1,25-DHCC from the kidney is increased by low plasma calcium concentration. As the active form is the derivative of a steroid it is thought to act by a cytoplasmic receptor and synthesis of new protein. Together with parathyroid hormone, 1,25-DHCC increases gut absorption and reabsorption from bone and kidney. It is also necessary for new bone formation. The serum concentration of 1,25-DHCC increases greatly during periods of growth; an effect controlled ultimately by the pituitary.

Parathyroid hormone. PH is secreted from the chief cells of the parathyroid glands (two pairs) which are always close to the thyroids in the neck (Fig. 11.15). The parathyroids are derived from gut and so—as in the general rule—the hormone structure is a polypeptide; the first 27–34

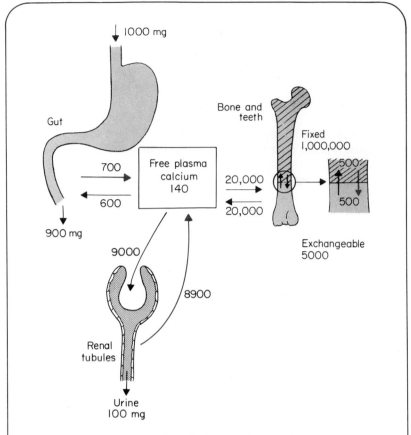

Fig. 11.42. Quantitative outline of calcium exchanges. The figures (in milligrams) represent the turnover per day or the total size of the various pools. The lines in red indicate the reactions affected by the hormones.

Calcium exchanges also take place with other tissues and with that bound to the plasma albumin (100 mg); some 20–300 mg/day is also lost by sweat. The sizes of the bone calcium compartments shown are proportional to the log of the calcium contents; the linear proportions are 1:200. Survival is possible on calcium intakes as low as 200 mg/day; in this case changes occur in the uptake mechanism from the gut.

amino acids contain all the activity. The release of PH is controlled by the plasma calcium concentration (Fig. 11.44) acting directly on the parathyroid; low plasma calcium concentration increases its secretion. There is evidence that it is made as a prohormone and a part of it is cleaved off before secretion. The hormone acts on the surface of the target cell producing the intracellular messenger cAMP. PH acts on bone on two time scales: an early effect (<20 minutes) leading to calcium release and a slower effect on bone turnover and remodelling.

In addition to its action on calcium movements in bone, gut and kidney it also increases renal loss of phosphate. This action means that the product of Ca × phosphate in the blood is not increased by its action and so precipitation does not occur.

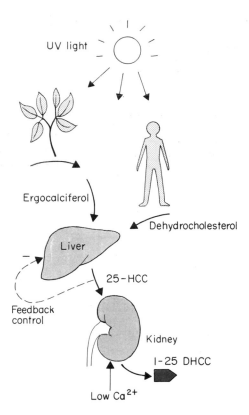

Fig. 11.43. Steps in the production and activation of the vitamin D group of substances (cholecalciferol or CC). Our bodies can produce the animal form of vitamin D by skin irradiation or from the plant form. Certain other animals cannot use the plant form.

25-HCC (25-hydroxycholecalciferol) is long lived because it is bound to an α globulin. 1,25-DHCC (1,25-dihydroxycholecalciferol) production in the kidney requires protein synthesis, and is the active form of the hormone. With low serum [calcium] 1,25-DHCC is released; with high serum [calcium] the alternative form 24,25-DHCC, is released; this has no effect on Ca movements.

Calcitonin is secreted by the C-cells of the thyroid gland. It is a small polypeptide which probably acts on the surface of target cells to produce an intracellular messenger. The release of calcitonin is controlled by the plasma calcium concentration acting directly on the producing cells. High plasma calcium concentration increases its secretion. It acts more quickly than the other two hormones. The main action of calcitonin is to decrease the calcium coming out of the bone; therefore, the plasma concentration drops due to the continued entry of calcium into bone. In adult man its role is not too clear; it is probably an accessory fine control to PH and may be important in pregnancy.

Control of plasma calcium concentration. Calcitonin, PH, and vitamin D (1,25-DHCC) interact as shown in Fig. 11.44 to control the plasma calcium concentration. The cellular interactions of the hormones can be understood if you remember the scheme shown in Fig. 11.45, where vitamin D (1,25-DHCC) produces a transport protein whose action is allowed by PH but ineffective in the presence of calcitonin. This scheme,

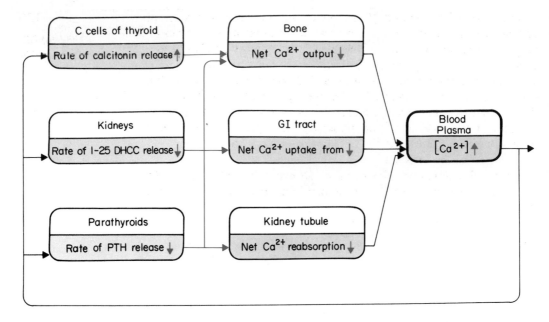

Fig. 11.44. Hormonal control of plasma calcium concentrated (simplified). A disturbance of plasma [Ca²⁺] (shown as an increase) is followed by the consequences shown (arrows in the other compartments) which correct the disturbance. Ca²⁺ is the regulated parameter; the hormones act as regulatory agents.

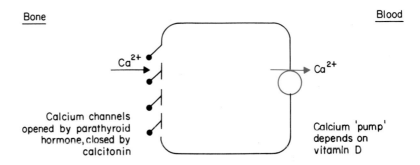

Fig. 11.45. Scheme for interaction of the three hormones affecting calcium mobilisation from bone. Vitamin D causes the production of a transport protein which pumps calcium into the blood from the lining cell. Parathyroid increases and calcitonin decreases the calcium permeability of the side of the cell facing bone, thus controlling the rate at which calcium enters the cell and hence the overall rate of transport from bone to blood. (Adapted from Catt.)

primarily developed for bone, is also applicable to gut and kidney if the action of calcitonin is excluded.

Other modulators of Ca
Mechanical stress on bone promotes the deposition of calcium and its absence causes resorption. This is a problem in prolonged bed rest and in space flights; a secondary complication which arises is deposition of calcium in the excretory pathways, causing renal stones.

Blood pH affects ionised plasma calcium, such that a rise in pH decreases the ionised calcium concentration and a fall in pH increases it. pH acts on plasma calcium by two mechanisms, both working in the same direction. If the pH is increased then: (1) more anionic sites appear in the plasma proteins and so more calcium is bound; and (2) calcium deposition in bone is favoured. Both mechanisms, therefore, decrease the ionised calcium. Lowering blood pH reverses both mechanisms.

Clinically these effects are of importance in hyperventilation, which raises the blood pH and so lowers the plasma calcium concentration and may give muscular spasms.

The detailed actions of these hormones are more complex than stated above, and still imperfectly understood. A fuller account is given in Tepperman.

Part 4
Reproduction

Chapter 12
Reproductive Physiology

The great diversity of species in the animal kingdom is mainly a result of the development of sexual reproduction, in which the genetic material from two individuals is combined in their offspring.

The sex of higher animals is determined by one pair of chromosomes, the sex chromosomes. The nucleus of body cells of a genetic female contains two similar sex chromosomes, the X chromosomes, while those of the genetic male contain one X and one smaller Y chromosome; the many aspects of sexual differentiation arise either directly or indirectly from this chromosome difference. In the normal human there are, in addition to the sex chromosomes, 44 non-sex chromosomes or autosomes.

Fertilisation occurs when a male germ cell (the spermatozoon)—containing half the normal number of chromosomes (i.e. 22X or 22Y)—combines with a female germ cell (the ovum) again containing half the normal number of chromosomes (i.e. 22X).

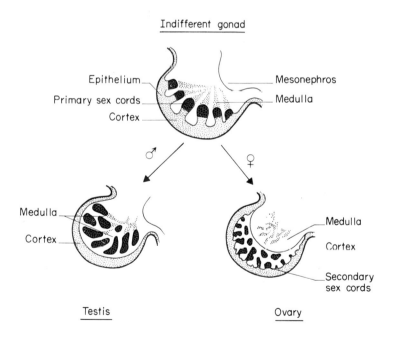

Fig. 12.1. Differentiation of the gonads. When first formed the embryonic gonad has the potential to develop into either a testis or an ovary; the course of development it takes depends upon the genetic constitution of the embryo. In a normal male the primary sex cords of the medulla develop to form the seminiferous tubules. In a normal female the primary sex cords regress and are replaced by secondary sex cords derived from the gonadal cortex (after Ganong).

The male and female reproductive systems both consist of three parts: 1, organs where the germ cells are made—the gonads (testes in the male, ovaries in the female); 2, ducts for conveying germ cells; and 3, the external sexual apparatus. The development of these three parts occurs as follows.

1   The gonads develop under genetic control from a primordial ridge of tissue near the adrenal gland. Initially the primordial gonad is divided into an outer cortical part and an inner medullary part, identical in both

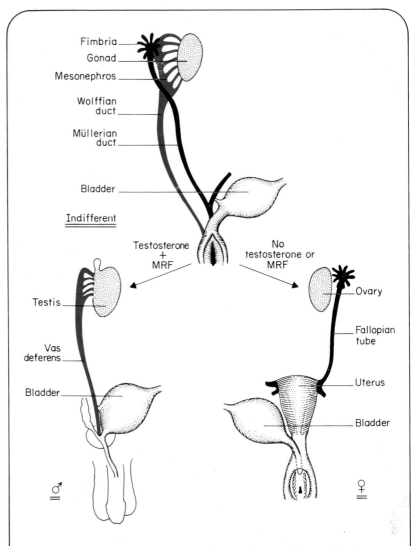

Fig. 12.2. Differentiation of the genital ducts. The genital ducts of the fetus initially have the potential to develop into those of either sex. If testosterone and Müllerian regression factor (MRF) are both secreted by the fetal testis, a male duct system is formed. In the absence of these substances female ducts alone persist (after Ganong).

sexes. In the male the medulla enlarges to form the testes, in the female the cortex becomes the ovary (Fig. 12.1). Once the testis develops it starts to produce the male hormone—testosterone—which then influences the subsequent development of the fetal reproductive apparatus.

2 Before the gonads differentiate to an ovary or testis, they are connected to the body surface by two sets of ducts; after sexual differentiation one or the other set survives, depending on the sex of the child (Fig. 12.2). In the normal male, during development, the testis produces both testesterone and a factor (Müllerian regression factor, MRF) which causes degeneration of the female ducts. Together, testosterone and Müllerian regression factor cause the development of the male duct system.

3 Testosterone alone is required for male external sexual apparatus to develop; in its absence the female ducts and external apparatus develop.

The reproductive apparatus therefore develops through a sequence of steps. Initially the genetic information carried in the fetal cells controls gonadal development; in the male testicular hormones induce subsequent male development. Female development takes place in their absence. Aberrant sexual development can therefore result from genetic or hormonal abnormalities.

FEMALE REPRODUCTIVE SYSTEM

The general structure of the female reproductive system is shown in Fig. 12.3. The ovaries have the dual function of producing germ cells and of secreting female sex hormones. The fetal ovary contains many primordial germ cells, which divide mitotically and develop into a larger type of cell (oogonia). The oogonia themselves undergo further mitotic division and finally give rise to primary oocytes. Oocytes are the cells from which ova will be formed by meiotic (reduction) cell division. Multipli-

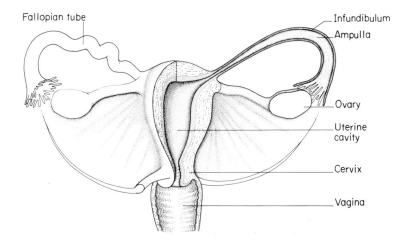

Fig. 12.3. Human female reproductive system. A single egg is normally released from one of the ovaries about every 28 days; it is then picked up by a Fallopian tube, where it may be fertilised before passing on to the uterus. If fertilisation occurs the embryo implants into the uterine wall; if not the egg degenerates.

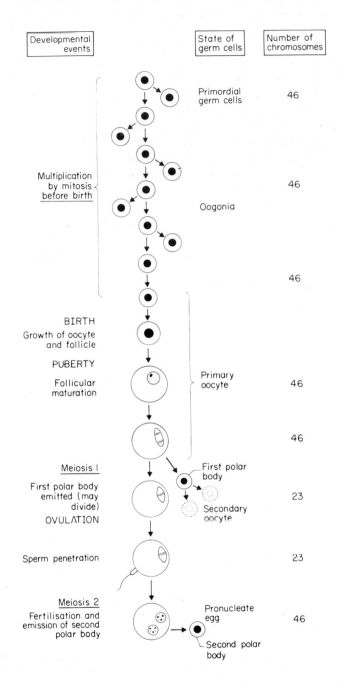

Fig. 12.4. Development of female germ cells. Before birth there is an immense proliferation of germ cells in the fetal ovary. By the time of birth this phase of multiplication is complete; the germ cell remains in a state of arrested development until shortly before ovulation, when they undergo the first meiotic (reduction) cell division. The second meiotic division takes place after a spermatozoon penetrates the oocyte membrane (after Austin & Short).

cation of female germ cells is completed before birth, so that if they are destroyed at this stage they cannot be replaced. Primary oocytes remain in a state of arrested development from the time at which they are formed (before birth) until just before ovulation, when they divide by meiosis. Meiosis consists of two cell divisions; one in which the number of chromosomes is halved, and a second non-reductive division. The second meiotic division of the oocyte does not actually take place until fusion occurs between a sperm and the oocyte surface. At both meiotic divisions, there is an uneven distribution of cytoplasm to the daughter cells; those which receive least cytoplasm are termed polar bodies and will eventually degenerate. Although the female germ cell is usually termed an ovum after ovulation, strictly speaking it remains an oocyte until after the second division at fertilisation. Development of the female germ cell is shown in Fig. 12.4.

Oocytes in the fetal ovary become surrounded by a single layer of cells termed the membrana granulosa; together, the oocyte and the surrounding cell layer constitute a primordial follicle. At puberty in the human, each ovary contains approximately 100,000 to 500,000 primordial follicles; this represents a great over-production, since only a few hundred ova will ever be released.

Ovarian and uterine cycles

Release of female germ cells is cyclical and intermittent; this is reflected by corresponding cyclic structural and functional changes throughout the female reproductive system. These changes are dependent on two interrelated cycles: the ovarian cycle, and the uterine (menstrual) cycle. Both these cycles, although variable, last approximately 28 days. The menstrual cycle is controlled by the ovarian cycle via sex hormones.

Ovarian cycle

Fig. 12.5 represents diagrammatically the structures which may be seen within the ovary at different phases of the ovarian cycle. At no one time are all those stages present together. At the beginning of each cycle several follicles start to enlarge. Normally one continues to enlarge, while the others regress and are said to become atretic. The number of follicles that develop is determined by the circulating levels of pituitary gonadotrophic hormones (see below). During the development of the follicle, the granulosa cells proliferate and form a layer several cells thick. A layer of flattened cells (termed the theca) then forms around the granulosa cells. As the follicle develops further the granulosa cells start to secrete a fluid which forms a space in the follicle—the antrum. After further enlargement the mature (Graafian) follicle is ready to be released from the ovary (ovulation). This normally takes place at about 14–16 days after the start of the preceding menstrual period. (Both the ovarian and menstrual cycles are said to begin on the first day of the menstrual period.)

Ovulation probably occurs in the following way. The enlarging Graafian follicle stretches the capsule surrounding the ovary, so locally restricting blood flow through the capillaries. This causes local degeneration and weakening of ovarian tissue, which, apparently together with

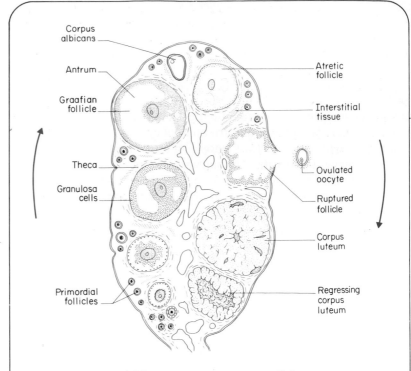

Fig. 12.5. Diagram of the ovary. At no one time are all the structures illustrated present. At the beginning of the ovarian cycle several primordial follicles start to develop; one of these will mature to form a Graafian follicle, while others regress. Near the middle of the cycle the mature follicle releases its oocyte (ovulation) and then differentiates to become a corpus luteum; this starts to regress towards the end of the cycle and finally forms an avascular scar (corpus albicans) (after Turner).

enzymic tissue breakdown, causes the ripe follicle to break through the ovary wall. The oocyte is released along with many granulosa cells, some of which remain attached to the oocyte and some which form a cloud of cells around the oocyte (the 'cumulus oophorus'). After ovulation the oocyte (at this stage often loosely called an ovum) is picked up by the finger-like projections at the mouth of the Fallopian tube. Cells of the Graafian follicle which are left in the ovary after ovulation persist and form a folded structure the cavity of which becomes filled with blood (the corpus haemorrhagicum). The granulosa cells at this stage begin to proliferate and soon replace the blood clot; the whole structure is then known as a corpus luteum. In an infertile cycle the corpus luteum only lasts for about 10 days before it starts to regress, finally forming an avascular scar (a corpus albicans). If pregnancy does occur the corpus luteum is maintained for several months.

Uterine cycle
The uterus has three layers: a thin layer in contact with the body cavity, a thick muscular layer (myometrium) and a mucous membrane bound-

ing the uterine cavity (endometrium). During the course of the uterine cycle the superficial layers of the endometrium undergo cyclic structural changes which can be divided into three phases: the menstrual phase, during which time the superficial layers are lost; the proliferative (or follicular) phase, during which the destruction resulting from menstruation is repaired by proliferation of endometrial cells; and the secretory (or progestational) phase, during which glands in the endometrial wall become more complex in structure and start to secrete mucus. In this last phase the uterine lining is prepared for implantation of the fertilised ovum. The endometrial changes which occur during the course of one cycle are shown diagrammatically in Fig. 12.6.

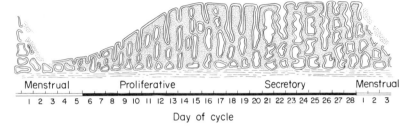

| Menstrual | Proliferative | Secretory | Menstrual |

| 1 2 3 4 5 6 7 8 9 10 11 12 13 14 15 16 17 18 19 20 21 22 23 24 25 26 27 28 1 2 3 |

Day of cycle

Fig. 12.6. Uterine (menstrual) cycle. After a menstrual period the uterine endometrium rapidly thickens as a result of the cellular proliferation (proliferative phase). In the second half of the cycle the structure of the glands in the endometrium becomes more complex and they start to secrete mucus (secretory phase). Finally, the endometrium breaks down (menstrual phase) (after Ganong).

Uterine cavity

Gland

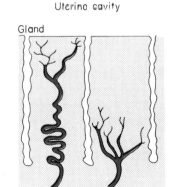

Fig. 12.7. Vascular supply to the uterine endometrium. The layers of the endometrium nearest the uterine cavity are supplied by spiral arteries. deeper layers receive blood from the basal arteries (after Ganong).

Spiral arteries   Basal arteries

Shedding of the outer layers of the endometrium during menstruation probably takes place as follows. Spiral arteries which supply the superficial layer of the endometrium (Fig. 12.7) constrict and cause local ischaemia and weakening of the capillary walls. The spiral arteries then re-open and blood leaks out of the blood vessels, so detaching the outer layers of the endometrium and producing the menstrual flow.

Ovarian hormones

The ovary produces two major types of steroid hormone: oestrogens and progesterone. In the human the major oestrogen is 17β-oestradiol. In addition to other important functions, these ovarian steroids are of key importance in maintaining both the ovarian and menstrual cycles. During the early part of the ovarian cycle, oestradiol is synthesised by the developing follicle so that the plasma level of this hormone increases progressively and reaches a peak approximately 2 days before ovulation (Fig. 12.8); oestradiol is responsible for the proliferative (follicular) phase

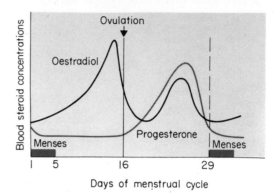

Fig. 12.8. Blood concentrations of 17β-oestradiol and progesterone during the course of the ovarian cycle. The blood concentration of oestradiol rises as the secretory activity of developing ovarian follicles increases; the level starts to fall just before ovulation. The peaks in progesterone and oestradiol concentrations which occur after ovulation result from secretion of these hormones by the active corpus luteum (after Austin & Short).

of endometrial development. The corpus luteum, formed after ovulation, starts to synthesise both 17β-oestradiol and progesterone (Fig. 12.8). Progesterone induces the secretory (progestational) phase of the uterine cycle; that is to say, it prepares the uterine endometrium for implantation of the fertilised egg. To be maintained in its secretory phase, the endometrium requires the hormonal support of oestrogen and progesterone. If fertilisation does not take place, the corpus luteum starts to regress towards the end of the uterine cycle and so secretes decreasing amounts of oestrogen and progesterone; the circulating levels of these hormones fall, and so the uterine endometrium is shed.

Anterior pituitary and hypothalamus

So far we have seen that cyclic changes in the ovary hormonally control corresponding changes in the uterine endometrium. The ovarian cycle is, however, itself controlled by the anterior pituitary, which, in turn, is under the influence of the hypothalamus.

The anterior pituitary produces three protein hormones (gonadotrophins) which are of primary importance in reproductive function; they are follicle stimulating hormone (FSH), luteinising hormone (LH) and prolactin. The secretion of gonadotrophins is dependent on the activity of the hypothalamus, which is one reason why female reproductive cycles and fertility can be greatly affected by emotional influences. The hypothalamus controls the anterior pituitary by the release of substances from the terminals of specialised neurosecretory neurones into the hypophyseal blood portal system (see Fig. 9.49). The secretion of both FSH and LH is controlled by a single hypothalamic gonadotrophin-releasing hormone (GnRH) which is sometimes referred to as luteinising hormone releasing hormone or factor (LHRH or LRF).

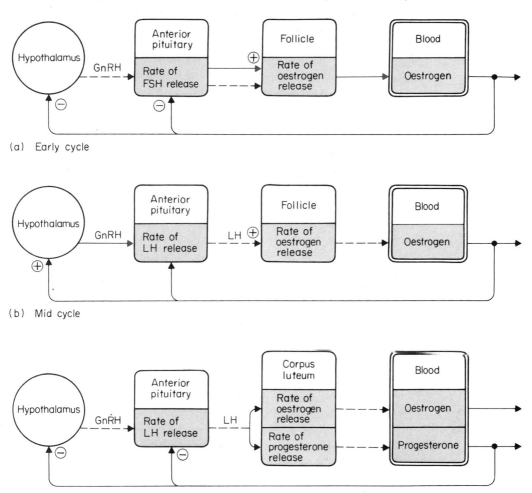

(a) Early cycle

(b) Mid cycle

(c) Late cycle

Fig. 12.9. Hormonal control of ovarian function. At the beginning of the ovarian cycle follicular development is triggered by a rise in level of follicle stimulating hormone (FSH) secreted from the anterior pituitary gland under the influence of gonadotrophin-releasing hormone (GnRH) from the hypothalamus; the developing follicles start to secrete increasing amounts of oestrogens. These oestrogens exert two feedback effects on pituitary gonadotrophin secretion; firstly, they inhibit secretion of FSH (A), and secondly, they enhance secretion of luteinising hormone (LH), which, itself further increases oestrogen secretion (B). The resulting peak in plasma LH concentration triggers ovulation. After ovulation LH stimulates the corpus luteum to secrete both oestrogens and progesterone; progesterone depresses LH secretion by a negative feedback effect (C).

Ovarian steroids probably exert their feedback effects upon gonadotrophin secretion both by altering secretion of GnRH from the hypothalamus and by modulating the sensitivity of the pituitary to this releasing hormone.

Prolactin secretion is controlled by prolactin inhibitory factor (PIF), which, when released from the hypothalamus, inhibits secretion of prolactin from the pituitary. Therefore, prolactin is secreted in the absence of PIF. GnRH is a decapeptide, while PIF is the catecholamine dopamine (see p. 234).

Control of the ovarian cycle

At the onset of the ovarian cycle the blood concentrations of both LH and FSH are relatively low (Fig. 12.9). As the cycle progresses their concentrations start to increase. The rise in FSH is primarily responsible for triggering follicular development and oestradiol secretion while later stages of follicular development are dependent on both FSH and LH. Oestradiol mediates two types of feedback effect upon the secretion of pituitary gonadotrophins; at low plasma concentrations it produces negative feedback and so depresses FSH secretion. More prolonged exposure to relatively high plasma concentrations of oestradiol results in a positive feedback which enhances gonadotrophin secretion. This positive feedback effect is responsible for the mid-cycle peak in plasma LH which actually triggers ovulation (Fig. 12.9b). After ovulation the concentration of LH remains slightly elevated and this hormone, probably together with prolactin, is responsible for the formation and maintenance of the corpus luteum. However, LH secretion is apparently inhibited towards the end of the cycle as a result of negative feedback exerted by a rising concentration of progesterone (Fig. 12.9c). The mechanism responsible for the regression of the corpus luteum at the end of the human ovarian cycle has not been resolved. In some animals, such as sheep, one of a group of substances termed prostaglandins is involved; this, however, is apparently not the case in man. After regression of the corpus luteum, and the fall in levels of the ovarian steroids, oestradiol and progesterone, the concentrations of FSH gradually increase again.

Oestradiol and progesterone probably mediate their feedback effects both by modulating GnRH release from the hypothalamus and by altering the pituitary sensitivity to this releasing hormone.

The relative time course of various events associated with the ovarian and uterine cycles are summarized in Fig. 12.10.

Oral contraceptives

Although it was known for some time that oestrogens and progesterone could theoretically influence fertility, they are not effective as contraceptive agents because they are rapidly inactivated in the liver after oral administration. Synthetic compounds have been developed which have the biological activity of oestrogens or progesterone, but which are resistant to enzymic degradation. One of the most commonly used oral contraceptive formulations is the 'combined pill', which contains synthetic oestrogen and a progestogen (a compound with a progesterone-like action) and is taken from approximately the fifth to the twenty-fifth day of the menstrual cycle. This type of oral contraceptive acts by blocking the release of gonadotrophins from the pituitary, thus preventing ovulation. It is also thought that they have an additional anti-fertility action at the site of the uterine cervix and endometrium.

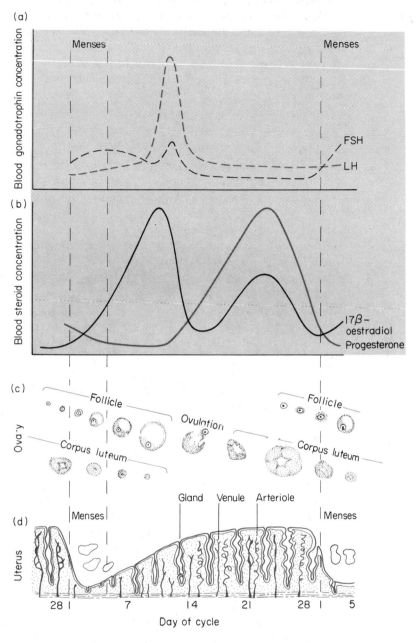

Fig. 12.10. Summary diagram of the major events which take place during the course of the ovarian and uterine cycles (modified from Segal).

(a) Blood concentration of follicle-stimulating hormone (FSH) and luteinising hormone (LH).
(b) Blood concentrations of ovarian steroids.
(c) Development of ovarian follicles and the corpus luteum.
(d) Changes in the uterine endometrium.

Pregnancy

Fertilisation normally takes place in the widest part (ampulla) of the Fallopian tube (Fig. 12.11). After a short delay the fertilised egg (zygote) starts dividing and forms a ball of of cells termed a morula. As the cells continue to divide they start to form a fluid-filled cavity, the blastocoele. A single layer of cells, the trophoblast, forms around the blastocoele: after implantation these trophoblast cells will form the placenta and embryonic membranes. At one pole of the blastocoele there is an agre-

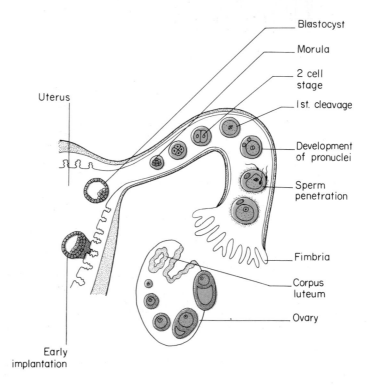

Fig. 12.11. Fertilisation of the egg and initial development of the zygote. The oocyte is released into the body cavity from the ripe ovarian follicle; it then enters the Fallopian tube, where it is fertilised. The zygote then starts to divide mitotically, becoming first a morula, then developing into a blastocyst; during implantation the outer layer of the blastocyst burrows into the uterine endometrium (modified from Tuchmann-Duplessis *et al.*).

gate of cells, the inner cell mass, which later forms the actual fetus. At this stage the embryo is called a blastocyst (Fig. 12.12).

The fertilised egg takes about 2–4 days to travel from the ampulla to the uterus (as a blastocyst). It then implants into the uterus (usually on the posterior wall) through the burrowing action of the trophoblast layer. After implantation the cell membranes of the trophoblast cells start to break down to form a syncytium. In the human placenta the trophoblast tissue disrupts the walls of blood vessels in the uterine wall to become directly bathed in maternal blood; as the placenta enlarges the trophoblast layer develops into the chorion.

The embryo is immunologically different from the mother and might

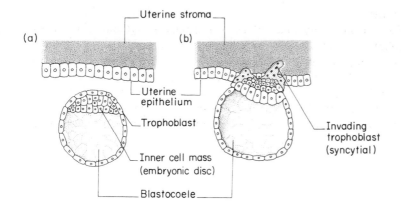

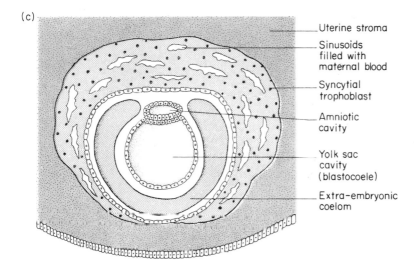

Fig. 12.12. Early stages in the development of an embryo (after Austin & Short).
(a) The blastocyst is composed of the trophoblast layer and the inner cell mass; the inner cell mass will later form the actual embryo, while the trophoblast layer forms the embryonic membranes and the fetal part of the placenta.
(b) As the blastocyst implants into the uterine wall the membranes of the invading portions of the trophoblast develop into a syncytium.
(c) Finally, the syncytial trophoblast breaks down the walls of the blood vessels in the uterine lining and becomes bathed in maternal blood.

be expected to be rejected as would be a graft from an incompatible donor. At present the way in which the embryo avoids rejection is unclear.

Hormones in pregnancy

One of the first problems to be overcome by the fertilised ovum is to prevent menstrual loss of the uterine endometrium and so prevent abortion. To maintain the uterine lining it is essential to prevent the fall in the blood progesterone concentration which occurs at the end of an infertile cycle as a result of luteal regression. Although the mechanism is not entirely clear, the corpus luteum is probably maintained by the secretion of two protein hormones from the trophoblast of the blastocyst. These hormones are known as human chorionic gonadotrophin (HCG) and human chorionic somatomammotrophin (HCS, also known as human placental lactogen). Thus, in a sense, the placenta takes over the role of the pituitary in maintaining hormone production by the corpus luteum at the beginning of pregnancy. In early pregnancy, as in the ovarian cycle, the corpus luteum is the major source of progesterone and oestrogen; however, as pregnancy progresses the placenta itself begins to take over some of the hormonal functions of the ovary by producing oestrogen and progesterone. In fact, by the sixth week of gestation the corpus luteum is no longer necessary. Fig. 12.13 shows

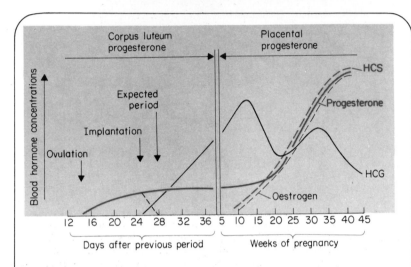

Fig. 12.13. Concentration of hormones in the maternal blood during the course of pregnancy. If fertilisation occurs the blood concentration of progesterone does not fall, but is maintained mainly through the action of human chorionic gonadotrophin (HCG) on the corpus luteum. By about the fifth week of gestation the placenta starts to take over progesterone synthesis from the corpus luteum; after the initial stages of pregnancy the placenta is responsible for synthesising progesterone, oestrogens, HCG and human chorionic somatomammotrophin (HCS).

the concentrations of human chorionic gonadotrophin, somato-mammotrophin and progestgerone during the course of pregnancy.

The ovary is able to synthesise progesterone from acetate (the starting point for the synthesis of all steroids) and can convert this to oestrogens; the placenta, on the other hand, lacks some of the enzymes necessary both to convert acetate to progesterone and progesterone to oestrogens. It therefore extracts the steroid cholesterol from the maternal blood supply and converts it to progesterone. Some of this progesterone passes to the fetus, where the fetal liver and adrenal cortex convert it to precursors which, in turn, can be used by the placenta to synthesise oestrogens. Thus, steroids are shuttled between the placenta and the fetus allowing the biosynthesis of oestrogens; this cooperation has given rise to the term 'feto-placental unit'. Of course, it must be remembered that the mother is essential in these syntheses, since she initially supplies

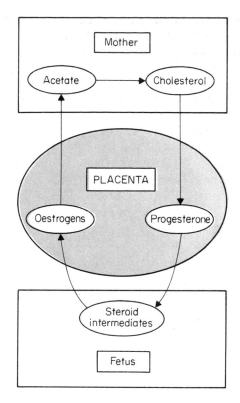

Fig. 12.14. Biosynthetic cooperation between the feto-placental unit and the mother. The placenta lacks the ability to perform some of the necessary steps in the synthesis of progesterone and oestrogens from acetate; the placenta therefore extracts cholesterol from the maternal blood and from it synthesises progesterone. Some of the progesterone produced in the placenta passes into the fetus, where it is converted into steroid intermediates, which, on returning to the placenta, can be used to form oestrogens.

the necessary cholesterol. Fig. 12.14 shows diagramatically the transfer of materials between the mother and the 'feto-placental unit'.

Pregnancy diagnosis

Modern pregnancy diagnosis depends on the detection of human chorionic gonadotrophin (HCG) in the urine of pregnant women by an immunological method. The urine sample to be tested is mixed with antiserum containing antibodies to HCG. The antiserum plus urine is then mixed with latex particles coated with HCG. If the urine contains no HCG the antiserum will react with the HCG on the latex particles and produce a precipitate, indicating a negative diagnosis. If the urine sample contains HCG the antibodies on the serum will have been used up and so cannot react with the HCG on the latex particles so there is no precipitation; this indicates a positive diagnosis. The general principles underlying immunological tests are discussed on p. 331 in Chapter 11.

Parturition

At present the mechanism by which parturition is started is not fully understood. Until recently the mother was thought to be entirely responsible for the control of parturition, while the fetus played a passive role. The onset of parturition was thought to be almost entirely controlled by the relative concentrations of progesterone and oestrogens in the maternal bloodstream, since uterine contractility is depressed by progesterone and enhanced by oestrogens.

It now seems that the fetus can control the onset of parturition. Certain undefined stimuli (possibly stress) can act upon the fetal hypothalamus to trigger the release of adrenocorticotrophic hormone (ACTH) from the fetal pituitary (Fig. 12.15). The ACTH, in turn, stimulates the fetal adrenal to secrete corticosteroids. These steroids cause a fall in the placental progesterone concentration (and hence increase uterine contractility) and an increase in the secretion of a prostaglandin. (Prostaglandins can induce powerful uterine contractions.) Although this view of parturition is based mainly on work with sheep, it is probably also applicable to humans.

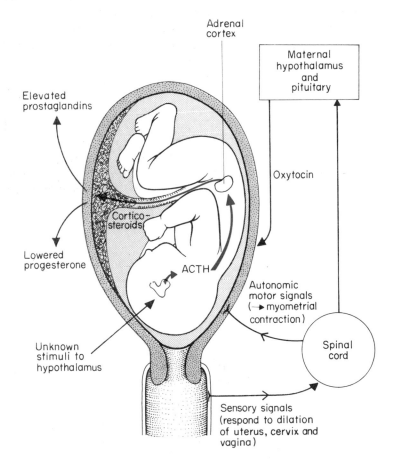

Fig. 12.15. Control of parturition. Stimuli (probably including stress) act upon the fetal hypothalamus and cause the release of adrenocorticotrophic hormone (ACTH) from the anterior pituitary gland; this in turn causes the secretion of corticosteroids from the fetal adrenal glands. Corticosteroids lower progesterone secretion and increase the secretion of prostaglandins by the placenta, so inducing uterine contractions. Once started uterine contractions are reinforced both by autonomic reflexes and through the release of oxytocin from the posterior pituitary gland (after Research in Reproduction).

Once parturition has begun uterine activity itself enhances uterine contraction; as the uterus contracts it dilates the cervix and stimulates stretch receptors in the cervical wall. Nerve impulses travel into the spinal cord, and cause uterine contraction via autonomic reflexes. Impulses from stretch receptors also travel up the spinal cord to the brain, and act on specialised hypothalamic neurones which send their axons through the pituitary stalk to terminate in the posterior lobe of the pituitary (see Fig. 9.49). When they are stimulated they release an octapeptide hormone (oxytocin) from their terminals into the blood. Oxytocin produces strong uterine contractions. The events which take place during parturition are summarised in Fig. 12.15.

Lactation

The mammary gland consists of two types of tissue: glandular tissue or parenchyma, and supporting tissue or stroma. The glandular tissue consists of small sacs of alveoli, which are lined with a single layer of glandular epithelium. The alveoli are surrounded by a layer of contractile myoepithelial tissue. Milk is formed by the cells of the glandular epithelium; it is conveyed from the alveoli to the nipple by a system of ducts (Fig. 12.16).

Lactation has two phases: milk secretion and milk removal. In milk secretion milk is secreted by glandular cells into the lumen of the alveoli. This process is controlled by a number of hormones. In the rat both prolactin and ACTH (and hence, presumably, corticosteroids) are necessary. Since both these hormones normally influence mammary development, after parturition their action must be switched so that they initiate milk secretion. Probably lactogenesis is blocked during gestation by the direct action of progesterone on the gland cells; when the progesterone

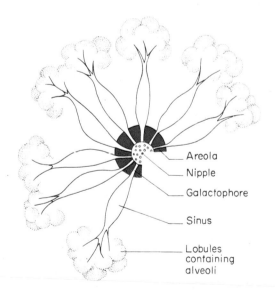

Areola

Nipple

Galactophore

Sinus

Lobules containing alveoli

Fig. 12.16. Structure of the human breast. Milk is secreted by sacs (alveoli) arranged into a number of lobules. Milk passes from the alveoli to a number of sinuses and is finally carried to the nipple via galactophores (after Austin & Short).

level falls at parturition, the protein hormones are able to induce lacto-genesis.

Milk removal is the transfer of milk from the alveoli to the nipple, where it is available to the infant. This is brought about in the following way by the milk ejection reflex. Sensory receptors in the nipple, stimulated by the suckling infant, send impulses up the spinal cord to the hypothalamus, where they trigger oxytocin release from the posterior pituitary (see p. 289). In addition to its action on the uterine myometrium (see p. 395) oxytocin causes contraction of the myoepithelial cells

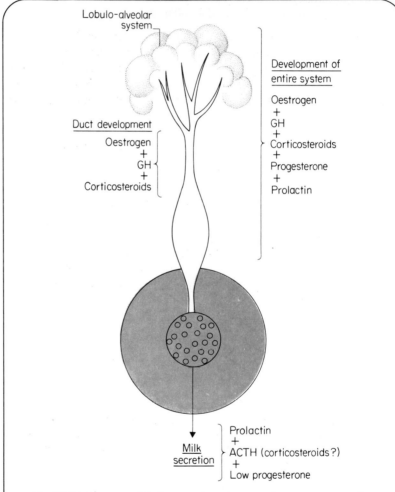

Fig. 12.17. Summary of the hormonal requirements for mammary growth and function. Hormones from the ovaries, pituitary gland and adrenal glands are necessary for normal breast function: duct development may take place in the presence of oestrogens, growth hormone (GH) and corticosteroids; the lobulo-alveolar system will develop if, in addition, progesterone and prolactin are available.

The hormonal requirements for milk secretion are not known with certainty, although it seems that prolactin and adrenocorticotrophic hormone (ACTH) or corticosteroids must be supplied; it also seems that the blood progesterone concentration must be relatively low.

surrounding the alveoli of the breast. This forces milk from the alveoli into the ducts and sinuses of the mammary gland and makes the milk available.

The hormonal control of mammary function is extremely complex and is dependent on hormones from the ovary, the anterior pituitary, the adrenal cortex and, during pregnancy, from the placenta; the requirements for duct development are different from those for development of the lobulo-alveolar system (Fig. 12.17).

MALE REPRODUCTIVE SYSTEM

The general structure of the male reproductive system is shown in Fig. 12.18. Spermatozoa formed in the testis are mainly stored in the epididymis. As they pass along the reproductive tract at ejaculation the secretions of the accessory glands (ampullary glands, seminal vesicles, prostate, urethral glands and bulbo-urethral glands) are added to the spermatozoa to form the semen.

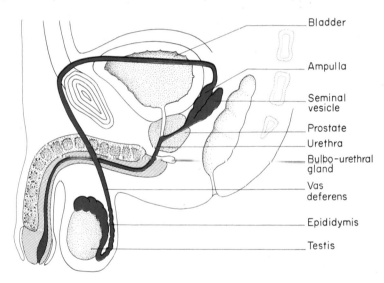

Bladder
Ampulla
Seminal vesicle
Prostate
Urethra
Bulbo-urethral gland
Vas deferens
Epididymis
Testis

Fig. 12.18. Human male reproductive system. Spermatozoa formed in the seminiferous tubules of the testis are mainly stored in the epididymis. At ejaculation they are expelled from the male reproductive tract in the semen, which consists of spermatozoa suspended in the secretions of the accessory glands.

Spermatogenesis

The testis consists of a number of convoluted seminiferous tubules in which spermatogenesis takes place. Germ cells in various stages of development can be seen in cross-sections of seminiferous tubules (Fig. 12.19). Clusters of spermatozoa are attached to Sertoli cells; although the function of Sertoli cells is not certain they give the spermatozoa mechanical support and may nourish them. Between the seminiferous tubules are found the Leydig (interstitial) cells which are responsible for the secretion of testosterone, the major male hormone produced by the testis. The formation of spermatozoa is shown diagrammatically in Fig.

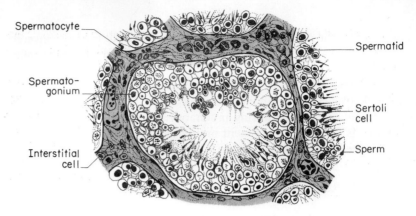

Fig. 12.19. Cross section of a human testis showing seminiferous tubule. Within the tubule germ cells may be seen at different stages of development; Sertoli cells are found among the germ cells. Interstitial (Leydig) cells are dispersed in the spaces between the seminiferous tubules (after Ganong).

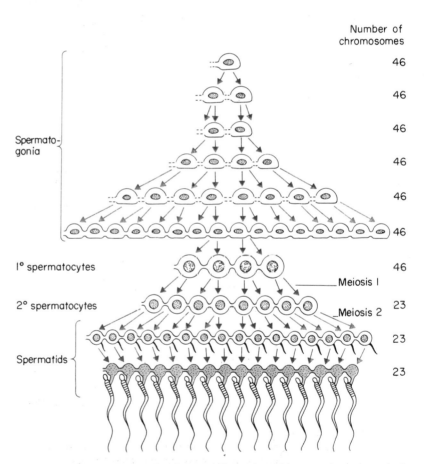

Fig. 12.20. Spermatogenesis. All male germ cells are derived from spermatogonia which undergo a number of mitotic divisions, finally forming primary spermatocytes; secondary spermatocytes are formed by a reduction division of primary spermatocytes and so have 23 chromosomes. Spermatids are produced by a second (non-reductive) meiotic cell division and develop into spermatozoa without further cell division. This whole process continues throughout the reproductive life of the male (after Research in Reproduction).

12.20. A major difference between germ-cell production in the male and female is that in the male germ cells continue to multiply in adult life while in the female this process is complete before birth.

The formation of spermatozoa from spermatids takes place by a complex process of differentiation without further cell division. This process is termed spermiogenesis and is illustrated in Fig. 12.21, while the structure of the mature human spermatozoon is shown in Fig. 12.22.

Formation of the spermatozoa from spermatogonia takes a number of weeks and can only take place at a temperature slightly below that of the body; for this reason the testes are held in the scrotum outside the abdominal cavity.

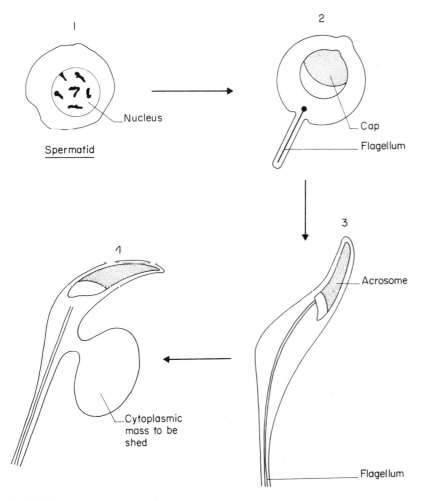

Fig. 12.21. Spermiogenesis.
This is the development of spermatozoa from spermatids and does not involve cell division. Development of the rat spermatozoon is illustrated (after Clermont).
1 The spermatid initially bears no resemblance to a spermatozoon.
2 Initially a flagellum starts to form and a cap develops over one pole of the nucleus.
3 The flagellum lengthens and the nucleus takes on its characteristic shape; by this time the nuclear cap has developed into an acrosome. Mitochondria migrate towards the base of the flagellum.
4 A cytoplasmic droplet forms on one side of the sperm head and is finally cast off.

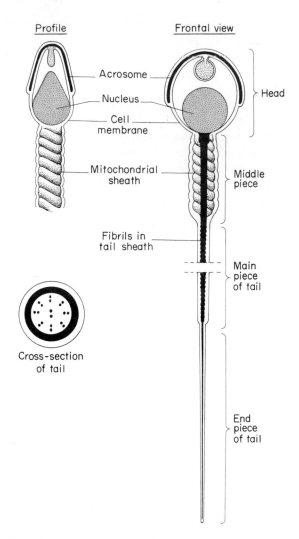

Profile   Frontal view

Acrosome
Nucleus
Cell
membrane
} Head

Mitochondrial
sheath
} Middle
piece

Fibrils in
tail sheath

Main
piece
of tail

Cross-section
of tail

End
piece
of tail

Fig. 12.22. Mature human spermatozoon (after Ganong).

Control of testicular function

Although the primary function of the testis is spermatogenesis it is also an endocrine organ, being the source of the androgenic hormone testosterone. Testosterone itself exerts an effect upon spermatogenesis, as we shall see later, but it is also responsible for a wide variety of male characteristics, such as certain aspects of behaviour, increased muscle mass, laryngeal changes which cause the voice to 'break' at puberty and balding in older men. Testosterone also causes the development of the accessory glands of the male reproductive tract.

Testosterone is synthesised by Leydig cells which are found in the interstitial spaces between the seminiferous tubules. Secretion of testosterone is controlled by LH* released from the anterior pituitary. Circu-

---

* In the male LH is often termed interstitial cell stimulating hormone (ICSH), since one of its primary actions is upon the Leydig (interstitial) cells; however, it is chemically identical with LH.

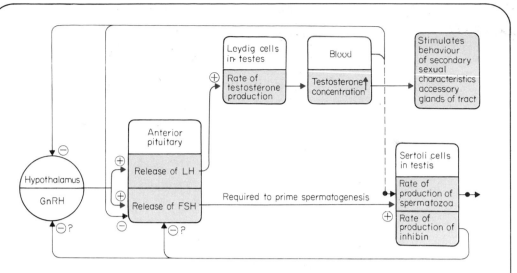

Fig. 12.23. Diagram showing the hormonal control of testicular function. Luteinising hormone secreted by the anterior pituitary controls testosterone synthesis by the Leydig cells which lie between seminiferous tubules; testosterone, in turn, exerts negative feedback on LH secretion. Testosterone controls the rate of spermatogenesis by acting upon Sertoli cells, but only appears to operate after the Sertoli cells have been primed by the action of FSH; secretion of this hormone appears to be controlled by negative feedback exerted by a protein known as 'inhibin' which is secreted by Sertoli cells in an amount related to the rate of spermatogenesis.

lating testosterone in turn limits LH secretion via negative feedback upon the hypothalamic–pituitary axis in a manner similar to that exerted by progesterone in the female. Testosterone maintains spermatogenesis by acting on receptors in Sertoli cells but is itself unable to initiate the process; for spermatogenesis to be initiated, Sertoli cells first must be primed by the action of FSH from the anterior pituitary.

There is evidence to suggest that Sertoli cells secrete a protein hormone known as 'inhibin' which exerts a negative feedback upon FSH output. The rate of 'inhibin' secretion apparently reflects the rate of spermatogenesis. The hormonal control of testicular function is summarised diagrammatically in Fig. 12.23.

Fertilisation

During coitus semen is deposited in the vagina by the contractions of the muscular walls of the male reproductive tract which occur at orgasm. The volume of the ejaculate is approximately 3 ml; it normally contains about 100 million spermatozoa per ml (Fig. 12.24). The migration of spermatozoa from the vagina to the Fallopian tube is mainly brought about by contractions of the muscles of the female genital tract.

Spermatozoa need to remain in the female genital tract for several hours before they are capable of fertilising an ovum (capacitation);

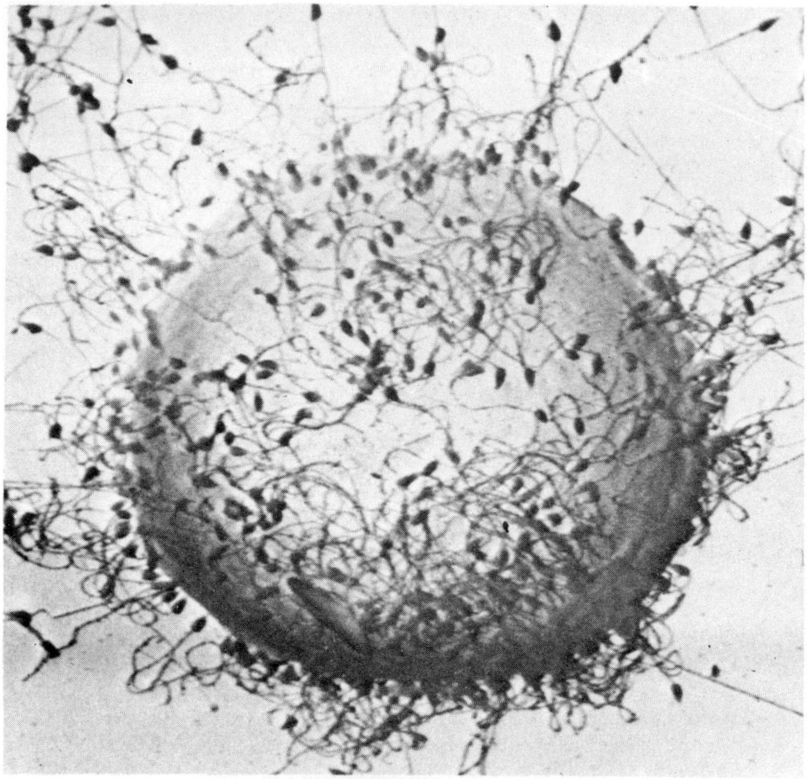

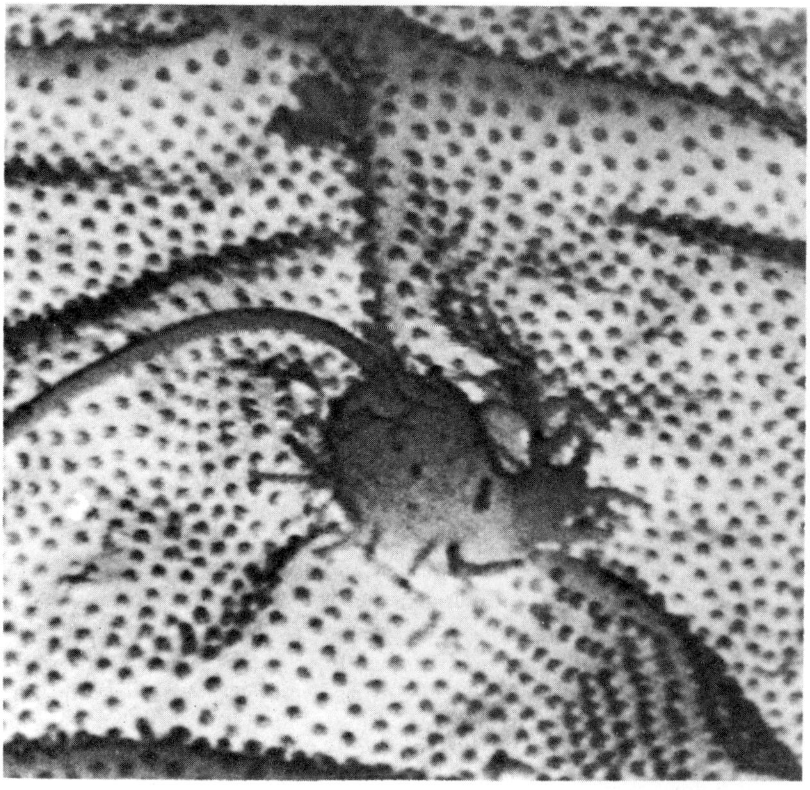

thereafter they retain this ability for about another 24 hours and so have a limited lifespan. When a sperm reaches the ovum changes take place in the surface membrane of the sperm head over the acrosome and in the adjacent acrosome membrane so that enzymes in the acrosome can digest a channel in the layers surrounding the ovum and allow the sperm to come into contact with the surface membrane of the ovum. These changes in the sperm head constitute the 'acrosome reaction'. Further changes take place in the acrosome, allowing fusion of the surface membranes of the ovum and the sperm. The fusion triggers changes in the ovum surface which prevent fertilisation by other spermatozoa. It also triggers the second meiotic cell division of the ovum. For some time the nuclei of the ovum and the sperm remain separate within the ovum cytoplasm. After about 12 hours, having undergone a number of changes including a marked increase in size, the genetic components from each parent become truly mixed.

PUBERTY

During the period of adolescence a number of dramatic changes take place. These include development of the gonads, genitalia and secondary sexual characteristics and changes in body size and proportions. The attainment of potential reproductive capacity is reached with puberty. In females this is achieved with the onset of menstruation (menarche).

There are two major theories on the mechanism by which the onset of puberty is controlled. In the first theory the hypothalamic–pituitary axis is considered to be very sensitive to inhibition by gonadal steroid hormones; as a result of this, gonadotrophin release from the pituitary is inhibited by extremely low circulating concentrations of gonadal hormones. Synthesis of sex hormones is in turn dependent on the presence of pituitary gonadotrophins in the blood (see Figs 12.9 and 12.23), so their secretion is kept in check. As puberty approaches the sensitivity of the feedback loop decreases and allows the concentrations of gonadal steroid to rise, producing the hormone-dependent changes seen during adolescence.

According to the second theory, before puberty, neural circuits in the hypothalamus are incompletely developed and therefore cannot bring about the release of gonadotrophin releasing factors. Thus, the development of this capability controls the onset of puberty. Neither theory is completely adequate and more information will be required before we can gain a complete understanding of the mechanisms by which puberty is controlled.

Fig. 12.24. Egg surrounded by many spermatozoa. (a) A scanning electron micrograph showing a sea urchin egg to which are attached an enormous number of spermatozoa; this illustrates the vast over-production of spermatozoa and the importance of mechanisms allowing only one spermatozoon to fertilise the egg. (b) Surface of the egg shown at a higher magnification to show the complexity of the egg membrane (photographs taken by Dr G.P. Schatten).

Part 5
Integration

Chapter 13
Hydrogen-ion Regulation

Most bodily processes are very sensitive to the hydrogen-ion concentration of the fluid which bathes them; a sensitivity due mainly to the marked effect of hydrogen-ion concentration on enzyme activity. Normally the hydrogen-ion concentration is very closely regulated in body fluids, a good example of the 'constancy of the internal environment' noted in earlier sections. In this section we shall consider the main mechanisms concerned in this regulatory process.

Basic ideas

The hydrogen ion (or proton) is a hydrogen atom with its only electron removed. Although usually written as H^+ it does not actually exist as such in body fluids, but rather is combined with water to form the hydronium or hydroxonium ions (OH_3^+). It is convenient, however, to use the conventional symbols. Another convention is to quote the hydrogen-ion concentration in negative logarithm units—the pH—rather than in moles of hydrogen ion:

$$pH = \log_{10} \frac{1}{[H^+]}$$

This convention is so widely used that we also use it. It does, however, have great disadvantages for students starting the subject: thus (1) a change in concentration of 10 times becomes a change in pH of 1 unit, and (2) increasing $[H^+]$ leads to decreasing pH (Fig. 13.1).

Many compounds dissociate reversibly, when dissolved in water, to produce negatively charged ions (anions) and positively charged ions (cations) of which the hydrogen ion is one:

$$\underset{\text{carbonic acid}}{H_2CO_2} \rightleftharpoons H^+ + \underset{\text{bicarbonate}}{HCO_3^-}$$

the degree to which a substance ionises depends both on the substance and on the conditions. A substance which liberates hydrogen ions is called an acid, one which accepts hydrogen ions is called a base; thus a bicarbonate ion is a base since it can bind a hydrogen ion. The term acidity is used to describe the concentration of free hydrogen ions present; a solution which has a greater acidity than another has a lower pH.

Strong and weak acids

A strong acid is one which dissociates completely when dissolved in water, thus producing the maximum number of hydrogen ions possible;

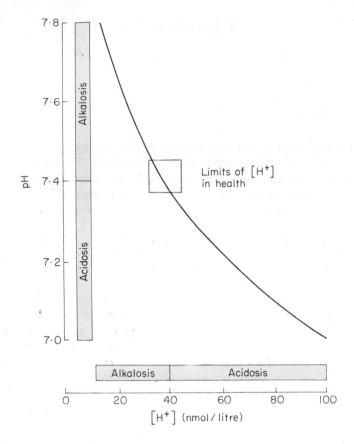

Fig. 13.1. Comparison of hydrogen-ion concentration and pH units, showing the range of each compatible with life. A change from a pH of 7·35 to 7·45 is equivalent to a hydrogen ion change from 45 to 35 nmol/; a 25 percent alteration (from Passmore & Robson).

hydrochloric acid is a strong acid and so dissociates completely, leaving little or no HCl present.

$$HCl \rightleftharpoons H^+ + Cl^-$$

A weak acid is one which only slightly dissociates when dissolved in water thus producing less hydrogen ions per unit of acid than a strong acid; carbonic acid is a weak acid and remains largely undissociated in solution. If the components of a strong and a weak acid could be collected and separated they would appear as in Fig. 13.2; strong acids are strongly biased towards ionisation, weak acids towards no ionisation. This characteristic of weak acids forms the basis of the action of buffers, substances which minimise changes in hydrogen-ion concentration.

Buffers
These consist of a weak acid and a salt of that acid (Fig. 13.3); the weak acid partially ionises whereas the salt ionises completely. The (undissociated) weak acid can be considered as a store of hydrogen ions (protons) which can be called upon to supply them if required, the (dissociated) base can be regarded as a store of base ready to combine with any free

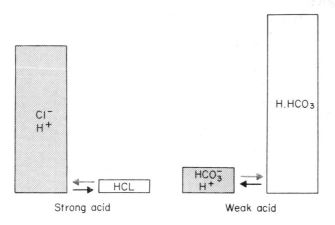

Fig. 13.2. Diagrammatic representation of strong and weak acids; in this and the following diagrams the heights of the columns represent concentration of the components shown. Strong acids are completely dissociated, weak acids are hardly dissociated at all (the preferred direction is shown by the red arrows).

A weak acid dissociates thus:

$$H^+ + HCO_3^- \rightleftharpoons H \cdot HCO_3$$

At equilibrium this can be expressed as

$$[H^+] = K_a \cdot \frac{\overset{\text{acid}}{(H \cdot HCO_3)}}{\underset{\text{base}}{(HCO_3^-)}}$$

Where K_a is the dissociation constant and the brackets indicate concentrations. If pH notation is used:

$$pH = pK_a + \log \frac{\overset{\text{base}}{(HCO_3^-)}}{\underset{\text{acid}}{(H \cdot HCO_3)}}$$

thus the pK_a of a weak acid is equal to the pH when the concentration of base equals that of acid.

hydrogen ions added; thus these two stores regulate the hydrogen-ion concentration of the solution they are in. A strong acid added to water produces vast changes in hydrogen-ion concentration and therefore pH; when added to an appropriate buffer system it only produces a small change in hydrogen-ion concentration and so of pH. For example, 1 ml of molar HCl added to 40 ml of water produces a change in pH of 5·4 units, added to 40 ml of a buffer consisting of equal molar volumes of carbonic acid and its salt it would produce a change of only 0·071 pH— the buffer has reduced the change by about 500 times.

Production of hydrogen ions in the body
Hydrogen ions are produced in the body in two main ways: (a) during metabolism of various substances; and (b) as a result of CO_2 production by the tissue cells.

(a)   The phosphorus and sulphur present in many biological molecules

pH 7·4

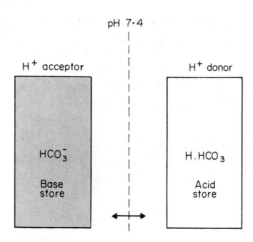

Fig. 13.3. Buffers consist of a weak acid, which acts as a H⁺ donor, in the presence of its base, which acts as a H⁺ acceptor. When H⁺ are added to or removed from the system the ratio of base/acid tends to remain constant keeping the H⁺ concentration constant.

The system arises from the two reactions:

weak acid $\boxed{HCO_3^-}$ $+ H^+ \rightleftharpoons H \cdot HCO_3$

salt of weak acid $\boxed{HCO_3^-}$ $+ Na^+ \rightleftharpoons Na \cdot HCO_3$

base acts as
a 'common ion'

In water the salt ionises completely and the acid hardly at all (the type of arrow indicates the bias) thus forming an acid or a base store. For any one system the pH is proportional to the log of the ratio of base/acid:

$$pH = pK_a + \log \frac{\text{(base, i.e. proton acceptor)}}{\text{(acid, i.e. proton donor)}}$$

are metabolised to phosphoric and sulphuric acids and these, on ionisation, produce hydrogen ions. Many organic acids, for example lactic and fatty acids, are produced which again liberate hydrogen ions on ionisation. Some 60 mmol of hydrogen ion is produced in this way every day.

(b) CO_2 produced by the tissues eventually combines with water (mainly in the erythrocytes) liberating hydrogen ions; in the lungs this reaction is reversed and CO_2 eliminated. Local regulation of these pH changes is discussed on p. 197; here we need only comment that buffering is largely due to the remarkable properties of the haemoglobin molecule and that normally there is no net gain of hydrogen ion by the body during the process. Some 20,000 mmol of CO_2 is excreted in this way every day.

Physiological buffers in the body

The normal hydrogen-ion concentration of body fluids is about 40 nmol/l (pH = 7·4); during the course of one day some 60 mmol of hydrogen ion is added to it which, if not buffered, would raise the concentration of the extracellular fluid (12 litres say) to 5 mmol/l. This would change the hydrogen-ion concentration by over 10,000 times (to a pH of 2·3) if not buffered in any way. This does not happen because most of these extra hydrogen ions are taken up by the various buffers found in body fluids. Ultimately, of course, these excess hydrogen ions and associated bases need to be excreted by the kidneys (making it acid), but initially they combine with buffers to minimise change in pH.

Bicarbonate–CO_2 buffer. The most important buffers in the body are the bicarbonate–CO_2 system, the large anion complexes such as plasma proteins and phosphates and haemoglobin in cells. In all of these the essential reaction is:

$$H^+ + buffer^- \rightleftharpoons H\text{-buffer}$$

If the hydrogen ion concentration increases then it combines with the buffer, if it decreases some hydrogen ions are released from the H-buffer complex; in this way the body fluids are stabilised against hydrogen-ion change. Here we shall be concerned only with the bicarbonate–CO_2 system for it shows some special features which must be discussed; haemoglobin also shows special features.

Quantitatively the bicarbonate–CO_2 system is a very good buffer in the body, yet in a test-tube the system is not exceptional; there are two reasons for this (Fig. 13.4):

1   Mammalian body fluids contain much dissolved CO_2, for they are in equilibrium with alveolar gas which contains 5 percent CO_2, rather than

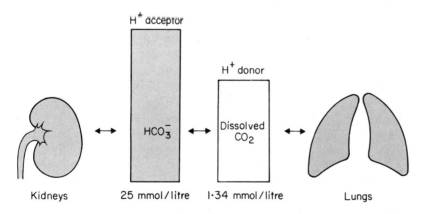

Fig. 13.4. The bicarbonate–CO_2 buffer system. Bicarbonate acts as the H^+ acceptor and the dissolved CO_2 acts as the H^+ donor; in addition the concentration of bicarbonate is determined by the action of the kidneys and the concentration of dissolved CO_2 is determined by the rate of CO_2 excretion through the lungs. This system gives a very tight control of the hydrogen-ion concentration of the blood. The pH is determined by

$$pH = \underset{\substack{pK_a \text{ of} \\ \text{carbonic acid}}}{6.1} + \log \frac{(25)}{(1.34)} = 7.4$$

The ratio of dissolved HCO_3^- to CO_2 is 25 to 1·34, i.e. nearly 19 to 1, not as shown in the diagram.

with air which contains practically none. As a buffer it therefore behaves as:

$$H^+ + buffer^- \rightleftharpoons H\text{-}buffer \rightleftharpoons dissolved\ CO_2$$
$$[HCO_3^-] \qquad [H_2CO_3]$$

The acceptor of hydrogen ions is the buffer base (HCO_3^-) as usual: the donor is the weak acid (H_2CO_3) which is in equilibrium with the dissolved CO_2; as the amount of CO_2 dissolved far exceeds the amount of carbonic acid present the dissolved CO_2 can be considered as the proton donor. The equation then becomes:

$$H^+ + HCO_3^- \qquad \rightleftharpoons \qquad dissolved\ CO_2$$

$$\text{proton acceptor} \qquad\qquad \text{proton donor}$$
$$(25\ mmol/l) \qquad\qquad (1 \cdot 34\ mmol/l)$$

From the concentrations given, the pH is 7·4, i.e. that of the body.

2   The bicarbonate–CO_2 system by itself can minimise changes in hydrogen-ion concentration, but must always allow some change to occur. In the body this limitation is overcome by systems which adjust the base (HCO_3^-) and the dissolved CO_2 to keep them at a constant value; kidney function fixes the concentration of base (p. 223), lung function that of dissolved CO_2 (see p. 201; both the CO_2 concentration itself and the hydrogen-ion concentration control respiration).

If pH disturbance occurs then the whole system responds in three stages:

1   blood buffers (very rapid, < second);

2   respiratory CO_2 system (rapid, seconds);

3   renal base system (slow, mins–hours), e.g. if hydrogen ions enter the blood then they combine with base (HCO_3^-) to form carbonic acid and CO_2; the CO_2 and original rise in hydrogen ions stimulate respiration to exhale more CO_2. The base associated with the hydrogen ions is then excreted. You will see that if a large amount of an acid is added to the blood the immediate compensation is by the blood buffers, then there is an excretion of CO_2 with reduction in the plasma bicarbonate; the slower compensation is that the acid is excreted by the kidneys which replaces bicarbonate in the blood (p. 221). Removal of hydrogen ions from blood—as for example occurs during HCl secretion into the stomach—is compensated for by retaining CO_2 and forming more base (in the short term). In both cases the respiratory system makes rapid buffering possible, the renal system supplies the long term buffering.

Chapter 14
Regulation of Body Fluids and Electrolytes

Some 60 per cent of the body is water, with potassium as the main intracellular ion and sodium as the main extracellular ion (Fig. 1.5, p. 11). The total body osmolality is proportional to the ratio of the potassium plus sodium to the total body water. The main control systems summarised here are those which regulate water, sodium and potassium balances in the body; those regulating the other ions are either dealt with elsewhere, for example, iron (p. 71), calcium (p. 371), or omitted.

The overall interaction between the circulation, kidneys, body fluid and electrolyte compartments is very complex and can only be outlined here. In considering these factors it is useful to have a scheme of the system in mind. Fig. 14.1 shows one devised for the computer simulation mentioned in the references; after you have mastered the points made in this chapter it would be helpful for you to obtain access to this model.

WATER BALANCE

The three sources of water for the body are : (1) water ingested; (2) water in the food; and (3) water formed in the tissues by the oxidation of the hydrogen present in food (metabolic water). Water is lost from the body in three ways: (1) in the urine; (2) in the alimentary tract (small amounts); and (3) by evaporation (large).

In normal sedentary life in a temperate climate the water excreted in the urine roughly balances that drunk at about $1\frac{1}{2}$ litres, for the other sources and losses of water are relatively equal (Fig. 14.2). Vigorous exercise or life in a hot climate, both of which lead to increased sweating, greatly increases the water lost by evaporation and consequently that drunk.

There is no reserve of water in the human body, despite the large amount present. A normal adult contains about 40 kg of water. The consequences of a reduction in water depends on what happens to sodium. A pure loss of water of 2 kg (5 percent) makes him very thirsty; a reduction of 4 kg (10 percent) makes him very ill; a reduction of 8 kg (20 percent) will kill him. With no water to drink obligatory water losses continue (about 1 litre/day) so that death will occur in 4–6 days, even in a cool climate. Consequently, a minimum intake of 1 litre/day is required for survival.

Control of body water
Body water is adjusted by altering the water intake—the thirst mechanism—or the water excretion by the kidney. The overall control mechanisms do so with an accuracy much greater than 5 percent, perhaps to 0·5 percent. It is convenient to outline the effector mechanisms before discussing the stimuli which control them.

413

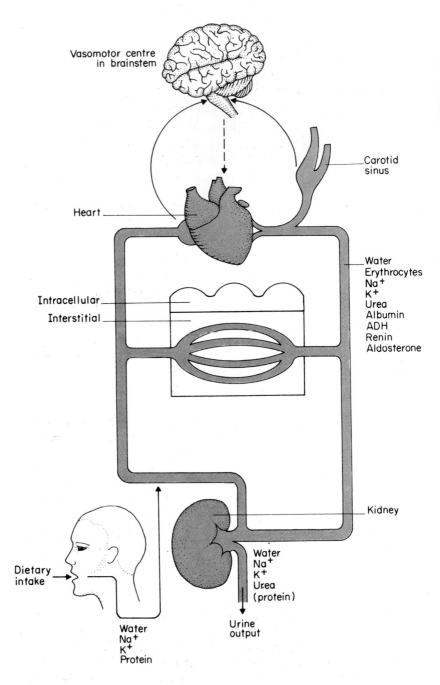

Fig. 14.1. Model of the fluid compartments, renal and vascular system of man. The neural pathways shown are greatly simplified (after MacPee).

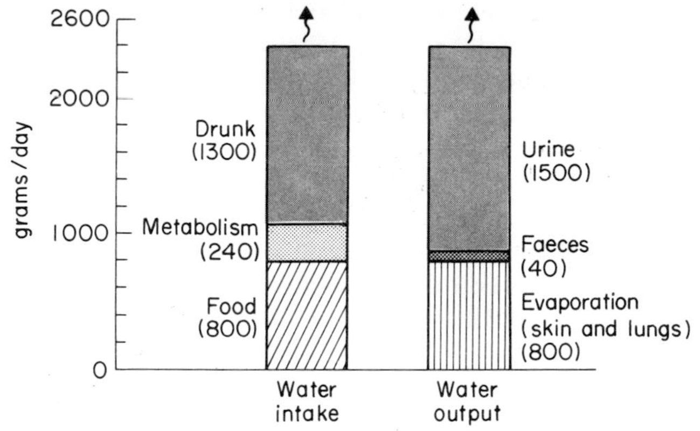

Fig. 14.2. Water balance in a young man leading a sedentary life in a temperate climate; the water excreted in the urine roughly balances that drunk (shown in red).

Thirst is an appetite mechanism under the control of the hypothalamus (pp. 174, 289). Lesions in the appropriate part of the hypothalamus diminish or abolish fluid intake, whereas stimulation causes drinking; both effects are unaccompanied by changes in food intake. Thirst is like appetite (p. 173) in that several other factors influence it, for example distension of the stomach decreases it, dryness of the mouth and throat increase it. The 'thrist' area varies in position in different animals, but is distinct from the osmoreceptor regions of the hypothalamus.

Water excretion. The excretion of water by the kidneys is controlled by the antidiuretic hormone (ADH) concentration in the blood, which in turn is controlled by the activity of osmoreceptors in the hypothalamus and elsewhere in the brain (p. 341). The presence of ADH causes an increased water permeability the collecting duct system in the kidney with a consequent reabsorption of water (p. 216). The output of water in the urine is reduced, and the concentration of solutes in the urine rises correspondingly. The total excretion of solutes is not changed, however, by ADH action. In normal life ADH is always present in the plasma in low concentrations and so may be increased or decreased as required.

Stimuli causing thirst and ADH release. The main stimulus for activating both thirst and ADH secretion is an increase in the osmotic pressure of the body fluids, usually brought about by a water deficit. The mechanism of thirst and antidiuresis together constitute an effective defence against water deficit by increasing water input and lowering output. Defence against a water load is less efficient, for the 'absence of thirst' does not necessarily prevent drinking; and some ADH secretion can occur under other influences (see below) even though its inhibition would be called for by a positive water balance. So the natural stress of water deprivation is better guarded against than the dangers, often artificial, of water loading.

In normal daily life, there are small and transient changes in tonicity of the body fluids, e.g. if not enough water is drunk, the body becomes slightly hypertonic, due to the continued loss of water via lungs, skin and obligatory renal losses. This state of hypertonicity is corrected by the release of ADH and the production of a state of thirst which leads to increased drinking. If excess water is drunk, this water rapidly becomes distributed throughout all the fluid compartments of the body; the lower osmolality which results causes a reduced production of ADH and so the plasma ADH concentration decreases due to liver inactivation and a diuresis ensues (Fig. 14.3).

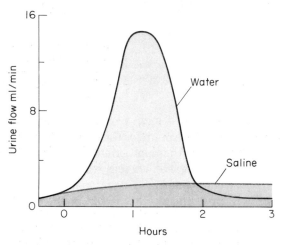

Fig. 14.3. The effect of drinking a litre of water or isotonic saline on the subsequent urine flow. Drinking water causes a diuresis which excretes the excess water within 3 hours or so. Drinking saline causes only a slight diuresis; this is because the ECF is expanded isotonically so there is no stimulation of the osmoreceptors. (after Schmidt & Thews)

The essential function of these mechanisms is to maintain osmotic constancy within the body. So if there is a primary change in the solute (mainly NaCl) content of the body this will tend to cause an equivalent but secondary change in the water content. This has been called the 'companionship' of water and electrolytes in the organisation of body fluids.

There are, however, other effective stimuli for these mechanisms. Both thirst and ADH secretion are stimulated by a decrease in blood volume (p. 341), but this response is more sluggish and less sensitive than the response to the pure water deficit discussed above (Fig. 14.4). ADH secretion is stimulated by pain, apprehension and exercise, and decreased by alcohol. Dryness of the mouth and throat causes a sensation of thirst in ourselves (p. 174).

SODIUM BALANCE

The body contains about 4 mol of sodium of which about one-third is held in bone (p. 371); the remainder is present in the extracellular space and is readily exchangeable. In addition to its specific functions in the action potential, etc. the sodium content of the body largely determines

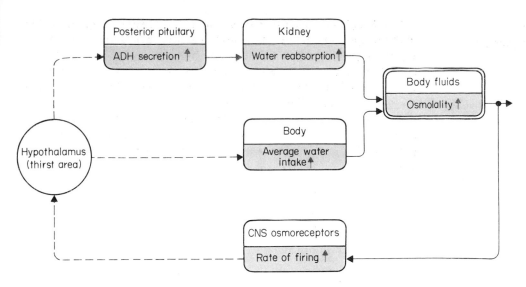

Fig. 14.4. The control of body water. A disturbance of osmolality (shown as an increase) is followed by the consequences shown (arrows in other compartments) which restore the disturbance. This control system interacts with that shown on Fig. 11.28 and should be considered in conjunction with it.

the volume of the extracellular space; severe depletion leads to 'shock' with a marked fall in blood pressure. As all land animals have evolved in an environment which is poor in sodium, it is found that the kidneys of such animals are capable of protecting the salt content of the body, even by secreting urine which contains no sodium. Conversely, the kidney can adjust its output to balance the large sodium intakes common to Western man, provided that sufficient water is available to excrete with it.

The dietary intake of sodium varies from 70 to as much as 350 mmol/day, corresponding to 4–20 g of NaCl (Fig. 14.5). As soil does not contain much salt most of the dietary intake comes from that added during or

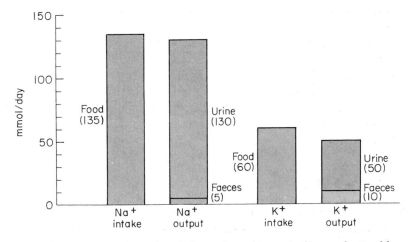

Fig. 14.5. Daily sodium and potassium balance of a young man leading a sedentary life on a normal diet. The losses of sodium and potassium from the skin were negligible.

after cooking; in almost all primitive societies salt is an important and sought-after commodity. We cannot, however, live on sea-water for it is too hypertonic (about 1 M-NaCl) and contains a high concentration of magnesium.

Control of body sodium
Total body sodium in man is regulated by mechanisms acting on its excretion through the kidneys, rather than on its intake. In some animals, however, there is a well-marked 'salt appetite' mechanism operated by areas in the hypothalamus and responding to the CSF concentration of sodium. Perhaps our inability to live on low sodium diets is a lingering effect of this mechanism. The renal mechanisms are largely dependent on the concentration of aldosterone in the renal plasma (pp. 216, 355), although there is evidence for the involvement of newly discovered hormones, such as atrial natriuvetic factor (ANF).

Stimuli causing aldosterone release
The current view is that there are two integrated mechanisms for the regulation of aldosterone release: one mechanism having the characteristics of a reflex, thus depending on volume receptors, nervous pathways and central networks; the other humoral, based on the renin–angiotensin system and so having its trigger in the kidneys (p. 356). The rest of this section will be devoted to an outline of these volume receptors.

Regulation of the volume of the extracellular fluid
The volume of the extracellular fluid is largely determined by the sodium content of this space, for given a particular amount of sodium present the osmoreceptor–ADH mechanism will adjust the water content to give a normal osmotic pressure. With any particular volume of extracellular fluid its partition between the plasma and the interstitial space depends on the Starling mechanism (p. 128) involving plasma proteins and blood pressure. The current view is that the volume of the extracellular space is kept constant by a mechanism which measures the 'fullness of the bloodstream'* in the chest and then adjusts the intake and output of water and the excretion of sodium (Fig. 14.6, see also Fig. 11.27). Examples of stimuli which affect this system are: (1) bleeding reduces the 'fullness' and so leads to a reduction of sodium and water excretion. (2) standing upright leads to a movement of fluid from the chest to the legs reducing this 'fullness' and so causes a reduction in sodium and water excretion; (3) lying down or immersion in water to the neck reverses this process and leads to a rapid sodium and water excretion.

Volume receptors and associated effector mechanisms. It seems likely that there are several such receptor areas and several mechanisms which bring about these changes, although knowledge about them is, at present, fragmentary. The receptor areas are thought to reside (a) in the atria and great veins near the heart (low-pressure receptors [p. 126]), and (b) in the arteries near the thyroid gland in the neck and in the

---

* i.e. the ratio of the blood volume to the volume of the vessels containing it.

juxtamedullary apparatus of the kidney (high-pressure receptors [p. 356]). It is not clear how the receptors in the chest and neck control excretion, a current suggestion is that they act reflexly via the sympathetic system on the renal vessels; the juxtamedullary apparatus exerts its control via the renin–angiotensin–aldosterone system on renal excretion (p. 216).

The interactions of sodium and water balance are very complex and can only be outlined in this account. Here we shall summarise the effects of 'pure' salt and water deprivation. In salt deprivation there is a reduced

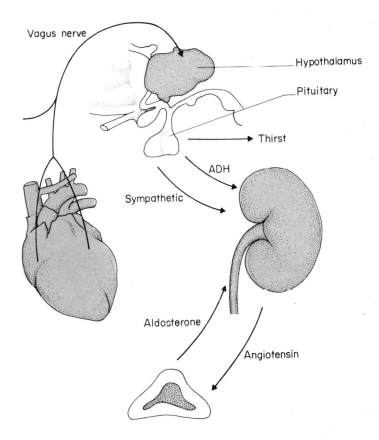

Fig. 14.6. Postulated system for volume control. Body sodium is measured indirectly by the 'fullness of the bloodstream' (after Passmore & Robson).

amount of sodium (Fig. 14.7) available for the extracellular space, water is excreted to preserve the osmotic pressure and so the extracellular volume decreases; the intracellular volume may be normal or even increased. With a mild water deprivation the osmotic pressure of body fluids increases while their volume diminishes; this diminution being spread evenly over the sub-divisions of body fluid. With more severe water deprivation, the diminished blood volume causes the secretion of aldosterone which leads to sodium retention and so the extracellular space expands at the expense of the intracellular space (Fig. 14.7). It is thought that this has a survival value. Water deprivation therefore leads to an increase in the osmotic pressure of the body fluids, which may be

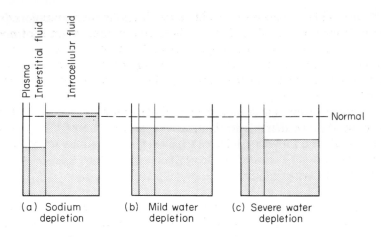

Fig. 14.7. Effect of sodium depletion, and of mild and severe water depletion, on the partition of fluid between the extracellular and intracellular fluids. The partial preservation of extracellular fluid at the expense of intracellular fluid in severe water depletion is probably due to increased secretion of aldosterone. (From Black.)

detected clinically by an increase in the sodium concentration of the plasma.

POTASSIUM BALANCE

The body contains about 3·5 mol of potassium of which some 95 percent is present as the main cation in cells. Plant cells also contain large amounts of potassium and so foods are never deficient in it. Potassium deficiency commonly arises from the increased losses associated with vomiting and diarrhoea, and sometimes the excessive use of purgatives. A typical daily balance is shown in Fig. 14.5.

Both the total amount of potassium in the body and the concentration in the extracellular fluids are less well controlled than that of sodium.* Moderate potassium depletion, say up to 10 percent of the body total, produces no symptoms in normal people but their excretion of potassium and their plasma concentration both fall. If these people are not deprived of sodium as well their extracellular volumes expand and their intracellular volume shrinks so that no loss of body weight occurs. If the body potassium is depleted to 30 percent some weakness and thirst appear, with oedema and changes in the ECG. This amount of depletion eventualy leads to death, perhaps by changes in enzyme activity and membrane potentials of the cells.

Control of body potassium

The problem of potassium regulation is, in a way, more complicated than that of sodium in that the bulk of potassium is intracellular so that absorption and excretion are at a remove from the major potassium mass. It has also been more difficult to measure in intact animals.

*80 percent of the population have serum sodium concentrations within $\pm 4\%$ of the mean, but serum potassium concentrations within $\pm 12\%$.

The total body potassium could be regulated by mechanisms acting on its intake or excretion. So far as is known, intake is not controlled and excretion is only rather slugglishly and poorly regulated. It is now known that the urinary potassium is determined mainly by the rate of potassium secreted into the tubular fluid, for that filtered is mostly absorbed (see p. 216).

Mechanisms regulating potassium balance are unclear, though they do exist, thus an increase in the intake of potassium is followed by an increased excretion, and a decreased intake by a decreased excretion. Both responses occur with a time course of 10–14 days. There is a daily rhythm of potassium excretion, which only adapts rather slowly to time zone changes.

Alderosterone causes an excretion of potassium as well as a retention of sodium, but is unlikely to be the means by which body potassium is regulated. It has been suggested that the potassium secreted into the urine comes from the inside of the renal tubule cells and so is intracellular rather than extracellular in origin. If renal cells showed a similar pattern of potassium content to other cells in potassium excess or depletion this mechanism might be adequate in controlling body potassium.

Chapter 15
Temperature Regulation

The temperature of the body depends on the amount of heat stored within it; this in turn depends both on the rate at which heat is being produced by the bodily processes and the rate at which heat is being lost to the surroundings.

Normal body temperature
Warm-blooded (homeothermic) animals like ourselves maintain a high and relatively constant body temperature, despite large fluctuations in that of the environment. In man the deep body (or core) temperature—i.e. that of the heart, lungs, viscera and brain—is kept within a narrow

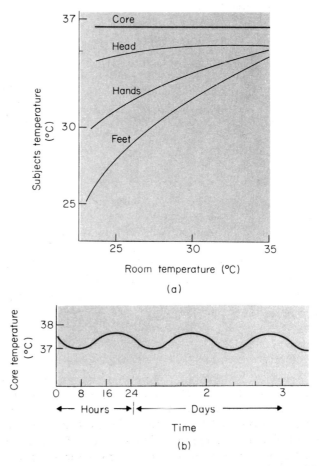

(a)

(b)

Fig. 15.1. (a) The temperature of various parts of the (naked) human body at different environmental temperatures. (b) The normal daily rhythm of core temperature; note the low temperature during the early hours of the morning, and peak in the afternoon.

range around 37°C (98·6°F). Surrounding the central organs there is a 'shell' of peripheral tissues and skin, whose temperature fluctuates within quite a wide range according to the environment (Fig. 15.1a).

It is useful to think of the body as having a 'set-point' temperature, around which the regulatory mechanisms operate. If the core temperature is greater than this set-point then cooling mechanisms ensue; if the core temperature is less than the set-point then heating mechanisms or behavioural changes occur. This set-point varies with a 24-hour cycle (Fig. 15.1b) causing the core temperature to vary with it. In fever the set-point is raised and regulation occurs around this new value. In exercise the set-point is (probably) unchanged, but the core temperature rises because heat is produced more quickly than it can be dissipated.

The core temperature is often estimated by the rectal temperature; other measures of it may be obtained in the oesophagus, the external meatus and nasopharynx. The latter is an index of brain temperature. The oral temperature (below the tongue) is sufficiently accurate for clinical purposes, but it is about $\frac{1}{4}$°C below the core temperature. It is difficult to measure the mean body temperature and in practice various formulae are used to estimate it.

THERMAL BALANCE

The heat produced by metabolic processes is lost by evaporation, radiation and convection to the environment. These quantities are related in the heat equation (Fig. 15.2). The units of heat used are kcal/h or the watt (W), where 1 kcal/h = 1·16 W.

The metabolic heat production (M) is measured directly in a calorimeter or by the O_2 consumption (p. 167). It is partly determined by muscular activity, and can be increased by shivering or by carrying out physical work. Although a small increase in metabolic heat production occurs after a meal and some chemical regulation can be made, these factors play a negligible part in maintaining body temperature in man under normal conditions. Small animals (for example, the rat) have a much higher ratio of surface area to volume than man and consequently have a greater problem in maintaining their body temperature; perhaps for this reason they are able to increase their metabolic heat production by several times in cold conditions.

Evaporative heat loss
The evaporative heat loss (E) is due to the loss of heat which occurs when water evaporates. At rest we lose about 30 g of water by evaporation from the skin and respiratory passages per hour;* as the vaporisation of each gram of water removes about 0·6 kcal (2·43 kJ) of heat this gives a loss of about 20 W. Evaporative heat loss is subject to physiological control and can be increased by sweating. This heat loss is only obtained if the sweat evaporates; sweat dripping from the body may look impressive, but does not cool it. (To cool quickly after exercise you

---

* Dogs can increase their respiratory loss of heat very greatly by panting.

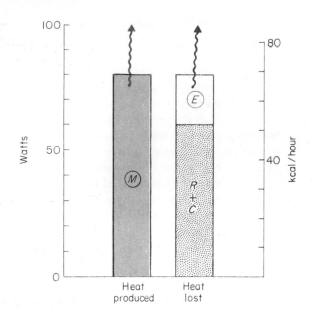

Fig. 15.2. The thermal balance at rest in a room at 21 °C.
The gain and loss of heat can be expressed by the heat equation:

$$M - E \pm (R + C) = \pm S$$

where M is the metabolic heat production, E the evaporative heat loss, $R + C$ the loss by radiation and convection and S the heat stored in the body. Increases in the metabolic rate (M) (mainly during exercise) are balanced by increases in sweating (E); during severe exercise (p. 433) the sweat rate may rise to 1 kg/h causing a heat loss of 700 W, an increase of thirty fold; ($R + C$) is fairly constant but the direction of heat flow depends on the environmental temperature. The arrows show those components which can be varied.

should spread your sweat over your skin, not mop it with a towel.) The sweat glands are under nervous control and sweating is the main physiological adjustment to an increased heat load (Fig. 15.3). It is, of course, less effective in a humid environment. In man sweat for cooling purposes is produced by the eccrine glands; the apocrine glands (temple, armpits and pubic regions) produce sweat in response to emotional stress.

The heat exchanges due to *radiation and convection* ($R + C$) are determined by physical factors, mainly the temperature gradient from the surface of the body to the environment, the amount of clothing worn and the amount of air movement. Normally heat is *lost* by this means, but if the environment is hotter than the body then heat is *gained*. Changes in behaviour (for example, curling up or stretching out your limbs), and skin blood supply (see below) modify the rate of transfer of heat to some extent but cannot, for example, protect the body from the inevitable losses which occur in a cold environment.

The *heat storage* (S) of the body can be positive or negative for short periods of time. As the specific heat of the body is 3·55 J/g (0·85 cal/g) a rise in mean body temperature of 1 °C in 1 hour in a 70 kg man requires about 70 W, i.e. a heat production almost twice that of normal.

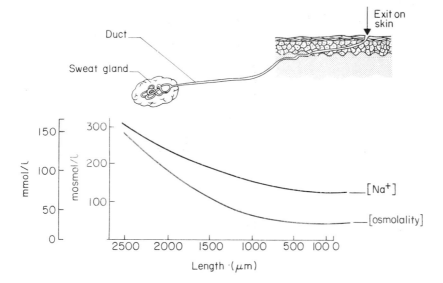

Fig. 15.3. The composition of human sweat at different points along a sweat gland.
Eccrine sweat glands in man are distributed over the whole skin. They are innervated by cholinergic sympathetic nerve fibres. They work by producing an isotonic salt solution which is modified by reabsorption of NaCl and water as it proceeds to the skin surface. The final sweat produced is hypotonic and the final concentration depends on the salt balance of the individual and rate of production being controlled by aldosterone (after Schulz *et al.*).

The insulation of the body
The 'shell' of tissues surrounding our bodily 'core' can be considered as a layer of insulation between the constant temperature of the core tissues (heart, liver, etc.) and the variable temperature of our surroundings. It has a similar function to the layers of insulation around a hot-water cylinder, in that it decreases heat loss to the environment. In our bodies the insulating properties of this layer can, to some extent, be altered to suit the conditions. This is done by varying the blood flowing from the core tissues to the skin; if the blood flow is high then more heat is brought to the skin surface and then dissipated to the environment; if low, little heat is transferred with consequently less heat dissipation. In cold conditions the counter-current heat-exchange systems in the limbs (Fig. 5.16) also decrease heat loss by cooling the blood reaching the hands and feet. Tissue insulation, however, only plays a small part (compared to clothing) in stopping heat loss in the cold, though the decreased tissue insulation which occurs in a hot environment significantly helps heat loss (Fig. 15.4).

The insulation of a clothed body resists heat flow from the tissues to the surrounding air. It is usually considered as being made up of three components: (1) that due to the tissues, which affects the flow of heat from the core of the body to the skin surface; (2) that due to the clothing; and (3) that due to the air near the body. An arbitrary unit of insulation (the clo) is used to measure the amount of insulation of the body and of clothing; 1 clo is the amount of insulation (from clothing) required by a person to be comfortable while sitting quietly in a room at 21°C (it is

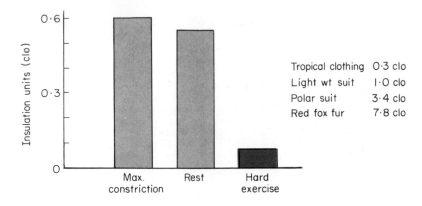

Fig. 15.4. Human tissue insulation at rest; with maximum vasoconstriction; and maximum vasodilatation. The figures on the right show the insulation values of typical clothing.

equivalent to $0.155°C\ m^2\ W^{-1}$). At rest in still air more insulation is contributed by clothing than by the tissues.

The temperature of the air surrounding us is not a satisfactory measure of the environmental heat load; for when indoors we gain or lose heat by radiation from the surfaces around us, and outdoors we are affected by radiation from the surrounding surfaces, the wind speed and humidity as well as solar radiation. The term 'operative' temperature is used to describe the combination of air and radiant temperature. Unclothed man can maintain thermal equilibrium at operative temperatures around 30–34°C at rest, but at much lower temperatures (say 20–10°C) during active work.

Temperature control

Body temperature is regulated by a combination of behavioural and physiological responses; in man the behavioural responses are much the most important, the physiological ones acting as a fine control system. Thus, when hot we move to a cooler place and remove clothing; more blood circulates through our skins and we start to sweat. When cold we move to a warmer place, curl ourselves up to reduce our surface area and put on more clothes; the circulation of blood to our skin decreases and we may shiver. The motor pathway for these physiological events is the sympathetic system, which controls the diameter of the blood vessels and the activity of the sweat glands. The hypothalamus plays a cardinal role in controlling these activities and damage there (in the pre-optic region) can put the regulating system out of action. The sensory input signals come from fibres in the skin (Fig. 15.5), from cells in the hypothalamus itself which sense the temperature of the blood perfusing them and compare it with the 'set-point' of the system, and perhaps from cells situated in other regions of the central nervous system which do likewise. The current suggestion is that the hypothalamic temperature-sensitive cells activate defences against a temperature rise, i.e. lead to sweating and increased peripheral blood flow while an interaction between central and skin receptors guards against a temperature fall. Thus cooling responses are switched on at a lower core temperature

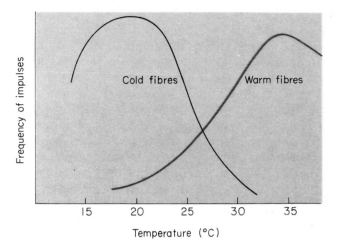

Fig. 15.5. Average frequency of static discharge of populations of single warm and cold fibres of mammalian skin as a function of temperature. The receptors are both naked nerve endings and organised receptor structures.

if the skin is hot rather than cold. This system can be thought of as a protective one to guard against temperature rises, and an attempt to obtain a mean body temperature measurement as the temperature falls.

When we suffer from a fever—the oldest hallmark of disease—the temperature regulatory system behaves as though the 'set-point' had been reset at a level higher than normal. The hypothalamus detects the difference, we feel cold, and temperature-raising mechanisms (shivering, vasoconstriction, etc.) occur to raise our temperature. When our 'thermostat' is reset again (naturally, or with aspirin) the opposite events occur and so we sweat, have a peripheral vasodilatation and feel hot.

On keeping a cool head
Temperature changes applied to the head have a much greater effect on heat regulatory systems than the same changes applied to the body. In man the venous blood from the face, mouth, etc. returns via the internal

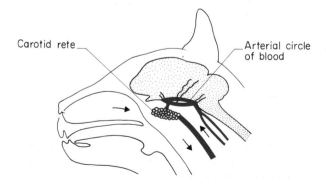

Fig. 15.6. Section of a cat head showing the heat exchanger between the arterial blood supply to the brain and the blood returning from the nasal mucosal and respiratory tract (adapted from Baker).

jugular vein which runs closely along the common carotid artery. It seems that this system acts as a counter-current heat exchanger so that cool blood from the head affects the arterial blood going to the brain, which then organises appropriate responses by the rest of the body. In this way the brain temperature is kept remarkably constant (changes of 0·01°C there cause thermoregulatory effects). Hot people sponge their faces, hot rodents smear saliva over theirs; thus both achieve a greater cooling response than they otherwise would. In various animals there is a heat-exchange system between the arterial blood supply to the brain and venous blood returning from the nasal cavity (Fig. 15.6). Perhaps the reputation of hot tea as a cooling drink originates from some such effect; the hot tea heats the brain and thus sets in motion or amplifies sweating, leading to a lower skin temperature.

Central mechanisms regulating body temperature

The integrating centre in the anterior hypothalamus compares the 'core' temperature with the 'set-point' temperature; if they differ, a thermo-regulatory response is made. This response is modified by information from skin and spinal cord temperature sensors, and other stimuli, for example osmotic and barostatic inputs; but these inputs are of secondary importance. These neurones are close to others in the hypothalamus

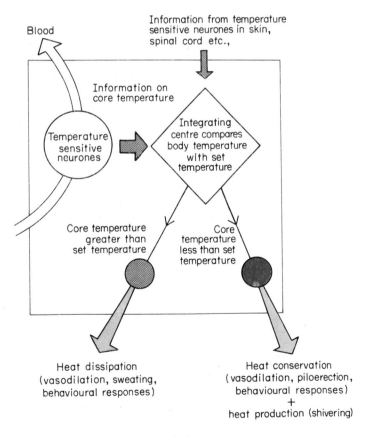

Fig. 15.7. Central mechanisms regulating body temperature.

which regulate vasomotor activity, and also send connections to areas subserving behaviour and consciousness (Fig. 15.7).

The threshold for sweating during exercise at different skin temperatures
During vigorous exercise the core temperature rises, when it exceeds a threshold value (inflection on Fig. 15.8) sweating starts; above this value the rate of sweating is proportional to the core temperature (bold line in Fig. 15.8). At lower skin temperatures the core temperature needs

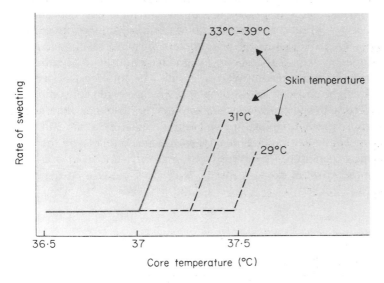

Fig. 15.8. The threshold for sweating during exercise at different skin temperatures (after Benzinger).

to rise to a higher level before sweating starts, but, once started, it has the same proportionality to core temperature. After exercise, which raises the core temperature by 1°C, sweating brings it back to the set-point in about 30 minutes.

These results are explained by supposing that sweating is primarily determined by the difference between the core temperature and the set-point temperature, but that the starting point for sweating is also influenced by the skin temperature. The set-point in the subject of Fig. 15.8 was about 37°C.

Chapter 16
Exercise

Work load and energy expenditure during exercise

The physiological responses to exercise in man will depend on the frequency, duration and intensity of the work being performed, as well as age, environment, etc. For example, a fit person will have a higher maximal working capacity and a greater endurance at submaximal intensities than someone who is less fit. The total cost of performing work can be obtained indirectly by measuring oxygen consumption and converting this to kilocalories or kilojoules. In metabolic work it is usual to ignore protein metabolism and take a value of 5 kcal (20 kJ) as the caloric equivalent per litre of oxygen uptake. Examples of the average energy expenditure in various types of activity are given in Fig. 16.1.

At the onset of exercise there is a gradual increase in oxygen con-

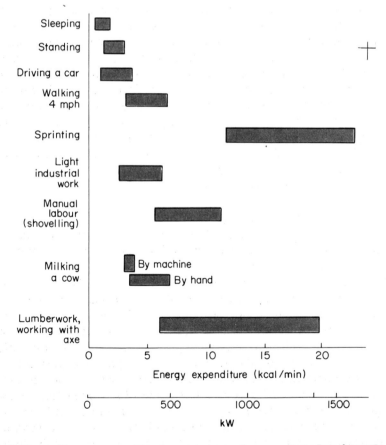

Fig. 16.1. Aerobic energy expenditure in various types of human activity. Bars show range of values found.

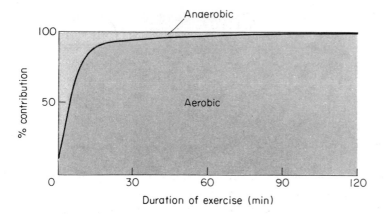

Fig. 16.2. The relative contribution of aerobic (shown in red) and anaerobic metabolism during severe exercise lasting for different durations. Thus during the 100 metre sprint, which lasts around 10 s, over 80 percent is anaerobic; during the marathon, which lasts for 2–4 h, most is aerobic (after Åstrand & Rodahl).

sumption which reaches a steady state value within a few minutes. Provided the energy demand does not exceed that which can be supplied aerobically then the level of oxygen consumption will be proportional to the work performed. Measurements of oxygen consumption provide a useful reference point both in assessing fitness and in describing the intensity of effort for different types of exercise. Fitness testing is often performed with reference to bicycle exercise expressed as kilogram metres/minute. The length of time a person can perform at a given work load is inversely proportional to the severity of the exercise. Thus work at 90 percent of maximum oxygen consumption can be kept up for about 30 minutes, whereas below about 60 percent of maximum the exercise may continue for hours, and often there is not an easily defined point of exhaustion. The relative contribution of aerobic and anaerobic metabolism in supplying energy depends both on the intensity and the duration of the exercise (Fig. 16.2).

Respiratory adjustments during exercise
Pulmonary ventilation is finely adjusted to the work performed during exercise; it can increase from around 5–6 l/min at rest to a maximum of 100–150 l/min in exercise.

This rise in ventilation could be achieved both by a rise in the tidal volume and in the frequency of breathing, or (at least for moderate exercise) by a change in one of these only. The situation in man is complex because it varies between individuals and also with the level of exercise; it is also different from that in other animals. In addition, with exercise involving repeated movements, for example, cycling, running, etc., respiratory rate is often adjusted to be in time with the movements. For moderate exercise (say up to seven times the resting level) ventilation increases mainly by an increase in tidal volume with only a small increase in respiratory rate; thus a fourfold increase in tidal volume is accompanied by a twofold increase in rate. A further peculiarity of this is that during moderate exercise the inspiratory time remains constant but the expiratory time decreases, so that the increase in respiratory rate occurs solely as a result of a faster expiration. This can be remembered as being due to expiration becoming an 'active' rather than a passive event. At very high rates of ventilation the inspiratory time does shorten and further increases in ventilation are mainly due to increases in the respiratory rate. During exercise, as at rest, it seems likely that tidal volume and rate are adjusted to give the minimum 'work of breathing' at any rate of ventilation.

As the level of exercise increases, ventilation increases in proportion until anaerobic metabolism becomes a large fraction (see Fig. 16.2) at which point ventilation often increases at a somewhat higher rate. This point is often referred to as the anaerobic threshold. Athletes are characterised by having higher anaerobic thresholds than untrained individuals (Fig. 16.3).

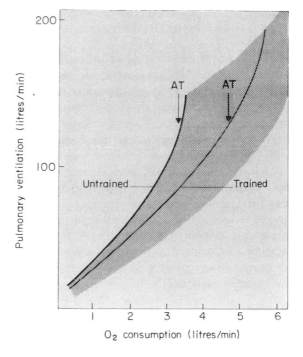

Fig. 16.3. Pulmonary ventilation at different levels of oxygen consumption for trained and untrained individuals doing bicycle exercise. Ventilation increases in proportion to oxygen uptake until levels of effort are reached which require a large contribution from anaerobic metabolism. At this point ventilation often increases at a somewhat greater rate (the anaerobic threshold, AT). The shaded area indicates the wide range of values obtained for different individuals (from Åstrand & Rodahl).

Although the level of respiration is closely related to the level of energy expenditure it is not clear how this occurs; for although Po_2 and Pco_2 of the venous blood returning to the heart are vastly different during severe exercise than at rest, so efficient is the ventilatory process in the lungs that the corresponding concentrations for arterial blood are virtually unchanged by even the severest exercise. At present it is thought that a combination of stimuli lead to the rise in ventilation, but that individually these vary in importance according to the duration and intensity of the exercise.

The following five factors (see also respiratory section) probably account for the ventilatory rise in exercise.

1   Neural stimuli spread from higher centres and from joint and muscle inputs to give the rise in ventilation which occurs at the start of exercise.

2   CO_2 production increases in exercise and this may (especially initially) raise the arterial Pco_2 and thus stimulate respiration. Later in the exercise the response to CO_2 is sensitised by a rise in temperature and hypoxia, so that respiratory stimulation occurs at levels of Pco_2 which would be 'normal' at rest.

3   The lactic acidosis which occurs during exercise stimulates ventilation, particularly as the exercise proceeds and lactic acid levels increase. This effect is additive with the others.

4   Hypoxia may occur during exercise, though usually the arterial Po_2 is kept remarkably normal even during severe exercise. There seems, however, to be some hypoxic drive during exercise in that increasing the inspired Po_2 reduces ventilation during exercise but not at rest. It seems possible that breath to breath fluctuations of Po_2 and/or Pco_2 during exercise are detected by the chemoreceptors and acted upon. The timing of the arrival of impulses from the chemoreceptors at the respiratory neurones seems to be important; those arriving during inspiration have a greater effect than those arriving during expiration.

5   These factors do not, however, seem sufficient to account for the rapid rise in ventilation at the start of exercise. An attractive suggestion is that regulation is by 'feedforward' means (p. 13), the CNS changes ventilation at the same time as it instructs the muscles to act.

Cardiovascular adjustments in exercise

The increased oxygen demand of the muscles during exercise is achieved partly by an increased extraction of oxygen from the blood and partly by an increase in cardiac output. During exercise cardiac output increases linearly in step with the increase in oxygen consumption (Fig. 16.4).

This increase in cardiac output is achieved through a small increase in stroke volume and a large increase in heart rate which at maximal rates of work can reach around 200 strokes/min; there is, of course, a corresponding increase in the venous return to the heart (p. 131). There is a small increase in systolic blood pressure due to the increased cardiac output and a drop in diastolic blood pressure due to a large decrease in the peripheral resistance; this is largely due to the dilation of the vessels in the exercising muscles. At the same time blood is diverted away from other organs—including the skin—by vasoconstriction (see CVS section and Fig. 16.5). An exception is cerebral blood flow which remains relatively constant during exercise. As the body temperature rises, vasoconstriction of the vessels in the skin is relaxed to aid in heat dissipation by evaporative cooling (see section on the control of body temperature). The increase in muscle blood flow in response to muscular contraction is extremely rapid, so that within 0·3 second of a brief contraction of

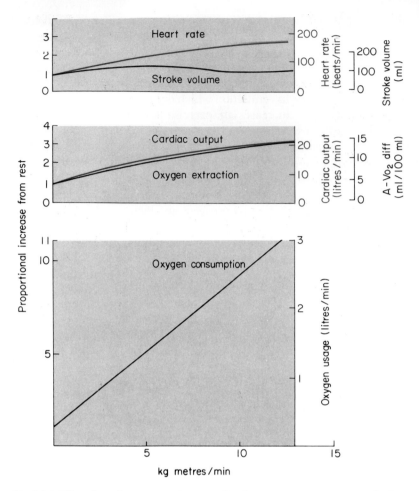

Fig. 16.4. Effect of steady-state exercise on cardiac function. Increasing exercise leads to a proportional increase in oxygen consumption (lower trace) which is carried by almost equal increments in cardiac output and oxygen extraction from the blood (middle trace). The increase in cardiac output is largely due to a proportional increase in heart rate with a small increase in stroke volume (upper trace). In 'horizontal' exercises (for example, swimming) the stroke volume is higher and does not change much with exercise level. (Note that the stroke volume falls at higher outputs due to lesser filling.) Data: Asmunsen & Neilson, 1952. The left scale shows the proportional increase in the parameter compared to rest, the right scales show the actual values.

In patients with transplanted hearts, which are of course denervated, increased cardiac output is due to increased stroke volume with little change in heart rate.

forearm muscles it is possible to measure an increase in blood flow. This initial increase does not appear to involve sympathetic vasodilator activity but is mediated through local mechanisms such as the liberation of metabolites which act directly on the blood vessels or sensory nerves (see p. 116 and Fig. 16.6).

In addition to the changes in systemic circulation described there is also a small rise in pulmonary arterial pressure and a small decrease in pulmonary resistance.

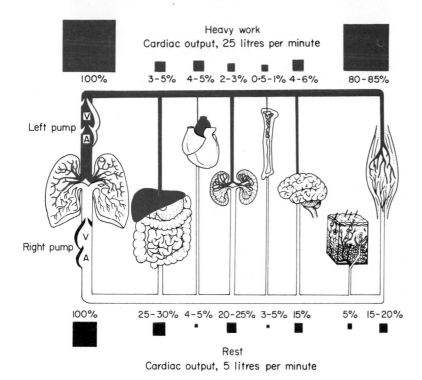

Fig. 16.5. Distribution of blood to the various organs at rest (lower scale) and during exercise (upper scale). The areas shaded are proportional to the minute volume of blood flow (the figures show the percentage distribution) (from Åstrand & Rodahl).

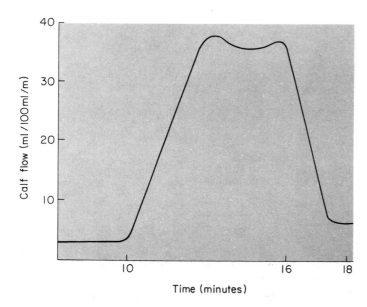

Fig. 16.6. Increased blood flow (ml/100 ml/min) in the calf muscle following rhythmic exercise consisting of short periods of contraction interspersed with a few seconds of relaxation between contractions (adapted from Barcroft & Swan).

The regulation of energy supply to the muscles during exercise

The fuels available for supplying energy to working muscles are the same as those normally available to the body (p. 357 et seq.). The largest energy store is in the form of fat from the adipose tissue. Carbohydrate forms a lesser store with about 55–90 g glycogen available in the liver and about 12 g/kg in the individual muscles. Although a normal man contains around 10 kg of protein none of this is used directly as an energy source for muscle. The preferred fuel of the muscle fibres varies not only according to the particular fibre type (p. 55) but also according to the severity of the exercise. At low levels of exercise involving oxidative muscle fibres most of the energy comes from fat (Fig. 16.7). In

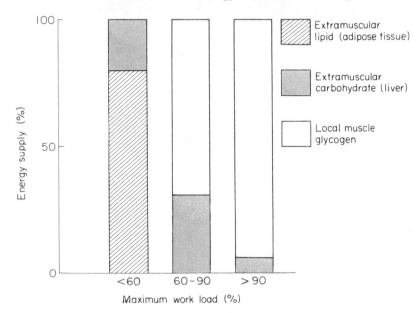

Fig. 16.7. Relative contribution of fat and carbohydrate from liver and local stores in supplying energy at various work loads in a muscle. Part of the explanation for this pattern of energy store utilisation is the recruitment of different types of fibre at different work loads (see text).

contrast, at high levels of exercise almost all of the energy demand is met by stores of glycogen within the muscle fibres themselves (Fig. 16.7). At low levels of exercise (walking, crawling, etc.) most of the power can be provided by the recruitment of aerobic slow-twitch fibres (p. 57). As the level of effort increases an increasing proportion of aerobic fast-twitch fibres are recruited.

The increase in lipolysis and glucose production by the liver is mainly due to an increase in sympathetic activity, although circulating catecholamines, growth hormone, glucocorticoids and insulin may also be involved. The entry of free fatty acids (FFA) from plasma to muscle cells is probably passive and is normally limited by the rate of blood supply; so it increases with the increase in muscle blood supply occurring during exercise (Fig. 16.6). During supramaximal (anaerobic) exercise the entry of FFA into plasma may be inhibited by the high lactate levels prevailing, thus sparing the body's fat stores.

Finally, for really high power outputs the supply of oxygen and nutrients by the blood cannot keep pace with the energy demand. Really fast contractions (sprinting, etc.) are therefore achieved by recruiting anaerobic fast-twitch fibres. These fibres are able to provide ATP at a very high rate by anaerobic glycolysis using local glycogen stores. Usually exhaustion coincides with a depletion of muscle glycogen stores, except at very high levels of effort when respiratory acidosis (resulting from accumulated anaerobic end products) may be a limiting factor.

Human muscles contain appropriate mixtures of these three fibre types according to the particular function they are required to perform. Adoption of this 'three-geared system' allows the most efficient use to be made of the available fuels during different levels of muscular activity (p. 55). It should be noted, however, that this classification of fibre types is somewhat artificial and that there is really a continuous gradation of properties between each type (p. 55).

Effects of training

The type of adaptations seen in the respiratory and cardiovascular systems and muscles in response to training will depend on the nature, frequency and duration of the training programme employed.

Endurance training leads to the following adaptations:

1   An increase in the maximal blood supply to the muscles and an increase in the maximal cardiac output.

2   The ability of muscles to use oxygen increases; this is due partly to the increased blood flow and partly to an increased extraction of oxygen (so the venous PO_2 falls).

This is shown particularly by slow- and fast-red-twitch fibres. These show the following changes: an increased number of (a) capillaries, (b) mitochondria, (c) myoglobin concentration and (d) oxidative enzymes. They also have an increased ability to use endogenous triglycerides.

3   Individuals trained for endurance obtain more of their energy from fat and less from carbohydrates. This is reflected in a slower depletion of glycogen in their muscles with a slower build up of lactate.

4   Endurance training leads to less fatigue. As fatigue is partly due to depletion of the glycogen stores, these biochemical observations may explain this.

A different type of adaptation is seen in training designed to build up the strength of the muscle such as weight-lifting exercises. In this case there is a hypertrophy of the muscle fibres rather than an increase in their capacity for aerobic metabolism.

Chapter 17
Circadian Rhythms

Many functions of our bodies vary throughout the day. These variations are known as diurnal or circadian rhythm (from *circa dies*, around the day); the latter has a more precise meaning so we shall use it.

It is common experience that we tend to get up around the same time each day, to feel tired and go to bed around the same time, to feel hungry at certain times of the day, to pass less urine at night etc. If we measure various aspects of bodily function then we find that blood pressure (Fig. 17.1), temperature, cortisol production (Fig. 11.23) etc. varies accord-

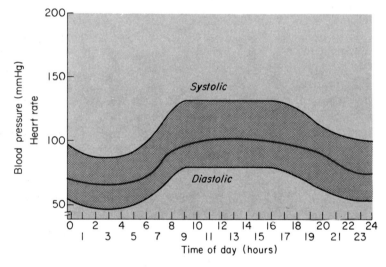

Fig. 17.1. Daily variations in the systolic and diastolic pressures and heart rates (red line) in five normal subjects (mean age 29 years). The records were taken by an indwelling catheter over 2 days, while the subjects worked normally. (Note (1) heart rate and mean BP well correlated; this is probably because the CVS 'centre' is being reset throughout the day; (2) pulse pressure remains constant (after Millar-Craig *et al.*).)

ing to a daily pattern. At first sight these phenomena might be thought to be due to simple cause-and-effect reasons, e.g. we get up because we have 'slept enough', we go to bed because it is dark outside, we pass less urine at night because we drink less etc. The simplest of investigations show that this direct causal relationship is not the correct one. Many of you will know this for yourselves from working on night-duty or from 'jet-lag'; in both cases the body tries to follow its previously normal pattern, which may be inappropriate to the new conditions. Fig. 17.2 shows that when a man is kept in constant light, the sleep pattern follows an approximate 24 hour, or circadian, rhythm. The important

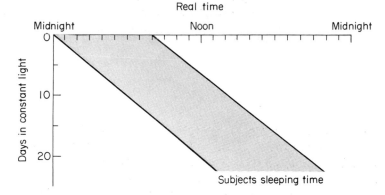

Fig. 17.2. Sleeping times of a human subject kept in a constantly lit environment. The subject's sleeping times lag further behind real time each day, showing that the 'sleep oscillator' runs slower than external time (data from Passmore & Robson).

point to notice is that sleep does not become random under these conditions.

The current suggestion is to suppose that man and other animals have developed an internal model of external time, i.e. a biological clock. This clock runs at about the same rate as that of normal clocks (actually slightly slower). In normal life our internal clock is synchronised with external time every day, so that it runs synchronously with it. If the normal 'cues' from the environment are removed, then our internal clock runs more slowly than external time, so the two get progressively out of step (Fig. 17.2). The main cue in man is light input from the eyes, with feeble cues from sound.

Site of clock. We can think of the clock as an oscillator which completes one cycle every 24 h (Fig. 17.3). The environment sends 'cues' to this

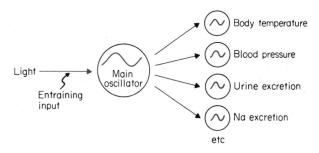

Fig. 17.3. A model of the internal clock arrangements explaining circadian rhythms. Multicellular organisms are made up of a population of circadian oscillators, coordinated together and connected to the external world by a master oscillator.

oscillator and so 'entrains' it. This oscillator then imposes a rhythm on the bodily processes. At present it is thought that the main oscillator drives a number of other oscillators, which then drives the BP, sleep pattern, urine output etc. It is thought that the main oscillator is in the suprachiasmatic nucleus (SCN), a collection of some 10,000 cells some 0.5 by 1 mm in size in the hypothalamus. This nucleus is entrained by

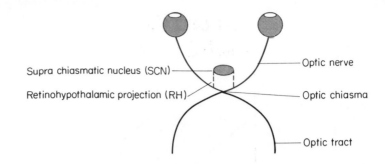

Fig. 17.4. Anatomical connections of the optic tracts, suprachiasmatic nucleus (SCN) and retinohypothalamic projection (RH). The SCN is in the hypothalamus.

a small pathway—the retinohypothalamic (RH) projection connecting it to the optic tract (Fig. 17.4), which anatomically lies just below it. This means that the RH pathway is the first branch from the optic tract; there is probably also a connection via the geniculate body.

Usefulness of circadian rhythms. In lower animals it is clearly useful to have an internal clock, for reasons such as long distance navigation, when to emerge from burrows to hunt etc. In man their usefulness is less clear, and they often do not fit well with our current lifestyle. For example, the performance of various tasks varies throughout the day: some are done better at night, others in the morning; sudden time shifts as in 'jet-lag' require some time (days to weeks) for all bodily functions to settle down etc. In this context it is perhaps as well to remember that man has followed the day/night cycle for the million or so years preceding the last few hundred years; during most of this time there were no clocks.

Appendix
Useful Data

The unit of *mass* is the mole (mol); it is defined as the number of atoms in 12 grams of carbon (6×10^{23} Avogadro's Number). The atomic weight of any element and the molecular weight of any substance expressed in grams contains this number of molecules.

> 1 mole (mol) contains 6×10^{23} particles
> 1 millimole (mmol) contains 6×10^{20} particles
> 1 micromole (μmol) contains 6×10^{17} particles

The most convenient unit of *volume* in biology is the cubic decimetre (dm^3), equivalent to one litre (l). The preferred smaller units are the cubic centimetre (cm^3) and the cubic millimetre (mm^3).

> dm^3 = litre (l)
> cm^3 = millilitre (ml)
> mm^3 = microlitre (μl)

Concentrations are expressed as moles per litre (mol/l), millimoles per litre (mmol/l) or micromoles per litre (μmol/l) and may be referred to as molar (M), millimolar (mM) or micromolar (μM) in strength, respectively. This system cannot be used for substances whose molecular weights are not known accurately; as most proteins are in this category it is therefore common to express them as grams per unit volume, usually grams per litre (g/l) or grams per decilitre or per 100 ml (g/dl or g/100 ml).

The *osmotic strength* of a solution may be expressed as osmoles per litre of volume (osmolarity) or osmoles per litre of water (osmolality). Remember that salts ionise to a greater or lesser extent in water so that a molar solution of NaCl has an osmotic strength of (nearly) two osmoles. The SI unit of *pressure* is the pascal (Pa) which is one newton per square metre ($N\ m^{-2}$). For general physiological purposes the kilopascal (kPa), which equals 7·5 mmHg, is more convenient. Currently, this system is used for gas tensions but not for blood-pressure measurement.

As 100 kPa = 750 mmHg ≑ atmospheric pressure, the percentage composition of a gas at sea level is approximately equal to the tension in kPa. Thus

> O_2 in air = 21% ≑ 21 kPa
> CO_2 in air = 0·03% ≑ 0·03 kPa
> O_2 in alveoli = 13% ≑ 13 kPa
> CO_2 in alveoli = 5% ≑ 5 kPa

Energy (or heat) is the capacity of a body for doing work. The SI unit is the joule (J), the previous but still commonly used unit is the Calorie.

> 1 kcal = 4·19 kJ

The calorie is the amount of heat required to raise the temperature of 1 g of water $1°C$ (actually from $15-16°C$). The 'Calorie' used in nutrition is really a kilocalorie (kcal) and will be so called here.

Force is that which when acting on a body, which is free to move, produces an acceleration in the motion of the body. The SI unit is the newton (N). A previous unit is the dyne ($= 10^{-5}$ N).

$$1 \text{ newton} = 1 \text{ joule acting over 1 metre (J m}^{-1})$$

Power is the rate of doing work. The SI unit is the watt (W).

$$1 \text{ watt} = 1 \text{ joule per second}$$

Frequency is now expressed as hertz (Hz) where 1 Hz is equal to 1 cycle per second.

The customary *temperature* is expressed as degree Celsius ($°C$).

For distances of the same order of magnitude as a chemical bond the ångstrom unit (Å) is still often used. This equals 10^{-10} m. 10 Å $= 1$ nanometre.

FORMULAE

Henderson–Hasselbalch equation

This equation is based on the fact that dissolved CO_2 combines with water to form H_2CO_3 which then dissociates to H^+ and HCO_3^-.

$$CO_2 + H_2O \rightleftharpoons H_2CO_3 \rightleftharpoons H^+ HCO_3^-$$

An equilibrium thus exists between dissolved CO_2 and H^+ and HCO_3^-. The dissociation or equilibrium constant, K', is thus:

$$K' = \frac{[H^+][HCO_3^-]}{[\text{dissolved } CO_2]}$$

The Henderson–Hasselbalch equation arises by taking the logarithm of both sides:

$$\log K' = \log H^+ + \log \frac{[HCO_3^-]}{[\text{dissolved } CO_2]}$$

or

$$pH = pK' + \log \frac{[HCO_3^-]}{[\text{dissolved } CO_2]}$$

In normal plasma $pK' = 6·1$; $[HCO_3^-] = 24$ mmol/l; and dissolved $CO_2 = 1·2$ mmol/l; hence $pH = 6·1 + \log \dfrac{24}{1·2} = 7·4$

NERNST EQUATION

For potassium ions in equilibrium across a semi-permeable membrane, the potassium equilibrium potential E_K is given by

$$E_K = \frac{RT}{ZF} \log_e \frac{[K_o^+]}{[K_i^+]}$$

where F is the Faraday constant (coulombs per mole of charge)

Z is the valency of potassium

R is the universal gas constant

T is absolute temperature

K_o^+, K_i^+ concentrations of potassium ions outside and inside the cell, respectively.

443

Appendix
Useful Data

Substituting and converting to logarithms to the base 10

$$E_K = 61 \log_{10} \frac{[K_o^+]}{[K_i^+]} \text{ mV at } 37°C.$$

If $E_K = 0$, the equilibrium condition for ions is reduced to that for neutral substances and the concentrations inside and outside the cell are equal. The Nernst equation can be written for any species of ion present.

Blood-flow measurement

Fick principle
O_2 uptake is found with Douglas bag or spirometer; arterial O_2 by arterial puncture; venous O_2 by catheterisation of right ventricle. Pulmonary blood flow is equal to cardiac output.

$$\text{Pulmonary blood flow} = \text{a–v } O_2 \frac{\text{ml } O_2 \text{ used/min}}{\text{difference in ml/l}}$$

Indicator dilution method
If a quantity of an indicator is injected into a stream with a steady flow and the mean concentration of the indictor at a point downstream measured over the transit time then:

$$\text{flow} = \frac{\text{amount injected}}{\text{mean concentration} \times \text{transit time}}$$

Respiratory quotient (RQ)

$$RQ = \frac{CO_2 \text{ expired/min}}{O_2 \text{ used/min}}$$

Michaelis–Menten equation

This equation occurs for example in enzymology where Y is the initial velocity of an enzyme-catalysed reaction. S is the substrate concentration, V_{max} is the maximum velocity (when $S \to \infty$) and K_m is the Michaelis constant (when $Y = \frac{1}{2}V_{max}$)

then
$$Y = \frac{V_{max} S}{K_m + S}$$

The most generally useful transformation to fit experimental data is:
$Y = V_{max} - K_m (Y/S)$.

Blood counts

Red cell count (RBC)

	Female	$4\cdot8 \pm 1 \times 10^{12}$/litre
	Male	$5\cdot5 \pm 1 \times 10^{12}$/litre

Haemoglobin (Hb)

	Female	$14 \pm 2\cdot5$ g/dl
	Male	$15\cdot5 \pm 2\cdot5$ g/dl

Haematocrit

	Female	$0\cdot42 \pm 0\cdot05$
	Male	$0\cdot47 \pm 0\cdot07$

Mean corpuscular volume (MCV)	85 ± 8 fl
Mean corpuscular haemoglobin (MCH)	$29\cdot5 \pm 2\cdot5$ pg
Mean corpuscular haemoglobin concentration (MCHC)	33 ± 2 g/dl
Reticulocytes ($0\cdot2$–$2\cdot0$%)	10–100×10^9/l
Leucocytes	$7\cdot5 \pm 3\cdot5 \times 10^9$/l

Differential count—

Neutrophils (40–75%)	$2\cdot0$–$7\cdot5 \times 10^9$/l
Lymphocytes (20–45%)	$1\cdot5$–$4\cdot0 \times 10^9$/l
Monocytes (2–10%)	$0\cdot2$–$0\cdot8 \times 10^9$/l
Eosinophils (1–6%)	$0\cdot04$–$0\cdot4 \times 10^9$/l
Basophils (< 1%)	$< 0\cdot1 \times 10^9$/l
Platelets	150–$400 + 10^9$/l

Blood constituents

(As measured in whole blood (B), plasma (P) or serum (S).

Bilirubin (S)	3–21 μmol/l
Calcium (S) (total)	$2\cdot1$–$2\cdot6$ mmol/l
Cholesterol (P)	$3\cdot2$–$8\cdot5$ mmol/l
Chloride (P)	98–108 mmol/l
Creatinine (S)	60–125 μmol/l
Folic acid (S)	12–35 nmol/l
Glucose (P)	$3\cdot3$–$5\cdot9$ mmol/l
Iodine (protein bound) (S)	$0\cdot28$–$0\cdot63$ μmol/l
Iron (S)	14–31
	11–29 μmol/l
	μmol/l
Lactic acid (B)	$0\cdot5$–$1\cdot2$ mmol/l
Nitrogen (non-protein) (S)	$10\cdot7$–$25\cdot0$ mmol/l
Osmolality (P)	278–295 mOsmol/kg
Phosphate (inorganic) (P)	$0\cdot8$–$1\cdot4$ mmol/l
Potassium (S)	$3\cdot9$–$5\cdot1$ mmol/l
pH (b)	$7\cdot35$–$7\cdot45$
PCO_2 (B)	$4\cdot7$–$6\cdot9$ kPa
PO_2 (B)	$11\cdot3$–$14\cdot0$ kPa
Protein (total) (S)	62–82 g/l

albumin (S)   33–48 g/l
globulin (S)   20–40 g/l
fibrogen (P)   1·8–4·2 g/l
Sodium (S)   131–145 mmol/l
Urea (P)   2·5–6·6 mmol/l

Cerebrospinal fluid
Calcium   1·1–1·3 mmol/l
Chloride   120–130 mmol/l
Potassium   2·4–3·2 mmol/l
Sodium   100–250 mmol/l
Glucose   3–4·5 mmol/l
Protein   20–30 mg/100 cm^3

Urine
Electrolytes
  calcium   2·5–7·5 mmol/24 h
  chloride   50–250 mmol/24 h
  potassium   40–120 mmol/24 h
  sodium   100–250 mmol/24 h
17-Ketosteroids
  males   10–30 mg/24 h
  females   5–15 mg/24 h
Reaction (pH)   4·5–7·5
Volume   5–30 dl/24 h
Urea   250–500 mmol/24 h
Urea clearance   50–100 ml blood/min
Protein   <150 mg/24 h

Faeces
Total fat   10–20% w/v
Free fatty acid   7·5–15% w/v
Neutral fat   2·5–5% w/v

Gastric secretion
Fasting volume   10–60 ml/h
  pH   0·9–1·5
  basal concentration   25 mmol/l
  maximum pentagastrin stim.   50 mmol/h

FRACTIONS AND MULTIPLES

Fraction	Prefix	Symbol	Multiple	Prefix	Symbol
10^{-1}	deci	d	10	deca	da
10^{-2}	centi	c	10^2	hecto	h
10^{-3}	milli	m	10^3	kilo	k
10^{-6}	micro	μ	10^6	mega	M
10^{-9}	nano	n			
10^{-12}	pico	p			
10^{-15}	femto	f			

CONVERSION TABLES

Pressure		Energy	
mmHg	kPa	kcal	kJ
1	0·13	1	0·0042
10	1·33	100	0·42
20	2·67	200	0·84
30	4·00	300	1·26
40	5·33	400	1·67
50	6·67	500	2·09
60	8·00	600	2·51
70	9·33	700	2·93
80	10·67	800	3·35
90	12·00	900	3·77
100	13·33	1000	4·19
110	14·67	2000	8·37
120	16·00	3000	12·56
130	17·33		
140	18·67		
150	20·00		

Reference Material

Further information on the various aspects of physiology may be obtained from:
1   Larger general textbooks, such as those listed below
2   Specialised textbooks, monographs and reviews, a selection of which are listed below
In addition to long reviews such as *Physiological Reviews* and *Annual Reviews of Physiology* there are many shorter review articles available. For example. *Scientific American* publish useful articles covering various aspects of physiology. These will be in your local library. Offprints of individual articles can be bought from WH. Freeman & Co., 660 Market Street, San Francisco, California 94104 or Werner House, Folkestone, Kent, England. *The Lancet* and *British Medical Journal* also publish many short topical reviews on medical aspects of Physiology.

TEXTBOOKS

Astrand P-O. Rodahl K. *Textbook of Work Physiology. Physiological basis of exercise.* 2nd ed. New York: McGraw-Hill Book Company, 1977.

Brobeck JR. ed. *Best and Taylor's Physiological Basis of Medical Practice.* Baltimore: Williams and Wilkins, 1973.

Davson H. Segal MB. *Introduction to Physiology.* London: Academic Press. Vol. 1 (1975). Vol. 3 (1976).

Eckert R. Randall D. *Animal Physiology.* San Francisco: Freeman, 1978.

Ganong WF. *Review of Medical Physiology.* Los Altes. California: Lafe Medical Publications, 1975.

Guyton AC. *Textbook of Medical Physiology.* 6th ed. London: Saunders, 1981.

Mountcastle VB. ed. *Medical Physiology* 13th ed. St Louis, Missouri: Mosby, 1974.

Passmore R. Robson JS. *A Companion to Medical Studies.* Vol. I. 2nd ed. Oxford: Blackwell Scientific Publications, 1976.

Prosser CL. *Comparative Animal Physiology.* Philadelphia: Saunders, 1973.

Schmidt RF, Thews G. *Human Physiology.* 20th ed. (English) Berlin, New York: Springer-Verlag, 1982.

Stryer L. *Biochemistry.* 2nd ed. San Francisco: WH. Freeman, 1981.

Vander AJ. Sherman JH. Luciano DS. *Human physiology: the mechanisms of body function.* 2nd ed. New York, 1975.

American Physiological Society. *Handbook of Physiology.* Washington: American Physiological Society, 1962.

REVIEW AND ORIGINAL ARTICLES

Chapter 1

Baker PF. The sodium pump in animal tissues and its role in the control of cellular metabolism and function. In Hokin LE, ed. *Metabolic Pathways:* Vol. VI, Metabolic Transport, pp. 243–268. New York: Academic Press, 1972.

Hinkle PC, McCarty RE. How cells make ATP. *Scientific American* 1978; **238**(3): 104–121.

Open University Course S321: *Physiology of Cells and Organisms.* Unit 1–Cells and their Ultrastructure. Units 3 and 4–Membranes and Transport. Milton Keynes: The Open University Press, 1974.

Robinson JR. *A Prelude to Physiology.* pp. 1–83. Oxford: Blackwell Scientific Publications, 1975.

Chapter 2

Katz B. *Nerve Muscle and Synapse.* New York: McGraw Hill, 1976.

Robinson JR. *A Prelude to Physiology.* pp. 67–102. Oxford: Blackwell Scientific Publications, 1975.

Chapter 3

Aidley DJ. *The Physiology of Excitable Cells.* pp. 191–294. Cambridge· Cambridge University Press, 1971.

Bagshaw CR. *Muscle Contraction.* London: Chapman and Hall, 1982.

Carlson FD, Wilkie DR. *Muscle Physiology.* New Jersey: Prentice Hall, 1974.

Hill AB. *First and Last Experiments in Muscle Mechanics.* Cambridge: Cambridge University Press, 1970.

Huddart H. *Comparative Physiology of Smooth Muscle.* Glasgow: Blackie, 1976.

Keynes RD, Aidley DJ. *Nerve and Muscle.* Cambridge: Cambridge University Press, 1981.

Open University Course S321: *Physiology of Cells and Organisms.* Unit 6–Muscles and Contractility. Milton Keynes: The Open University Press, 1974.

Wilkie DR. *Muscle.* 2nd ed. Institute of Biology Series. London: Edward Arnold, 1976.

Chapter 4

Roitt IM. *Essential Immunology.* 3rd ed. Oxford: Blackwell Scientific Publications, 1977.

Playfair JHL. *Immunology at a Glance.* Oxford: Blackwell Scientific Publications, 1980.

Weir DM. *Immunology—an outline for students of medicine and biology.* 4th ed. London: Churchill, Livingstone, 1977.

Chapter 5

Rushmer RF. *Structure and Function of the Cardiovascular System.* New York: Saunders, 1972.

Chapter 6

Duthie HL, Wormsley KG. *Scientific Basis of Gastroenterology.* Edinburgh: Churchill, Livingstone, 1979.

Durnin JBGA, Womersley J. Body fat assessed from total body density and its estimation from skin fold thickness. *British Journal of Nutrition*—1974; **32:** 77.

Gregory RA. Recent advances in the physiology of gastrin. *Proceedings of the Royal Society B* 1968; 170: 81.

Chapter 7

Cotes JE. *Lung Function.* 4th ed. Oxford: Blackwell Scientific Publications, 1979.

West JB. *Respiratory Physiology—the essentials.* Oxford: Blackwell Scientific Publications, 1974.

Chapter 8

De Wardener HE. *The Kidney. An outline of normal and abnormal structure and function.* London: Churchill, Livingstone, 1974.

Hladky SB, Rink TJ. *Body Fluid and Kidney Physiology.* London: Edward Arnold, 1986.

Valtin H. *Renal Function.* Boston: Little, Brown & Co, 1973.

Chapter 9

Kandel ER, Schwartz JH. *Principles of Neural Science.* New York, Amsterdam, Oxford: Elsevier/North Holland, 1981.

Netter FH. *The Ciba Collection of Medical Illustrations.* Vol. 1. Nervous System. New Jersey: Ciba, 1972.

Noback CR, Demarest RJ. *The Human Nervous System. Basic Principles of Neurobiology.* 2nd ed. Tokyo: McGraw-Hill, 1975.

Stein JF. *An Introduction to Neurophysiology.* Oxford: Blackwell Scientific Publications, 1982.

Chapter 10

Kuffler SW, Nicholls JG. *From Neurone to Brain.* Sunderland, Mass.: Sinauer, 1976.

Von Békésy G. The ear. *Scientific American* 1957; **197**(8): 66–78.

Werblin FS. The control of sensitivity in the retina. *Scientific American* 1973; **228**(1): 70–79.

Chapter 11

Catt KJ. *An ABC of Endocrinology*. London: The Lancet, 1971.

Tepperman J. *Metabolic & Endocrine Physiology*. Chicago: Year Book Medical Publishers, 1973.

Chapter 12

Austin CR, Short RV. eds. *Reproduction in Mammals*: Books 1–5. Cambridge: Cambridge University Press, 1973.

Hogarth PJ. *Biology of Reproduction*. Glasgow: Blackie, 1978.

Johnson MH, Everitt BJ. *Essential Reproduction*. Oxford: Blackwell Scientific Publications, 1980.

Parkes AS. *Patterns of Sexuality and Reproduction*. London: Oxford University Press, 1976.

Chapter 13

Robertson JR. *Fundamentals of Acid–Base Regulations*. 4th ed. Oxford: Blackwell Scientific Publications, 1972.

Chapter 14

Black DAK. *Essentials of Fluid Balance*. 4th ed. Oxford: Blackwell Scientific Publications, 1967.

Chapter 15

Open University Course S321: *Physiology of Cells and Organisms*. Unit 13–Adaptation to the External Environment: Temperature. Unit 16–Physiological Integration and Control. Milton Keynes: The Open University Press, 1974.

Precht H, Christophersen J, Heinsel H, Larcher W. *Temperature and Life*. Berlin, Heidelberg, New York: Springer-Verlag, 1973.

Chapter 16

Bourne GH. ed. *The Structure and Function of Muscle*. Vol. III. New York: Academic Press, 1973.

COMPUTER SIMULATIONS

The MAC suite of programs. These programs model various aspects of Human Physiology and are designed for interactive use by students and research workers. They consist of the following:

DOPE Program to simulate pharmacokinetics
MAN Program to simulate heart and blood circulation
PEE Program to simulate the systemic circulation, kidneys and body fluids
PUF Program to simulate gas exchange and circulation

Program details may be obtained from:

In North America: McMaster Medical School, Hamilton, Ontario L8s 4L8
In United Kingdom: The Department of Medicine (Computer Unit). The Dunn Laboratories, St Bartholomew's Hospital and Medical College, West Smithfield, London EC1

Sources of Figures*

Allweis C. Control system diagrams in Physiology, Biology and Medicine. *Israel J Med Sci* (supplement) 1971; **7**: 1–84.

Astrand P-O, Rodahl K. *Textbook of work physiology*. New York: McGraw Hill Book Company, 1977.

Austin CR, Short VR. eds. *Reproduction in Mammals*. Cambridge: Cambridge University Press, 1972.

Barcroft H, Swan HJC. *Sympathetic Control of Human Blood Vessels*. London: Arnold, 1953.

Bell GH, Davidson JN, Emslie-Smith D. *Textbook of Physiology and Biochemistry*. 8th ed. Edinburgh: Churchill Livingstone, 1972.

Benzinger TH. Heat regulation: homeostasis of central temperature in man. *Physiol Rev* 1969; **49**: 671–759.

Burton AC. Relation of structure to function of the tissues of the wall of blood vessels. *Physiol Rev* 1954; **34**: 619–642.

Clermont Y. *Arch Anat micr Morph* 1967; exp. 56: 7. Masson et Cie Paris.

Colquhoun D. *Lectures on Biostatistics*. Oxford: Clarendon Press, 1971.

Davis JO. What signals the kidney to release renin? *Circulation Res* 1971; **28**: 301–306.

Davson H. *A Textbook of General Physiology*. Vol. 1. 4th ed. London: J & A Churchill, 1970.

Davson H, Segal MB. *Introduction to Physiology*. Vol. 1. London: Academic Press, 1975 (After Greep).

Deetjen P. *Nierenphysiologie*. In: Gauer OH, Kramer K, Jung R. *Physiologie des Menschen*. Vol. 7. Niere und Wasserhaushalt. München, Berlin, Wien: Urban & Schwarzenberg, 1970.

Edwards RG. ed. *Research in Reproduction. The Control of Parturition*. London: International Planned Parenthood Federation, 1975.

Edwards RG. ed. *Research in Reproduction. The Testis*. London: International Planned Parenthood Federation, 1973.

Ehrlanger J, Gasser HS. *Electrical Signs of Nervous Acitivity*. Philadelphia: University of Pennsylvania Press, 1937.

Eisenberg E, Hill TL. A cross-bridge model of muscle contraction. *Prog Biophys Molec Biol* 1978; **33**: 55–82.

Eyzaguiree C. *Physiology of the Nervous System*. Chicago: Year Book Medical Publishers, Inc., 1969.

Finean JB, Coleman R, Michell RH. *Membranes and their Cellular Functions*. 2nd ed. Oxford: Blackwell Scientific Publications, 1978.

Garrow JS. Regular review: Weight penalties. *Br med J* 1979; ii: 1171–1172.

Gordon AM, Huxley AF, Julian FJ. The variation in isometric tension with sarcomere length in vertebrate muscle fibres. *J Physiol* 1966; **184**: 170–192.

Grimes AJ. *Human Red Cell Metabolism*. Oxford: Blackwell Scientific Publications, 1980.

Guyton, AC. *Structure and Function of the Nervous System*. Philadelphia, London, Toronto: WB. Saunders, 1972.

Harvey W. *An Anatomical Disputation Concerning the Movement of the Heart and Blood in Living Creatures*. Transl. Whitteridge G. Oxford: Blackwell Scientific Publications, 1976.

Hodgkin AL. *The Conduction of the Nerve Impulse*. Liverpool: Liverpool University Press, 1964.

Hodgkin AL, Huxley AF. Ionic Currents. *J Physiol* 1952; **117**: 500–544.

Hubel DH. The visual cortex of the brain. *Scientific American* 1963; **209**: 5.

Jamison RL. Recent advances in the physiology of Henle's Loop and the collecting tubule system. *Circulation Res* 1974; 34–35 (supplement).

*If not included in reference material list.

Katz B. Depolarisation of sensory terminals and the initiation of impulses in the muscle spindle. *J Physiol London* 1950; **111**: 261–282.

Kuffler SW, Nicholls JG. *From Neuron to Brain*. Sunderland, Mass.: Sinauer Associates, 1976.

Lote CJ. *Principles of Renal Physiology*. London: Croom Helm, 1982.

McDonald DA. *Blood Flow in Arteries*. 2nd ed. London: Edward Arnold, 1974.

Makowski L, Caspar DLD, Phillips WC, Goodenough DA. Gap junction structures, *J Cell Biol* 1977; **74**: 629–645.

Michel CC. Filtration coefficients and osmotic reflexion coefficients of the walls of single frog mesenteric capillaries. *Journal of Physiology* 1980; **309**: 341.

Millar-Craig MW, Bishop CN, Ra EB. Circadian variation of blood pressure. *Lancet* 1978; **i**: 795–997.

Netter FH. *Ciba Collection of Medical Illustrations*. Nervous System, Vol. 1. 1972.

Noback CR, Demarest RJ. *The Human Nervous System. Basic Principles of Neurobiology*. Tokyo: McGraw-Hill, Kogakusha, 1975.

Noble D. *The Initiation of the Heart Beat*. Oxford: Clarendon Press, 1975.

Paul DH. *The Physiology of Nerve Cells*. Oxford: Blackwell Scientific Publications, 1975.

Peachy LD. The sarcoplasmic and transport tubules of the frog's sartorius. *J Cell Biol* 1965; **25**: 209–32.

Penfield WG, Rasmussen T. *The Cerebral Cortex of Man*. London and New York: Macmillan, 1950.

Perutz MF. Fundamental research in molecular biology: relevance to medicine. *Nature* 1976; **262**: 499–453.

Perutz MF. Hemoglobin structure and respiratory transport. *Scientific American* 1978; **239**: 68–86.

Pickering BT. Lessons from a peptidergic neurone. *Nature, Lond.* 1980; **288**: 117.

Polak JM, Bloom SR. Regulatory peptides: key factors in the control of bodily functions. *British Medical Journal* 1983; **286**: 1461.

Rhodin JAG, Missier P del, Reid LC. The structure of the specialised impulse-conducting system of the steer heart. *Circulation* 1961; **24**: 349–367.

Ruch TC. In Stevens SS. ed. *Handbook of Experimental Psychology*. New York: Wiley, 1951.

Segal SJ. *Scientific American* 1974; **231**(3): 53–62.

Shipley RE, Study RS. Changes in renal blood flow, extraction of inulin, glomerular filtration rate, tissue pressure and urine flow with acute alterations of renal arterial blood pressure. *American Journal of Physiology* 1951; **167**: 676.

Smith H. *The Kidney*. New York: Oxford University Press, 1951.

Thompson D'Arcy W. *On growth and form*. Cambridge: Cambridge University Press, 1942.

Tuchmann-Dupleosis H, David G, Haegal P. *Illustrated Human Embryology*. Vol. 1. London: Chapman & Hall, 1971.

Turner CD. *General Endocrinology*. 5th ed. WB. Saunders Co., 1966.

Index